MORAL MINDS

ALSO BY MARC D. HAUSER

The Evolution of Communication

The Design of Animal Communication
(coauthor)

Wild Minds

From Monkey Brain to Human Brain
(coauthor)

People, Property, or Pets?
(coauthor)

MORAL MINDS

How Nature Designed

Our Universal Sense

of Right and Wrong

Marc D. Hauser

HarperCollins books may be purchased for educational,
business, or sales promotional use.
For information, please write: Special Markets Department,
HarperCollins Publishers, 10 East 53rd Street, New York, NY 10022.

Designed by Kate Nichols

Library of Congress Cataloging-in-Publication Data is available upon request.

ISBN-10: 0-06-078070-3
ISBN-13: 978-0-06-078070-8

06   07   08   09   10   RRD   10   9   8   7   6   5   4   3   2

To five generations of extraordinary females

with the last name Hauser:

———

My grandmother, Lucille,

My mother, Alberta,

My wife, Lilan,

My daughters, Alexandra and Sofia, and

My cat, Cleopatra

OF ALL THE DIFFERENCES between man and the lower animals, the moral sense or conscience is by far the most important . . . [I]t is summed up in that short but imperious word ought, so full of high significance. It is the most noble of all the attributes of man, leading him without a moment's hesitation to risk his life for that of a fellow-creature; or after due deliberation, impelled simply by the deep feeling of right or duty, to sacrifice it in some great cause.

—CHARLES DARWIN

———

MORALS EXCITE PASSIONS, and produce or prevent actions. Reason of itself is utterly impotent in this particular. The rules of morality, therefore, are not conclusions of our reason.

—DAVID HUME

———

WHY DOES EVERYONE take for granted that we don't learn to grow arms, but rather, are designed to grow arms? Similarly, we should conclude that in the case of the development of moral systems, there's a biological endowment which in effect requires us to develop a system of moral judgment and a theory of justice, if you like, that in fact has detailed applicability over an enormous range.

—NOAM CHOMSKY

ACKNOWLEDGMENTS

THIS BOOK represents the culmination of much needed support and feedback. For excellent bibliographic detective work, and critical analysis of my thinking, I thank my research assistants Prasad Shirvalkar, Frances Chen, and especially Fiery Cushman. Fiery not only found references, he read much of the book, commented in detail, and argued me into many corners; he ultimately joined me as a graduate student on this adventure, one that continues today. During the spring term of 2002, I taught an undergraduate course in evolutionary ethics at Harvard and used bits of my book as fodder. The students devoured and regurgitated; I learned a great deal from this digestive process. I also presented parts of this book as an invited speaker to departments of psychology, neuroscience, law, and anthropology, as well as at conferences at MIT, NYU, Princeton's Center for Human Values, the Tanner Lectures on Human Values at the University of Michigan, Yale, Rutgers, University College London, and the Santa Fe Institute. A special thanks to Steve Stich for allowing me to present some of this work to a group of hungry moral philosophers who, for at least one day, tolerated an interloping biologist.

I am indebted to my closest academic family: my lab. My students are the source of inspiration, admiration, and fun. The generations involved during the gestation of this book include, in no special order except chronology: Laurie Santos, Cory Miller, Asif Ghazanfar, Tecumseh Fitch, Ruth Tincoff, Roian Egnor, Joanna Bryson, Jeff Stevens, Alberto Palleroni, Josh McDermott, Brian Hare, Keith Chen, Justin Wood, Liane Young, Fiery Cushman, Sang Ah Lee, Jenny Pegg, Tim O'Donnell, and an amazing cast of undergraduates.

I was lucky to have friends, students, colleagues, and experts at other universities who were willing and able to give me critical feedback on content and style. Of this illustrious group, several answered my questions, or commented on specific topics including evolution, economics, neurobiology, psychology, philosophy, linguistics, development, law, and religion, including: George Ainslie, Evan Balaban, Jonathan Baron, Kim Beeman, Ned Block, James Blair, Paul Bloom, Cedric Boeckx, Christopher Boehm, Robert Boyd, Samuel Bowles, Susan Carey, Dov Cohen, Richard Connor, Leda Cosmides, Alan Dershowitz, Adele Diamond, Susan Dwyer, Ernst Fehr, Daniel Fessler, Larry Fiddick, Robert Frank, Norman Frohlich, Drew Fudenberg, John Galaty, Susan Gelman, Herbert Gintis, Ann Graybiel, Joshua Greene, Paul Griffiths, Jonathan Haidt, David Haig, Lilan Hauser, Joseph Henrich, Sarah Hrdy, Ray Jackendoff, Susan Johnson, Daniel Kahneman, Frances Kamm, Louis Kaplow, Tim Ketelaar, David Laibson, Alan Leslie, Matthias Mahlmann, John Mikhail, Earl Miller, Doug Mock, Shaun Nichols, Richard Nisbett, Joe Oppenheimer, David Premack, Jesse Prinz, John Rawls, Adina Roskies, Al Roth, Paul Rubin, Rebecca Saxe, Wolfram Schultz, Susanna Siegel, Peter Singer, Walter Sinnott-Armstrong, Elizabeth Spelke, Daniel Sperber, Steve Stich, Alan Stone, Judith Thomson, Robert Trivers, Fritz Tsao, Peter Unger, James Whitman, David Sloan Wilson, and Richard Wrangham. A special thanks to the late John Rawls for spending some time speaking to me about his views concerning the role of moral psychology in our conceptions of justice; to John Mikhail, whose thesis on Rawls's linguistic analogy greatly influenced my own thinking; and to Jon Haidt for boldly proposing that the field of moral psychology shift focus from reasoning to intuition. Jon and I disagree about when emotions jump in to our judgments, but share the intuition about moral intuition. Others read through drafts of selective chapters of the book, or the entire thing, and provided honest though often brutal criticism (what are friends for?): Noam Chomsky, Daniel Dennett, Susan Dwyer, Ray Jackendoff, Steven Pinker, and Peter Singer. Thank you all.

I spent a glorious sabbatical writing and thinking in Australia, Palau, and various countries in Africa. I love Harvard and Cambridge, but getting away was pure elixir. While in Australia, I had the good fortune of reuniting with four old friends, Anne and Alan Goldizen, Ove Hoegh-Guldberg, and Sophie Dove. They not only introduced me to many wonderful sites, but made my stay fun and productive. In Palau, I thank Lori Colin Bell for finding us an apartment on top of a hill overlooking the most gorgeous string of islands and turquoise water I have ever seen; I pity those on the show *Survivor* who had to struggle and could not profit from the majesty of this country and the beauty of its people. In Kenya, I reunited with old friends from my graduate-school days, facilitated by staying at Cynthia Moss's blissfully peaceful elephant camp in Amboseli National Park. This book is all the better for these experiences, showcasing both the universality of humanity and its interesting twists and turns.

I thank Harvard University for overall financial support, the J. S. McDonnell Foundation for a grant to work on issues concerning language, and the H. F. Guggenheim Foundation for a fellowship to finish this book while also collecting data from our Web site, several hunter-gatherer sites, and patients populations.

To my agents, John Brockman and Katinka Matson, thank you, thank you. Your enthusiasm, support, and guidance have been exceptional. Your dedication to bringing science to the public is worthy of several toasts.

For sharing the journey, providing significant editorial guidance, and supporting the book from its inception, I thank Dan Halpern at Ecco and Tim Whiting at Time Warner UK. I also thank Emily Takoudes at Ecco for lending a fine editing hand to my prose. Cutting major chunks of the book felt like surgically removing limbs, but the book is now all the more agile for these changes. Thanks to Ted Dewan for several wonderful illustrations; they are gems, making the book sparkle.

I know that many married writers single out a spouse for putting up with absentmindedness, singularity, and stress. I deny none of these attributes. My dear Lilan, how did you put up with me? I raise my glass in admiration, love, devotion, and everlasting passion.

CONTENTS

PROLOGUE:
RIGHTEOUS VOICES

———

THE CENTRAL IDEA of this book is simple: we evolved a moral instinct, a capacity that naturally grows within each child, designed to generate rapid judgments about what is morally right or wrong based on an unconscious grammar of action. Part of this machinery was designed by the blind hand of Darwinian selection millions of years before our species evolved; other parts were added or upgraded over the evolutionary history of our species, and are unique both to humans and to our moral psychology. These ideas draw on insights from another instinct: language.

The revolution in linguistics, catalyzed by Noam Chomsky in the 1950s[1] and eloquently described by Steven Pinker in *The Language Instinct*, was based on a theoretical shift. Instead of an exploration of cross-cultural variation across languages and the role of experience in learning a language, we should follow in the tradition of the biological sciences, seeing language as an exquisitely designed organ—a universal feature of all human minds. The universal grammar that lies at the heart of our language faculty and is part of our species' innate endowment provides a toolkit for building specific languages. Once we have acquired our native language, we speak and comprehend what others say without reasoning and without conscious access to the underlying rules or principles. I argue that our moral faculty is equipped with a *universal moral grammar,* a toolkit for building specific moral systems. Once we have acquired our culture's specific moral norms—a process that is more like growing a limb than sitting in Sunday school and learning about vices and virtues—we judge whether actions are permissible, obligatory, or forbidden, without conscious reasoning and without explicit access to the underlying principles.

At the core of the book is a radical rethinking of our ideas on morality, which is based on the analogy to language, supported by an explosion of recent scientific evidence. Our moral instincts are immune to the explicitly articulated commandments handed down by religions and governments. Sometimes our moral intuitions will converge with those that culture spells out, and sometimes they will diverge. An understanding of our moral instincts is long overdue.

The framework I pursue in *Moral Minds* follows in a tradition that dates back to Galileo, has been accepted by most physicists, chemists, and a handful of natural and social scientists. It is a stance that starts by recognizing the complexity of the world, admitting the futility of attempts to provide a full description. Humbled by this recognition, the best way forward is to extract a small corner of the problem, adopt a few simplifying assumptions, and attempt to gain some understanding by moving deeply into this space. To understand our moral psychology, I will *not* explore all of the ways in which we use it in our daily interactions with others. In the same way that linguists in the Chomskyan tradition sidestep issues of language use, focusing instead on the unconscious knowledge that gives each of us the competence to express and judge a limitless number of sentences, I adopt a similarly narrow focus with respect to morality. The result is a richly detailed explanation of how an unconscious and universal moral grammar underlies our judgments of right and wrong.

To show the inner workings of our moral instincts, consider an example. A greedy uncle stands to gain a considerable amount of money if his young nephew dies. In one version of the story, the uncle walks down the hall to the bathroom, intending to drown his nephew in the bathtub, and he does. In a second version, the uncle walks down the hall, intending to drown his nephew, but finds him facedown in the water, already drowning.

The uncle closes the door and lets his nephew drown. Both versions of the story have the same unhappy ending: the nephew dies. The uncle has the same intention, but in the first version he directly fulfills it and in the second he does not. Would you be satisfied if a jury found the uncle guilty in story one, but not in story two? Somehow this judgment rings false, counter to our moral intuitions. The uncle seems equally responsible for his actions and omissions, and the negative consequences they yield. And if this intuition holds for the uncle, why not for any moral conflict where there is a distinction between an action with negative consequences and an omission of an action with the same negative consequences?

Now consider euthanasia, and the American Medical Association's policy: "The intentional termination of the life of one human being by another—mercy killing—is contrary to that for which the medical profession stands and is contrary to the policy of the American Medical Association. The cessation of the employment of extraordinary means to prolong the life of the body when there is irrefutable evidence that biological death is imminent is the decision of the patient and/or his immediate family." Stripped to its essence, a doctor is forbidden from ending a patient's life but is permitted to end life support. Actions are treated in one way, omissions in another. Does this clearly reasoned distinction, supported by most countries with such a policy, fit our moral intuitions? Speaking for my own intuition: No.

These two cases bring three issues to light: legal policies often ignore or cover up essential psychological distinctions, such as our inherent bias to treat actions one way and omissions another way; once the distinctions are clarified, they often conflict with our moral intuitions; and when policy and intuition conflict, policy is in trouble. One of the best-kept secrets of the medical community is that mercy killings in the United States and Europe have risen dramatically in the last ten years even though policies remained unchanged. Doctors are following their intuitions against policy and the threat of medical malpractice.[2] In cases where doctors adhere to policy, they tend to fall squarely within the AMA's act-omission bias. For example, in June of 2004, an Oregon doctor explicitly opposed to his state's tolerance for mercy killings through drug overdose stated: "I went into medicine to help people. I didn't go into medicine to give people a prescription for them to die." It is okay to help a patient by ending his life support, but it is not acceptable to help the patient by administering an overdose. The logic rings false. As the American response to the Terry Schiavo case revealed in 2005, many see termination of life support as an act, one that is morally wrong. And for many in the United States, moral wrongs are equated with religious wrongs, acts that violate the word of God. As Henry Wadsworth Longfellow noted, echoing a majority voice concerning the necessity of religion as a guiding light for morality, "Morality without religion is only a kind of dead reckoning—an endeavor to find our place on a cloudy sea by measuring the distance we have run, but without any observation of the heavenly bodies." I will argue that this marriage between morality and religion is not only forced but unnecessary, crying out for a divorce.

It is clear that in the arena of medicine, as in so many other areas where moral conflicts arise, the policy wonks and politicians should listen more closely to our intuitions and write policy that effectively takes into account the moral voice of our species. Taking into account our intuitions does not mean blind acceptance. It is not only possible but likely that some of the intuitions we have evolved are no longer applicable to current societal problems. But in developing policies that dictate what people ought to do, we are more likely to construct long-lasting and effective policies if we take into account the intuitive biases that guide our initial responses to the imposition of social norms.

There is an urgency to putting this material together—in Martin Luther King's words, "the fierce urgency of Now." The dominant moral-reasoning view has generated failed policies in law, politics, business, and education. I believe that a primary reason for this situation is our ignorance about the nature of our moral instincts and about the ways they work and interface with an ever-changing social landscape. It is time to remedy this situation. Fortunately, the pace of scientific advances in the sciences of morality is so rapid that by the time you read these words, I will already be working on a new prologue, showcasing the new state of play.

MORAL MINDS

I

WHAT'S WRONG?

You first parents of the human race . . . who ruined
yourself for an apple, what might you have done for a
truffled turkey?

—BRILLAT-SAVARIN[1]

HUNDREDS OF SELF-HELP BOOKS and call-in radio stations, together with the advice of such American ethic gurus as William Bennett and Randy Cohen, provide us with principled reasons and methods for leading a virtuous life. Law schools across the globe graduate thousands of scholars each year, trained to reason through cases of fraud, theft, violence, and injustice; the law books are filled with principles for how to judge human behavior, both moral and amoral. Most major universities include a mandatory course in moral reasoning, designed to teach students about the importance of dispassionate logic, moving from evidence to conclusion, checking assumptions and explicitly stating inferences and hypotheses. Medical and legal boards provide rational and highly reasoned policies in order to set guidelines for morally permissible, forbidden, and punishable actions. Businesses set up contracts to clarify the rules of equitable negotiation and exchange. Military leaders train soldiers to act with a cool head, thinking through alternative strategies, planning effective attacks, and squelching the emotions and instincts that may cause impulsive behavior when reasoning is required to do the right thing. Presidential committees are established to clarify ethical principles and the consequences of violations, both at home and abroad. All of these professionals share a common perspective: conscious moral reasoning from explicit principles is the cause of our moral judgments. As a classic text in moral philosophy concludes, "Morality is, first and foremost, a matter of consulting reason. The morally right thing to do, in any circumstance, is whatever there are the best reasons for doing."[2]

This dominant perspective falls prey to an illusion: Just because we can consciously reason from explicit principles—handed down from parents, teachers, lawyers, or religious leaders—to judgments of right and wrong doesn't mean that these principles are the source of our moral decisions. On the contrary, I argue that moral judgments are mediated by an unconscious process, a hidden moral grammar that evaluates the causes and consequences of our own and others' actions. This account shifts the burden of evidence from a philosophy of morality to a science of morality.

This book describes how our moral intuitions work and why they evolved. It also explains how we can anticipate what lies ahead for our species. I show that by looking at our moral psychology as an instinct—an evolved capacity of all human minds that unconsciously and automatically generates judgments of right and wrong—that we can better understand why some of our behaviors and decisions will always be construed as unfair, permissible, or punishable, and why some situations will tempt us to sin in the face of sensibility handed down from law, religion, and education. Our evolved moral instincts do not make moral judgments inevitable. Rather, they color our perceptions, constrain our moral options, and leave us dumbfounded because the guiding principles are inaccessible, tucked away in the mind's library of unconscious knowledge.

Although I largely focus on what people do in the context of moral conflict, and how and why they come to such decisions, it is important to understand the relationship between description and prescription—between what *is* and what *ought* to be.

In 1903, the philosopher George Edward Moore noted that the dominant philosophical perspective of the time—John Stuart Mill's utilitarianism—frequently fell into the *naturalistic fallacy*: attempting to justify a particular moral principle by appealing to what is *good*.[3] For Mill, utilitarianism was a reform policy, one designed to change how people ought to behave by having them focus on the overall good, defined in terms of natural properties of human nature such as our overall happiness. For Moore, the equation of good with natural was fallacious. There are natural things that are bad (polio, blindness) and unnatural things that are good (vaccines, reading glasses). We are not licensed to move from the natural to the good.

A more general extension of the naturalistic fallacy comes from deriving *ought* from *is*. Consider these facts: In most cultures, women put more time into child care than men (a sex difference that is consistent with our primate ancestors), men are more violent than women (also consistent with our primate past), and polygamy is more common than monogamy (consistent with the rest of the animal kingdom). From these facts, we are not licensed to conclude that women should do all of the parenting while men drink beers, society should sympathize with male violence because testosterone makes violence inevitable, and women should expect and support male promiscuity because it's in their genes, part of nature's plan. The descriptive principles we uncover about human nature do not necessarily have a causal relationship to the prescriptive principles. Drawing a causal connection is fallacious.

Moore's characterization of the naturalistic fallacy caused generations of philosophers to either ignore or ridicule discoveries in the biological sciences. Together with the work of the analytic philosopher Gottlieb Frege, it led to the pummeling of ethical naturalism, a perspective in philosophy that attempted to make sense of the good by an appeal to the natural. It also led to an intellectual isolation of those thinking seriously about moral principles and those attempting to uncover the signatures of human nature. Discussions of moral ideals were therefore severed from the facts of moral behavior and psychology.

The surgical separation of facts from ideals is, however, too extreme. Consider the following example:[4]

FACT: The only difference between a doctor giving a child anesthesia and not giving her anesthesia is that without it, the child will be in agony during surgery. The anesthesia will have no ill effects on this child, but will cause her to temporarily lose consciousness and sensitivity to pain. She will then awaken from the surgery with no ill consequences, and in better health thanks to the doctor's work.

EVALUATIVE JUDGMENT: Therefore, the doctor should give the
child anesthesia.

Here it seems reasonable for us to move from fact to value judgment.
This move has the feel of a mathematical proof, requiring little more
than an ability to understand the consequences of carrying out an action
as opposed to refraining from the action. In this case, it seems reasonable
to use *is* to derive *ought*.

Facts alone don't motivate us into action. But when we learn about a
fact and are motivated by its details, we often alight upon an evaluative decision that something should be done. What motivates us to conclude that
the doctor should give anesthesia is that the girl shouldn't experience pain,
if pain can be avoided. Our attitude toward pain, that we should avoid it
whenever we can, motivates us to convert the facts of this case to an evaluative judgment. This won't always be the right move. We need to understand what drives the motivations and attitudes we have.

The point of all this is simple enough: Sometimes the marriage between
fact and desire leads to a logical conclusion about what we ought to do, and
sometimes it doesn't.[5] We need to look at the facts of each case, case by
case. Nature won't define this relationship. Nature may, however, limit
what is morally possible, and suggest ways in which humans, and possibly
other animals, are motivated into action. When Katharine Hepburn turned
to Humphrey Bogart in the *African Queen* and said, "Nature, Mr. Allnut, is
what we are put in this world to rise above," she got one word wrong: We
must not rise above nature, but rise with nature, looking her in the eye and
watching our backs. The only way to develop stable prescriptive principles,
through either formal law or religion, is to understand how they will break
down in the face of biases that Mother Nature equipped us with.[6]

THE REAL WORLD

On MTV's *Real World*, you can watch twentysomethings struggle with
"real" moral dilemmas. On the fifteenth episode of the 2004 season, a girl
named Frankie kissed a guy named Adam. Later, during a conversation
with her boyfriend, Dave, Frankie tried to convince him that it was a

mistake, a meaningless kiss given after one too many drinks. She told Dave that he was the real deal, but Dave didn't bite. Frankie, conflicted and depressed, closed herself in a room and cut herself with a knife.

If this sounds melodramatic and more like *Ersatz World*, think again. Although fidelity is not the signature of this age group, the emotional prologue and epilogue to promiscuity is distressing for many, and for thousands of teenagers it leads to self-mutilation. Distress is one signature of the mind's recognition of a social dilemma, an arena of competing interests.

But what raises a dilemma to the level of a moral dilemma, and makes a judgment a morally weighty one?[7] What are the distinguishing features of moral as opposed to nonmoral social dilemmas? This is a bread-and-butter question for anyone interested in the architecture of the mind. In the same way that linguists ask about the defining features of speech, as distinct from other acoustic signals, we want to understand whether moral dilemmas have specific design features.

Frankie confronted a moral dilemma because she had made a commitment to Dave, thereby accepting an obligation to remain faithful. Kissing someone else is forbidden. There are no written laws stating which actions are obligatory or forbidden in a romantic but nonmarital relationship. Yet everyone recognizes that there are expected patterns of behavior and consequences associated with transgressions. If an authority figure told us that it was always okay to cheat on our primary lovers whenever we felt so inclined, we would sense unease, a feeling that we were doing something wrong. If a teacher told the children in her class that it was always okay to hit a neighbor to resolve conflict, most if not all the children would balk. Authority figures cannot mandate moral transgressions. This is not the case for other social norms or conventions, such as those associated with greetings or eating. If a restaurant owner announced that it was okay for all clients to eat with their hands, then they either would or not, depending on their mood and attachment to personal etiquette.

To capture the pull of a moral dilemma, we at least need conflict between different obligations. In the prologue, I described a classic case of moral conflict framed in terms of two incompatible beliefs—we all believe both that no one has the right to shorten our lives and that we should not cause or prolong someone's pain. But some people also believe that it is permissible to end someone's life if he or she is suffering from a terminal disease. We thus face the conflict between shortening and not shortening someone else's life. This conflict is more extreme today than it was in our evolutionary past. As hunter-gatherers, we depended upon our own health for survival, lacking access to the new drugs and life-support systems that can now extend our lives beyond nature's wildest expectations. Thus, when we contemplate ending someone's life today, we must also factor in the possibility that a new cure is just around the corner. This sets up a conflict between immediately reducing someone's suffering and delaying their suffering until the arrival of a permanent cure. What kind of duty do we have, and is duty the key source of conflict in a moral dilemma?

To see how duty might play a role in deciding between two conflicting options, let me run through a few classic cases. Suppose I argue the presumably uncontroversial point that the moral fabric of society depends upon individuals who keep their promises by repaying their debts. If I promise to repay my friend's financial loan, I should keep my promise and repay the loan. This seems reasonable, especially since the alternative—to break my promise—would dissolve the glue of cooperation.

Suppose I borrow a friend's rifle and promise to return it next hunting season. The day before I am supposed to return the rifle, I learn that my friend has been clinically diagnosed as prone to uncontrollable outbursts of violence. Although I promised to return the rifle, it would also seem that I have a new duty to keep it, thereby preventing my friend from harming himself or others. Two duties are in conflict: keeping a promise and protecting others. Stated in this way, some might argue that there is no conflict at all—the duty to protect others from potential harm trumps the duty to keep a promise and pay back one's debts. Simple cost-benefit analysis yields a solution: The benefit of saving other lives outweighs the personal cost of breaking a promise. The judgment no longer carries moral weight, although it does carry significance.

We can turn up the volume on the nature of moral conflict by drawing upon William Styron's dilemma in *Sophie's Choice*. Although fictional, this dilemma and others like it did arise during wartime. While she and her children are kept captive in a Nazi concentration camp, a guard approaches Sophie and offers her a choice: If she kills one of her two children, the other will live; if she refuses to choose, both children will die. By forcing her to accept the fact that it is worse to have two dead children than one, the guard forces her into making a choice between her children, a choice that no parent wants to make or should ever have to. Viewed in this way, some might say that Sophie has no choice: in the cold mathematical currency of living children, $1 > 0$. Without competing choices, there is no moral dilemma. This surgically sterile view of Sophie's predicament ignores several other questions: Would it be wrong for Sophie to reject the guard's offer and let both of her children die? Would Sophie be responsible for the deaths of her two children if she decided not to choose?

Because it is not possible to appeal to a straightforward and uncontroversial principle to answer these questions, we are left with a moral dilemma, a problem that puts competing duties into conflict. Sophie has responsibility as a mother to protect both of her children. Even if she was constantly battling with one child and never with the other, she would still face a dilemma; personality traits such as these do not provide the right kind of material for deciding another's life, even though they may well bias our emotions one way or the other. Imagine if the law allowed differences in personality to interfere with our judgments of justice and punishment. We might end up convicting a petty thief to life in prison on the basis of his sinister sneer, while letting another petty thief off from a sentence because of his alluring smile.

Sophie chooses to sacrifice her younger and smaller daughter to save her older and stronger son. She loses track of her son and, years later, ridden by guilt, commits suicide.

In the cases discussed thus far, we first appear to generate an automatic reaction to the dilemma, and then critically evaluate what we would do if we were in the protagonist's shoes. We empathize with Sophie's conflict, feel that a choice is necessary, and then realize that without a firm basis for choice, we might as well flip a coin. Emotion fuels the decision to choose, while the lack of an emotional preference for one option over the other triggers the coin flip. When pushed to explain our decisions, we are dumbfounded. Although we undoubtedly feel something, how can we be sure that feeling caused our judgment as opposed to following from it? And even if our emotions sneak in before we deliver a verdict, we don't have evidence that the two are causally related, as when I stick a pin in someone and induce pain.

Neither we nor any other feeling creature can just *have* an emotion. Something in the brain must recognize—quickly or slowly—that this is an emotion worthy situation. Once Sophie decides to choose, her choice triggers a feeling of guilt. Why? Guilt represents one form of response to a social transgression—a violation of societal norms. Did Sophie transgress? Was her decision to choose morally permissible or reprehensible? If Sophie had never felt guilty, would we think any less of her? My hunch is that Sophie's act was permissible, perhaps even obligatory, given the choice between two dead children or one. Why then a guilty response? Most likely, this emotional response—like all others—follows from an analysis, often unconscious, of the causes and consequences of an agent's actions: Who did what to whom, why, and with what means and ends? This analysis must precede the emotions. Once this system cranks through the problem, it may trigger an emotion as rapidly and automatically as when our eyelashes detect pressure and snap shut. Understanding this process presents a key to explaining why Sophie felt guilty even though she didn't do anything wrong. Being forced to act on a choice may trigger the same kind of angst as when a choice is made voluntarily. The kind of emotion experienced follows from an unconscious analysis of the causes and consequences of action. This analysis, I argue, is the province of our moral faculty.

Arguing against the causal force of emotions are those who think that we resolve moral dilemmas by consciously reasoning through a set of principles or rules. Emotions interfere with clear-headed thinking. An extreme version of this perspective is that there are no moral dilemmas, because for every apparent conflict involving two or more competing obligations, there is only one option. A sign outside a church in Yorkshire, England, reads: "If you have conflicting duties, one of them isn't your duty." If we had a completely accurate theory of morality, there would be a precise principle or rule for arbitrating between options. Morality would be like physics, a system we can describe with laws because it exhibits lawlike regularities. Like Einstein's famous equation for understanding the relationship between mass and energy—$E = mc^2$—we would have a parallel and equally beautiful equation or set of equations for the moral sphere. With such equations in mind, we would plug in the details of the situation, crunch the numbers, and output a clearly reasoned answer to the moral choices. Dilemmas, on this view, are illusory. The feeling of moral conflict comes from the fact that the person evaluating the situation isn't thinking clearly or rationally, seeing the options, the causes, and the consequences. The person is using his gut rather than his head. Our emotions don't provide the right kind of process for arbitrating between choices, even if they tilt us in one direction once we have made up our minds.

To push harder on the challenge we face in extracting the source of our moral judgments, consider one final set of cases.[8] It is a set designed to make us think about the difference between actions and omissions—a distinction that I alluded to in the prologue when discussing euthanasia. You are driving along a country road in your brand-new convertible, outfitted with pristine leather interior. You see a child on the side of the road, motionless and with a bloody leg. As you approach, she yells out that she needs immediate medical attention and a lift to the hospital. You waver. Her bloody leg will ruin your leather interior, which will cost you $200 to repair. But you soon realize that these are insufficient grounds for wavering. A person's life is certainly worth more than a car's leather interior. You pick the child up and carry her to the hospital, accepting the foreseen consequences of your decision: a $200 repair bill.

Now consider the companion problem. You receive a letter from UNICEF asking for a contribution for the dying children of a poor Saharan country in Africa. The cause of death seems easy to repair: more water. A contribution of $50 will save twenty-five lives by providing each child with a package of oral rehydration salts that will eliminate dehydrating diarrhea and allow them to survive. If this statistic on dehydrating diarrhea doesn't grab you, translate it into any number of other equally curable causes of death (malnutrition, measles, vitamin deficiency) that result in over 10 million child fatalities each year. Most people toss these aid organization letters in the trash bin. They do so even if the letter includes a picture of those in need. The picture triggers compassion in many, but appears insufficient to trigger check-signing.

For those who care about the principles underlying our moral judgments, what distinguishes these two cases and leads most people to think—perhaps unconsciously at first—that we must stop and help the child on the side of the road whereas helping foreign children dying of thirst is optional? If reason drives judgment, then those who read and think critically about this dilemma should provide a principled explanation for their judgment. When asked why they don't contribute, they should mention things like the uncertainty associated with sending money and guaranteeing its delivery to the dying children, the fact that they can only help a token number of needy children, and that contributions like this should be the responsibility of wealthy governments as distinct from individuals. All of these are reasonable ideas, but as principles for leading a virtuous life, they fail. Many aid organizations, especially UNICEF's child-care branch, have exceptionally clean records of delivering funds to their target source. Even though a contribution of $50 helps only twenty-five children, wouldn't saving twenty-five be better than saving none? And although our governments could do more to help, they don't, so why not contribute and help save a few children? When most people confront these counter-arguments, they typically acquiesce, in principle, and then find some alternative reason. Ultimately, they are dumbfounded, and stumble onto the exhausted conclusion that they just can't contribute right now—maybe next year.

An appeal to our evolutionary history helps resolve some of the tension between the injured child and the starving children, and suggests an explanation for our roller-coaster reasoning and incoherent justifications in these and many other cases. In our past, we were only presented with opportunities to help those in our immediate path: a hunter gored by a buffalo, a starving family member, an aging grandfather, or a woman with pregnancy complications. There were no opportunities for altruism at a distance. The psychology of altruism evolved to handle nearby opportunities, within arm's reach. Although there is no guarantee that we will help others in close proximity, the principles that guide our actions and omissions are more readily explained by proximity and probability. An injured child lying on the side of the road triggers an immediate emotion, and also triggers a psychology of action and consequence that has a high probability of success. We empathize with the child, and see that helping her will most likely relieve her pain and save her leg. Seeing a picture of several starving children triggers an emotion as well, but pictures do not evoke the same kind of emotional intensity as the real thing. And even with the emotions in play, the psychology that links action with consequence is ill prepared.

We should not conclude from the discussion thus far that our intuitions always provide luminary guidance for what is morally right or wrong. As the psychologist Jonathan Baron explains, intuition can lead to unfortunate or even detrimental outcomes.[9] For example, we are more likely to judge an action with negative consequences as forbidden whereas we judge the omission of an action with the same negative consequences as permissible. This omission bias causes us to favor the termination of life support over the active termination of a life, and to favor the omission of a vaccination trial even when it will save the lives of thousands of children although a few will die of a side effect. As Baron shows, these errors stem from intuitions that appear to blind us to the consequences of our actions. Once intuitions are elevated to rules, mind blindness turns to confabulation, as we engage in mental somersaults to justify our beliefs.

Bottom line: Reasoning and emotion play some role in our moral behavior, but neither can do complete justice to the process leading up to moral judgment. We haven't yet learned why we have particular emotions or specific principles for reasoning. We give reasons, but these are often insufficient. Even when they are sufficient, do our reasons cause our judgments or are they the consequences of unconscious psychological machinations? Which reasons should we trust, and convert to universal moral principles? We have emotional responses to most if not all moral dilemmas, but why these particular emotions and why should we listen to them? Can we ever guarantee that others will feel similarly about a given moral dilemma?

Scholars have debated these questions for centuries. The issues are complicated. My goal is to continue to explain them. We can work toward a resolution by considering a recent explosion of scientific facts, together with the idea that we are equipped with a moral faculty—an organ of the mind that carries a universal grammar of action.[10] For those brave enough to leap, let us join Milton "Into this wild abyss, the womb of Nature."[11]

ILL LOGIC

In *Leviathan*, published in 1642, Thomas Hobbes wrote that "Justice, and Injustice are none of the Faculties neither of the Body, nor Mind." Translating: We start from a blank slate, allowing experience to inscribe our moral concepts. Hobbes defends this position by setting up the rhetorical wild-child experiment, arguing that if biology had handed down our moral reasoning abilities—thoughtful, reflective, conscious, deliberate, principled, and detached from our emotions or passions—then "they might be in a man that were alone in the world, as well as his Senses, and Passions."[12] Hobbes's outlook on our species generates the seductive idea that all bad eggs can be scrambled up into good ones, while good ones are cultivated by the wisdom of our elders. It is only through reason that we can maintain a coherent system of justice. Our biology and psychology are mere receptacles for information and for subsequently thinking about this database by means of a rational, logical, and well-reasoned process. But how does reason decide what we ought to do?

When we reason about what ought to be done, it is true that society hands down principles or guidelines. But why should we accept them? How should we decide whether they are just or reasonable? For philosophers dating back at least as far as René Descartes, there was at least one uncontested answer: Get rid of the passions and allow the process of reason and rationality to emerge triumphant. And from this rational and deliberately reasoned position, there are at least two possible moves. On the one hand, we can look at specific, morally relevant examples involving harm, cooperation, and punishment. Based on the details of the particular example, we might deliver either a *utilitarian* judgment based on whether the outcome maximizes the greatest good or a *deontological* judgment based on the idea that every morally relevant action is either right or wrong, independent of its consequences. The utilitarian view focuses on consequences, while the deontological perspective focuses on rules, sometimes allowing for an exception clause and sometimes not. On the other hand, we might attempt to carve out a general set of guiding principles for considering our moral duties, independent of specific examples or content. This is the path that the philosopher Immanuel Kant pursued, argued most forcefully in his categorical imperative.[13] Kant stated: "I ought never to act except in such a way that I could also will that my maxim should become a universal law." For Kant, moral reasons are powerful prods for proper action. Because they are unbounded by particular circumstances or content, they have universal validity. Said differently, only a universal law can provide a rational person with a sufficient reason to act in good faith.

If this is how the moral machinery works, then one of its essential design features is a program that enables it to rule out immoral actions. The program is written as an imperative, framed as a rule or command line: Save the baby! Help the old lady! Punish the thief! It is a categorical imperative in that it applies without exception. It is consistent with Kant's categorical imperative to support the Golden Rule: "Do unto others as you would have them do unto you." It is inconsistent with Kant's imperative to modify the Golden Rule with a self-serving caveat: "Do unto others as you would have them do unto you, but only if there are large personal gains or small personal costs."

Kant goes further in his universal approach, adding another imperative, another line of code: "Act in such a way that you always treat humanity, whether in your own person or in the person of any other, never simply as a means, but always at the same time as an end." It is categorical in describing a nonnegotiable condition, and it is imperative in specifying the condition: Never use people *merely as means* as opposed to ends—individuals with desires, goals, and hopes.

One way to think of Kant's categorical imperative is as a how-to manual:[14]

1. A person states a principle that captures his reason for action.
2. He then restates this principle as a universal law that he believes applies to all other similarly disposed (rational) creatures.
3. He then considers the feasibility of this universal law given what he knows about the world and its assemblage of other rational creatures. If he thinks the law has a chance of working, then he moves on to step 4.
4. He then answers this question: Should I or could I act on this principle in this world?

5. If his answer to step 4 is "yes," then the action is morally permissible.

One of Kant's central examples is of an unfaithful promise. Fred is poor and starving, and asks his friend Bill for a short-term loan so that he can buy food. Fred promises that he will pay Bill back, but actually has no intention of doing so. The promise is empty. Run the Kantian method. Step 1: Fred believes that because he is hungry and poor he is justified in asking for a loan and promising repayment. Further, Bill has the money, the loan will make a negligible dent in his finances, and their friendship will remain intact even if the loan is never repaid. The principle might go something like this: It is morally permissible for Fred to renege on his promise to Bill, since the benefits to Fred outweigh the costs to Bill. Step 2: Restate the case-specific principle as a universal law: It is morally justified for any rational creature to renege on a promise as long as the benefits to self outweigh the costs to others. Step 3: How feasible is the universal law? Is it possible to imagine a world in which promises are largely empty or, at least, largely unpredictable in terms of their truth or falsity? The answer seems clear: No! No step four or five. It is morally impermissible to offer unfaithful promises. As I pointed out in the last section, however, this doesn't mean that we are always obliged to keep our promises. It could be permissible to break a promise if the positive consequences of reneging on it outweigh the negative consequences of keeping it.

Kant's imperative shows that in a rational universe, the only path to a fair and universal set of principles is to guarantee that these principles apply to everyone, with no exceptions. The categorical imperative therefore blocks theoretically possible worlds in which stealing, lying, and killing are part of universal law. The reason is straightforward: The consequences of these laws harm even selfish individuals. Individuals may lose their own property, friends, or life.

Let us call individuals who deliver moral judgments based on conscious reasoning from relevant principles "Kantian creatures," illustrated below and throughout the book by the little character scratching his brain. Although I am using Kant as exemplary of this approach, let me note that even Kant acknowledged the role of our commonsense notions of right and wrong, and especially the role of our emotions in driving behavior. But for

Kant, and many others following in his footsteps, our emotions get in the way. We arrive at our ultimate moral judgments by conscious reasoning, a process that entails deliberate reflection on principles or rules that make some things morally right and others morally wrong.

To see how well the Kantian creature manages in a social world, consider the following example. What if I offered the principle that everyone must tell the truth all the time, because it leads to stable and more efficient relationships? Run this through the five-point method. It appears that the principle works, but perhaps too well. When my father was a young boy in German-occupied France, a kind young girl warned him that the Nazis were coming to the village, and said that if he was Jewish he could hide at her house. Although reluctant to announce that he was Jewish, he trusted the girl and went to her house. When the Nazis arrived and asked if they were hiding any Jews, the girl and her parents said "No," and, luckily, escaped further scrutiny. Both the girl and her parents lied. If they had been true Kantian creatures, it should have been obligatory for them to announce my father's whereabouts. I, for one, am delighted that Kantians can sometimes jettison their code.

Kantians run into a similar roadblock when it comes to harming another individual. They may want to hold everyone to the categorical imperative that killing is wrong because they can't will that individuals with good personal reasons can kill someone else. It also seems inappropriate for them to recommend killing as a morally permissible solution to saving the lives of several other individuals. Here, though the utilitarian calculus may push us to act, the deontological calculus should not: Killing is wrong, unconditionally.

Debates over substantive theories of moral judgment, as well as Kant's categorical imperatives, continue into the twenty-first century. This rich and interesting history need not concern us here. Of greater relevance is the connection between moral philosophy and moral psychology, and those who have followed in the conscious reasoning tradition championed by Kant.

Moral psychology—especially its development—has been dominated in the twentieth and twenty-first centuries by the thinking of Jean Piaget and Lawrence Kohlberg.[15] Both held the view that moral judgments are handed down from society, refined as a function of experience (rewards and punishments), and based on the ability to reason through the terrain of moral dilemmas, concluding with a judgment that is based on clearly defined principles. Kohlberg stated the position: ". . . moral principles are active reconstructions of experience."[16] Psychology therefore followed the philosophy of Plato and Kant, with conscious reasoning leading the charge toward moral judgment. The goal of child development was to matriculate to a perfectly rational creature, graduating with a degree in practical reasoning and logical inference.[17] In many ways, Kohlberg out-Kanted Kant in his view that our moral psychology is a rational and highly reasoned psychology based on clearly articulated principles.

Piaget and Kohlberg focused on problems of justice, defined the characteristics of a morally mature person, and attempted to explain how experience guides a child from moral immaturity to maturity. Piaget formulated three stages of moral development, whereas Kohlberg described six; the numerical difference is due to Kohlberg's attempt to distinguish each stage on the basis of more refined abilities. How does the child acquire these skills? Who, if anyone, gives them a tutorial on the distinction between right and wrong, enabling each child to navigate through the complex maze of actions that are morally relevant or irrelevant? In raising these questions, I am not doubting that some aspects of the child's moral psychology change. I am also not denying that children acquire an increasingly sophisticated style of conscious reasoning. The interesting issues are, however, what changes, when, how, and why?

Consider, for example, a story in which a girl named Sofia promises her father that she will never cross the big, busy street alone. One day, Sofia sees a puppy in the middle of the street, scared and not moving. Will Sofia save the puppy or keep her promise? Children under the age of six typically go with saving the puppy; when asked how the father will feel, they say "happy," justifying their response by stating that fathers like puppies, too. If these children learn that Sofia's father will punish her for running out into the street—breaking a promise—they explain that saving the puppy isn't an option. When asked how Sofia will feel about leaving the puppy, they answer "happy"; they think that because adherence to authority is good, that they, too, will feel good having listened to their father. Answers to these kinds of questions change over time. Children move away from answers that focus on smaller points, as well as the here and now, opening the door to more nuanced views about causes and consequences, and the difference in attitude that one must adopt in thinking about self and other. But what causes this change? Is it a fluid, choreographed walk from one stage to the next? Does everyone, universally, step through the same stages, starting at stage 1 anchored in the voice of authority and ending in stage 6, an ideal in which individuals have acquired principles for rationally choosing among options? How does our environment create a Kantian creature, a person who arbitrates between right and wrong by gaining conscious access to the relevant moral principles?

Assume, as did Piaget and Kohlberg, that children move through different stages of moral development by means of a growing capacity to integrate what their parents say. As Freud suggested, one can imagine that children map "good" or "permissible" onto what parents tell them to do, and map "bad" or "forbidden" onto what parents tell them not to do. Good things are rewarded and bad things are punished. In the same way that the animal-learning psychologist Burrhus Fred Skinner showed you can train a rat to press a lever for a reward or avoid pressing the lever if punished by a shock, parents can similarly train their children. Each stage of moral development puts in place different principles of action. Each stage is a prerequisite for advancing to the next stage. Early stages reveal the limits of the child's capacity to recognize the distinction between authority and morality, causes and consequences, and the importance of duties and responsibilities.

This theory of moral development plows right into a series of road-blocks.[18] Roadblock one: Why and how should authority matter? There is no question that rewards for appropriate actions and punishments for inappropriate actions can push a child to behave in different ways. But what makes a particular action morally relevant? Parents deliver numerous commands to their children, many of which have no moral weight. On the reward side, we have: do your homework, eat your broccoli, and take a bath. On the punishment side, we have: don't play with your food, run into traffic, or take medicine from the cabinet. The rewarded actions are certainly good from the perspective of pleasing one's parents and benefiting the child's self-growth. Similarly, the punished actions are bad from the perspective of triggering negative emotions in one's parents and harming self-growth. But what allows the child to distinguish this sense of good and bad from the sense of good or bad that comes from helping or hurting another person? Appealing to authority doesn't provide an answer. It pushes the problem back one step, and raises another question: What makes a parent's verdict morally good or bad?

A second and related roadblock concerns the mapping between experience and the linguistic labels of "good" and "bad," or the equivalent in other languages. The rich developmental literature suggests that some concepts are relatively easy for children to understand, because they are anchored in perceptual experiences. Others are more difficult, because they are abstract. Take, as an example, the words "sweet" and "sour." Although we might not be able to come up with satisfactory definitions, when we think about these labels, they tend to have a relatively direct relationship to what we have tasted or smelled. Sweet things trigger feelings of satisfaction, while sour things generally trigger aversion or feelings of withdrawal. The words "good" and "bad" lack this relationship to perception and sensation. Saying that "good" and "bad" provide convenient labels for what we like and dislike doesn't explain a thing. We must once again ask: Why do certain actions trigger feelings of like and dislike? The linguistic labels of "good" and "bad," together with their emotional correlates of positive and negative feelings, emerge after the mind computes the permissibility of an action based on its causes and consequences.

A third roadblock concerns stages of development. What criteria shall we use to place a child into one of the designated stages? Are the psychological achievements of a given stage necessary for advancement to subsequent stages, and is there a moral superiority to the more advanced stages? Consider Kohlberg's stage 1, a period that includes children as old as ten. At this stage, individuals see particular actions as inexorably good or bad, fixed and unchanging, and defined by parental authority. If parental authority provides the trump card, all the child gets from this interaction is a label. Eating mashed potatoes with your fingers is bad, and so is picking your nose and shoving things up it, hitting the teacher, kicking the dog, and peeing in your pants. This is a smorgasbord of cultural conventions, matters of physical safety, parental aesthetics, and morally prohibited actions. Saying that a child follows parental authority tells us little about her moral psychology. Children daily hear dozens of commands spoken with the voice of parental authority. How is the child to decide between social conventions that can be broken (you can eat asparagus with your hands, but not your mashed potatoes) and moral conventions that can't (you can't stab your brother with a knife no matter how much he annoys you)?

Matters worsen for the child moving up the moral ladder. Kohlberg recognized that the final stage was an ideal, but suggested that individuals pass through the other stages, confronting new arenas of conflict and engaging in a game of "moral musical chairs"—taking another's perspective. Kohlberg was right in thinking that conflict fuels the moral machinery. Shall I . . . keep the pie or share it? tell my mother that I skipped school or keep it a secret? save the puppy or keep my promise not to run out into the street? He was also right in thinking that perspective-taking plays a role in moral decisions. He was wrong, however, in thinking that every child resolves conflict in precisely the same way and by taking another's perspective. Kohlberg held this view because of his belief in the universality of the child's moral development: Each child marches through the stages by applying the same set of principles and achieving the same solution. Although empathy and perspective-taking are certainly important ingredients in building a moral agent—as I discuss in the next section—it is impossible to see how they might function as the ultimate arbiter in a conflict. I feel strongly that abortion is a woman's right. I meet people who disagree. I imagine what it must be like to hold their view. They do the same. Although we now have a better understanding of each other's position, it doesn't generate a unique solution. Compassions bias us, but they never judge for us.

Kohlberg's final stage is achieved by individuals who consciously and rationally think about universal rules, accepting many of Kant's principles, including his first: Never treat people as mere means to an end but always as an end in themselves. Kohlberg assessed an individual's moral development from a forty-minute interview that involved asking subjects to judge several moral dilemmas and then justify their answers. But there is an interpretive art to this kind of work. Consider: If I hire a cook and use him as the means to making my dinner, am I immoral? No, I have employed him with this goal in mind. He is the means to my ends, but the act doesn't enter the moral arena, because my actions are not disrespectful of his independence or autonomy. He accepted the job knowing the conditions of employment. There is nothing immoral about my request. Now, suppose that I asked my cook to bake a pork roast, knowing full well that he is a Hasidic Jew and this violates his religious beliefs. This is an immoral request, because it takes advantage of an asymmetry of power to use another as a mere means to an end.

Acceptance of Kant's principles as criteria for moral advancement immediately raises a problem. Although Kohlberg may support these principles, and use Kant's prowess in the intellectual landscape to justify his perspective, other philosophers of equal stature—Aristotle, Hume, and Nietzsche, to name a few—have firmly disagreed with Kant. This leaves two possible interpretations: either the various principles are controversial, or some of the greatest thinkers of our time never reached Kohlberg's final stage of moral development.

A final problem with the Piaget-Kohlberg framework is that it leaps from correlation to causation. We can all agree that we have had the experience of working through the logic of a moral dilemma, of thinking whether we should vote for or against stem-cell research, abortion, gay marriage, and the death penalty. There is no question that conscious reasoning is part and parcel of our capacity to deliver a moral verdict. What is at stake, however, is whether reasoning precedes or follows from our moral judgments. For example, a number of studies show that people who are against abortion hold the view that life starts at conception, and thus abortion is a form of murder; since murder or intentional harm is bad—morally forbidden—so, too, is abortion. For others, life starts at birth, and thus abortion is not a form of murder; it is morally permissible. Toward the end of 2004, a jury voted that Scott Peterson was guilty of two crimes, killing his wife and killing their unborn child: Laci Peterson was entering her eighth month of pregnancy. This appears to be a classic case of moving from a consciously explicated principle—abortion is the murder of a person—to a carefully reasoned judgment—murdering a person is forbidden. Though we end with a principle, and appear to use it in the service of generating a moral judgment, is this the only process? Here's an alternative: we unconsciously respond to the image of ending a baby's life with a negative feeling, which triggers a judgment that abortion is wrong, which triggers a post-hoc rationalization that ending a life is bad and, thus, a justification for the belief that life starts at conception. Here, too, we end with a principled reason, but reason flows from an initial emotional response that is more directly responsible for our judgment. Thus, even when children reach the most sophisticated stage of moral reasoning, none of Kohlberg's observations—or those of his students—settle the issue: reasoning first or later?

Piaget and Kohlberg deserve credit for recognizing the importance of studying the psychology of moral development, and for noting significant changes in the child's ability to reason through moral dilemmas. Like Kant's approach to moral judgment, it is clear that we can and often do engage in conscious moral reasoning based on expressible principles. It is also clear that this kind of reasoning does, in some circumstances, determine our moral judgments.

Acknowledging that we do engage in conscious, rational forms of reasoning is different from accepting that this is the one and only form of mental operation underlying our moral judgments. It is in this sense that Piaget and Kohlberg's assessment of the child's path was flawed, both conceptually and methodologically. Although the child's grasp of moral dilemmas may well change, assigning each individual to a particular stage is an art, and fails to explain how each stage is accomplished. Because the method of choice involved presentation of moral dilemmas followed by judgments and justifications, it is not possible to capture the child's full moral competence. As I explain below, even young children—well below the ages that would enter into Piaget and Kohlberg's moral stages—recognize the distinction between intentional and accidental actions, social and moral conventions, and intended and foreseen consequences. Many of their judgments are made rapidly, involuntarily, and without recourse to well-defined principles. And, importantly, adults make some of the same judgments, and are equally clueless about the underlying principles. Kantian creatures do not uniquely define our species' psychological signature.

PASSION'S WAY

Please rate each of the following scenarios on a ten-point disgust scale, where ten is "exceedingly disgusting."

1. You come home from work, and your daughter rushes up to you: "Dad! Just a few minutes ago I walked into the bathroom and found Mom crouched over the toilet bowl, licking ice cream off of the toilet seat. She looked up and said 'It's delicious,' and then asked if I wanted some."

The mother's licking ice cream from the toilet seat was:

   [*1—Not disgusting*] 1 2 3 4 5 6 7 8 9 10 [*10—Exceedingly disgusting*]

2. A brother and sister are on vacation together and decide that to enrich their wonderful relationship they should make love. Since he has been vasectomized and she is on the Pill, there is no risk of pregnancy. They make passionate love and it is a wonderful experience for both. They keep this as their secret, something that they will always remember and cherish.

The brother and sister's lovemaking was:

   [*1—Not disgusting*] 1 2 3 4 5 6 7 8 9 10 [*10—Exceedingly disgusting*]

My own emotional barometer reads about a 6 for the first scenario and an 8 for the second. When I first heard about these cases,[19] I had an immediate, unconscious response to them. Both are disgusting. But the two cases seem different. I can imagine being convinced, albeit with some coercing, of the appropriateness of licking ice cream (or any other food) from a toilet seat: it is a brand-new toilet seat and has been sterilized clean; it is the only food in the house and I haven't eaten a thing in two days; it was a dare from an old friend. I have a harder time imagining an argument in favor of incest. I can't imagine ever making love with my sister if I had one. Incest seems morally wrong, whereas toilet seat–licking seems simply gross! If someone worked out a newfangled, self-cleaning toilet seat, and the Board of Health told the public that it was now okay to eat off these seats, it probably wouldn't take too long before people converted. On the other hand, if our medical health boards decided that contraception was foolproof, thereby eliminating the conception risks of intercourse among siblings, most of us would presumably still balk at the thought, contraceptives or not. Is disgust the kind of promiscuous emotion that moves in and out of the moral sphere, *sans* passport? If so, what triggers a moral judgment in one case but not the other? What makes the act of intercourse among siblings disgusting and then morally wrong? Or is it morally wrong and therefore disgusting? There are now scientific facts that begin to answer these questions, but the ideas leading up to them are ancient.

If Kant deserves credit as the leading light of dispassionate moral reasoning, then the seventeenth-century philosopher David Hume deserves credit as the first secular modern philosopher, the primary architect of the idea that we alight on moral judgments by calling upon our emotions, and the first person to point out that moral behavior has "utility," designed for the greatest good. His entry into the nature of our moral judgments was to see them in the same light as we see our sensations and perceptions of the world. In the same way that we automatically and unconsciously see red, hear music, smell perfume, and feel roughness, we perceive helping as right because it feels good and cheating as wrong because it feels bad.

Hume's perspective directly challenged those who held that pure reason provides the only means to a virtuous life, a necessary antidote to our selfish core. Unlike these other theories, Hume saw his work in moral philosophy as central to a science of human nature. Unfortunately, at the time, Hume's major works on morality sold horribly, and the science never took off. His writings are now mandatory reading at most universities, and Hume's ideas have enjoyed a rebirth in the wake of new developments in the mind sciences.[20]

Hume's theory gets off the ground by looking at moral judgments through the lens of a three-party interaction: agent, receiver, and spectator. The idea is to understand the kinds of virtues and vices that motivate agents to act in particular ways, the manner in which an agent's actions directly influence the receiver's feelings, and how a spectator feels toward the agent and receiver upon observing their interaction. If an agent gives to a charity, this benevolent act maps onto a personality trait—a signature of virtue. Hume thought that some personality traits, such as benevolence, generosity, and charity, were innate, while others, such as justice, allegiance, and chastity, were acquired through the pedagogical guidance of culture. Although there wasn't a shred of scientific evidence to support these developmental distinctions, the primary force of his argument was that traits such as benevolence and justice were powerful motivators of action. Action then triggered reaction in both receiver and spectator. Giving to a charity makes receivers feel good and makes spectators sympathetically experience a good feeling by observing the receiver. When the spectator feels good, this triggers a judgment: moral approval of the agent's original act. Since the spectator vicariously feels good, he then converts this feeling into an appraisal of the agent's act, assigning it to the category of virtue, as opposed to vice. Approval is therefore like an aesthetic judgment of beauty, as opposed to a deduction or inference in mathematics. Judging the beauty of an orchid is automatic, carried out effortlessly and without thinking about reasons. Drawing an inference from a postulate or a string of numbers in mathematics is slow, deliberate, and thought out in the company of axioms and operations. Morally approving of a charity donation feels more like aesthetically acknowledging an orchid's beauty.

The logic of the three-party interaction led Hume to famously conclude that "Reason is and ought only to be the slave of the passions, and can never pretend to any other office than to serve and obey them." Thus was born the "Humean creature," equipped with an innate moral sense that provides the engine for reasoned judgments without conscious reasoning. Emotions ignite moral judgments. Reason follows in the wake of this dynamic. Reason allows us to think about the relationship between our means and our ends, but it can never motivate our choices or preferences. Our moral sense hands us emotional responses that motivate action, enabling judgments of right or wrong, permissible or forbidden. A Humean creature's moral sense is like its other instincts, part of nature's gift: "[The moral sentiments] are so rooted in our constitution and temper, that without entirely confounding the human mind by disease and madness, 'tis impossible to extirpate and destroy them . . . Nature must

furnish the materials, and give us some notion of moral distinctions."[21] Our moral sense is an inevitable outcome of normal growth, no different from the growth of an arm or eye.

Humean creatures are undoubtedly real. We experience countless moral dilemmas in our lifetime and frequently have the feeling that we resolve them quickly, unconsciously, and without any apparent reflection upon explicit laws or religious doctrine. Scientists have only recently recognized the omnipresence of Humean creatures. They reveal a fundamental illusion in our psychology: Conscious moral reasoning often plays no role in our moral judgments, and in many cases reflects a post-hoc justification or rationalization of previously held biases or beliefs.

If, as Hume suggests, we are equipped with a moral sense, then—like the sensory systems of seeing, hearing, tasting, touching, and smelling—it, too, should be designed with specialized receptors. For Hume, the moral sense was equipped with an evaluative mechanism, reading virtue or vice off of action, with sympathy providing a central motivating force. Any sensory organ feeding into the emotions functioned as a receptor for our moral sense. Thus, unlike the other senses with their dedicated input channels of sound, smell, touch, or sight, the moral sense is a free agent. I can see the man hit the woman, hear her cry upon contact, and feel her sorrow when she sobs on my shoulder. I can even imagine what it would be like to witness this man hitting this woman, running a video in my mind's eye. Events—the unfolding of a crime—and actions—a stabbing—directly trigger our emotions, moving us to approach or avoid, feel guilt or shame, hatred or sorrow. Sympathy promotes altruism and guards against violence by taking the receiver's perspective into account. Our emotions provide the code for what is right or wrong.

For Hume, our perception of a morally relevant action or event is like our perception of an object. When we see an object with a color, say red, there is an objective fact of its color: It is red. We believe it is red, and our belief is true. This color has to do with the light absorbed and radiated. Under different lighting conditions, we will perceive a different color, and once again, there will be a fact of the matter about its color and our understanding of it. Hume thought that our actions were similarly coded. In terms of permissible and forbidden actions, there is also a fact of the matter, a moral truth. This truth can change under different conditions, as can our perception of an object's color. Humean creatures may feel deep down that killing is wrong. Seeing an act of killing makes them feel anger, perhaps even hatred toward the killer. They may also feel guilty if they don't report the killer to the authorities, because killing is wrong. Seeing someone kill in the context of self-defense triggers a different set of emotions. Humean creatures are often in awe of those who have the courage to defend themselves and others against a killer, and are unlikely to feel guilt if they fail to report the individual acting in self-defense. There is, therefore, a moral fact of the matter about an action and its emotional surrogate.

Hume drew a distinction between an action's triggering effects and its potential to carry some objective moral value. Returning again to visual perception and our aesthetic judgments, an object is not in and of itself beautiful. We make judgments about an object's appearances, judging some as beautiful under certain conditions. Many of us are automatically awed by the beauty of the pyramids in Egypt but disgusted by their re-creation in the world's gambling capital, Las Vegas. As Hume stated, "Beauty is not a quality of the object, but a certain feeling of the spectator, so virtue and vice are not qualities in the persons to whom language ascribes them [i.e., agents], but feelings of the spectator." Neither a person nor an action is, in Hume's view, morally good or bad. "When you pronounce any action or character to be vicious, you mean nothing, but that from the constitution of your nature you have a feeling or sentiment of blame from the contemplation of it." When we pronounce that Eric Harris and Dylan Klebold were vicious because of their heinous attacks on the students and teachers of Columbine High School, we reveal a sentiment of blame. Now that we better understand the psychology underlying their attacks, with Klebold diagnosed as clinically depressed and Harris as psychopathic, the account triggers a different emotion in some of us. Although those who suffered such horrific losses during the shooting may not feel this way, many have now responded with sadness to both the losses at Columbine and to the troubled mental states of these two teenage murderers. Their actions are not morally excusable. We don't want a legal system that permits such actions. But we also don't want a legal system that mechanically assigns actions such as killing to specific moral categories, proclaims some as right and others as wrong, and ends the discussion as if these were moral absolutes. Like viewing an object under different lighting conditions, our moral sense triggers different emotions depending upon the conditions in which we perceive an action. Saying that our moral sense triggers different emotions doesn't, however, provide an explanation of how it does this. That we have different emotions is clear. What is unclear is the process that must come first, enabling nuanced emotional responses to different situations.

Kantian creatures challenge Humean creatures on two further points. First, Kantians think that you need good reasons for making a particular judgment. When a Humean creature is asked for a justification, all she can do is shrug and say "it feels right." As the philosopher James Rachels puts it, it's like asking someone to account for their aesthetic preferences:

> If someone says "I like coffee," he does not need to have a reason—he is merely stating a fact about himself, and nothing more. There is no such thing as "rationally defending" one's like or dislike of coffee, and so there is no arguing about it. So long as he is accurately reporting his tastes, what he says must be true . . . On the other hand, if someone says that something is morally wrong, he does need reasons, and if his reasons are sound, other people must acknowledge their force. By the same logic, if he has no good reasons for what he says, he is just making noise and we need pay him no attention.[22]

The Kantian's second point delivers a more devastating blow: If we decide what is right or wrong by recruiting our emotions, then how are we to achieve impartiality, an objective, universal sense of what is permissible or forbidden? Consider lying. While walking, I notice a man who has just dropped his wallet. I pick it up before he turns around. Inside, I see $300. I think about the leather coat that I want to buy and how this will cover the cost. The thought of wearing that coat makes me feel good. I have a sense, however, that it would be wrong to keep the wallet and money. On the other hand, I feel that anyone who can carry around this amount of cash must be doing quite well and would barely notice the missing money. When the man turns around and asks me if I saw a wallet, I immediately respond "no," and walk on. I justify my response by saying that it felt okay to lie, because the man wouldn't feel bad about missing his money. Emotions drive this story, but they are partial, biased toward selfish behavior. The emotionally driven intuition, if that is what it is, fails because it can't be impartial. These arguments do not negate the fact that emotions play a role in our moral actions. Rather, they point to a weakness in using emotions to develop general principles of moral action.

Hume's thoughts about moral behavior had little effect on the sciences until the developmental psychologist Martin Hoffman picked them up in the twentieth century. This gap between philosophy and science directly parallels the relationship between Kant and Kohlberg. In fact, Kant's cold, rational, and calculated morality is to Kohlberg's reasoning-based scheme for moral development as Hume's warm, intuitive, and emotional morality is to Hoffman's empathy-based scheme for moral development. Like Hume, Hoffman placed empathy at the center of his theory of morality and as the essence of Humean creatures. Empathy is "the spark of human concern for others, the glue that makes social life possible."[23] Unlike Kohlberg, Hoffman placed less emphasis on rigid stages of moral development. Developing into a Humean creature entails moving through early and late periods of emotional maturity. The early or primitive periods of moral development are rooted in the child's biology, with glints from her primate past, a combination of selfish and compassionate motives. Most of the child's earliest forms of empathy are automatic and unconscious, often triggered by her own uncontrollable imitative abilities. A newborn baby, barely able to see, can imitate the facial expressions of adults within one hour of delivery. By imitating facial expressions and other body gestures, the motor system feeds the emotional system. Thus, when the child unconsciously mimics another's facial expression of sadness or delight, she automatically creates a coupling between her expressions and her emotions. A consequence of this move is that when young children see others experiencing a particular emotion, they will simultaneously feel something similar. Empathy moves as a form of contagion, like a game of emotional tag. It is a form that never fully disappears, as evidenced by what the social psychologist John Bargh calls the "chameleon effect."[24]

When we interact with someone who gesticulates a lot during conversation, touches her face frequently, or speaks in a particularly distinctive dialect, we are more likely to gesture in the same way and speak with the same dialect, even if this is not our typical conversational signature. When confronted with information about someone's age or race, we unconsciously activate stereotypes or prejudices that subsequently influence, unconsciously as well, our behavior. When we see an elderly person, either in the flesh or in a photo, we are more likely to move slowly following the encounter. When we encounter an image of an ethnic group for which we carry unconscious negative or hostile attitudes, we are more likely to engage in aggressive behavior when provoked than when we encounter an image of our own ethnic group or a group that triggers positive attitudes. We are like chameleons, designed to try out different colors to match our social partner's substrate.

Hoffman describes this early period of empathy as foundational but simple, "based on the pull of surface cues and requiring the shallowest level of cognitive processing."[25] Here, then, is a distinction between a purely emotional Humean creature and one with some awareness of what she is doing and why. It is only in the more mature period of development that empathy couples with reflection and awareness. A mature Humean creature not only recognizes when someone is blissed out or in pain, but why she should feel moved to help, extend a kind hand, or join in a joyous moment. With maturity comes the capacity to take on another's perspective. Although the capacity to assume another's perspective, either in terms of their emotions or their beliefs, is not restricted to the moral domain, it plays a significant role. But the growth of the child's mind, like the growth of her lungs and heart, is on a maturational time course. Along the path of growth we witness the emergence of abilities that are necessary for moral judgments, but not specific to it. From a purely emotional form of empathy that is reflexive the child grows into a form of empathy that takes into account what others know, believe, and desire. We know that these abilities are necessary, because of developmental disorders such as autism in which the inability to take another's perspective has devastating consequences for understanding the moral sphere.

There is no doubt that Hoffman is right: empathy does play a role in our moral actions. It motivates all Humean creatures. Over the last twenty years, the mind sciences have provided a rich understanding of empathy, including its evolution, development, neural underpinnings, and breakdown in psychopathy. Although I will have much more to say about these findings, and the more general role of the emotions in shaping our moral behavior, I want to raise a cautionary flag. Our emotions can't explain how we judge what is right or wrong, and, in particular, can't explain how the child navigates the path between social norms in general and moral norms in particular. A child's experiences are insufficient to create the dividing line between generic social or conventional transgressions and specifically moral transgressions. A little girl hits a little boy because he won't let her play in the sandbox. The girl's father gets mad, tells her that hitting is wrong, and asks her to apologize to the little boy. From this experience, the little girl reads an emotion off of her father's face—anger—maps this onto the previous event, and concludes that hitting is bad. A few minutes later, the same girl picks up sand and puts it in her mouth. The father is again angry, slaps her hand, washes her mouth, and then tells her that she should never put sand in her mouth. Same emotion, different conclusion. Hitting has moral weight. Eating sand does not. How do the child's emotions send one action to the moral sense and the other to common sense? And how does the child come to understand that she can't hit another little boy but her father can sometimes hit her, slapping sand out of her hand? Emotions may guide what she does, but can't educate her on the difference between social and moral rules, and why certain actions are morally wrong in one context (hitting a boy out of frustration) but not in another (hitting a daughter out of concern for her safety).

What we need, therefore, is an understanding of the evaluative process that triggers emotion. And we need this knowledge for two reasons: whatever this system is, it is the first step in our moral analysis, and it may represent the locus of our moral judgments. Put differently, the part of our mind that evaluates intentions, actions, and consequences might be the center of moral deliberation, the piece of our psychology that delivers the initial verdict about permissible, obligatory, or forbidden behavior. If this view is right, then our emotions, including Hoffman's empathy, are downstream, pieces of psychology triggered by an unconscious moral judgment. Emotions play a role, as Hume and Hoffman argued. But rather than playing a role in generating a moral judgment, our emotions may function like weights, moving us to lean in one direction rather than another. When our emotions are too charged, or not working due to brain injury, our competence to judge moral situations may remain intact even if our capacity to do the right thing fails. Serial killers, pedophiles, rapists, thieves, and other heinous criminals may recognize the difference between right and wrong, but lack the emotional input to follow through on their intuitive deliberations.

LET'S TAKE STOCK. Moral dilemmas present us with conflict, typically between two or more competing duties or obligations. Confronted with such dilemmas, we deliver a judgment, a verdict of morally good or bad with respect to a person's character or the act itself. These judgments are always made, consciously or unconsciously, in reference to a set of culturally specific virtues and vices. Some judgments do arise like flashes of lightning—spontaneous, unpredictable, and powerful. Incest is bad, and so is torturing a child for fun. Helping an old man across the street is good, and so is nursing one's child and giving money to a charity. Other judgments emerge slowly and deliberately, alighting upon some executive decision after carefully weighing the pros and cons of each option. If we agree that a woman has the right to make decisions about her body, then abortion should be her right. But the fetus has some rights as well, even if it is unable to exercise them. Conflicts often arise when our split personalities go head to head, a clash between Kantian reasoning and Humean intuition: Abortion may be permissible, but it feels wrong because it is murder. What is central to the discussion ahead is that intuition and conscious reasoning have different design specs. Intuitions are fast, automatic, involuntary, require little attention, appear early in development, are delivered in the absence of principled reasons, and often appear immune to counter-reasoning. Principled reasoning is slow, deliberate, thoughtful, requires considerable attention, appears late in development, justifiable, and open to carefully defended and principled counterclaims. Like all dichotomies, there are shades of gray. But for now, we can start with these two contrasting positions, using them to spring forward into a third.

MORAL INSTINCTS

Consider the following scenarios:

1. A surgeon walks into the hospital as a nurse rushes forward with the following case. "Doctor! An ambulance just pulled in with five people in critical condition. Two have a damaged kidney, one a crushed heart, one a collapsed lung, and one a completely ruptured liver. We don't have time to search for possible organ donors, but a healthy young man just walked in to donate blood and is sitting in the lobby. We can save all five patients if we take the needed organs from this young man. Of course he won't survive, but we will save all five patients."

   Is it morally permissible for the surgeon to take this young man's organs?

2. A train is moving at a speed of 150 miles per hour. All of a sudden the conductor notices a light on the panel indicating complete brake failure. Straight ahead of him on the track are five hikers, walking with their backs turned, apparently unaware of the train. The conductor notices that the track is about to fork, and another hiker is on the side track. The conductor must make a decision: He can let the train continue on its current course, thereby killing the five hikers, or he can redirect the train onto the side track and thereby kill one hiker but save five.

   Is it morally permissible for the conductor to take the side track?

If you said "no" to the first question and "yes" to the second, you are like most people I know or the thousands of subjects I have tested in experiments.[26] Further, you most likely answered these questions immediately, with little to no reflection. What, however, determined your answer? What principles or facts distinguish these scenarios? If your judgment derives from religious doctrine or the deontological position that killing is wrong, then you have a coherent explanation for the first case but an incoherent explanation for the second. In the train case, it makes sense—feels right—to kill one person in order to save the lives of five people. In the hospital case, it feels wrong to kill one person to save five. You might explain the hospital case by saying that it is illegal to commit intentional homicide, especially if you are a responsible doctor. That is what the law says. That is what we have been raised to believe. Our culture inscribed this in our minds when we were young and impressionable blank slates. Now apply this bit of legalese to the train case. Here, you are willing to kill one person to save five. You are, in effect, willing to do something in the second case that you were unwilling to do in the first. Why the mental gymnastics? Something different is going on in the second scenario. For most people, the difference is difficult to articulate. In the hospital case, if there is no tissue-compatible person nearby, there is no way to save the five patients. In the train case, if the side track is empty, the conductor can still switch tracks; in fact, in this scenario, the conductor *must* switch tracks—an obligatory act—since there are no negative consequences to turning onto the side track.

We can unify and explain these ideas by appealing to the principle that it is permissible to cause harm as a by-product of achieving a greater good, but it is impermissible to use harm as a means to a greater good. In the train example, killing one person is a by-product—an indirect though foreseen consequence—of taking an action that saves five. The key act— flipping a switch to turn the train—has no inherent emotional value; it is neither positive nor negative, neither good nor bad. In the hospital example, the doctor harms one person as a means to save five. The act—ripping out organs—has an inherently negative feel; it is bad. These distinctions combine to provide an explanation known as the "principle of double effect." Philosophical detective work uncovered this principle, but only after years and years of debate and scrutiny of particular moral dilemmas.[27] Everyone listening to these dilemmas, however, judges them immediately, without any sense of thinking through the problems and extracting the underlying principles. Our answers seem reasoned, but we have no sense of reasoning. In fact, based on several studies that I discuss in chapter 2, few readers of these scenarios generate this principle as an explanation or justification for their judgments. This incapacity to generate an appropriate explanation is not restricted to the young or uneducated, but rather includes educated adults, males and females, with or without a background in moral philosophy or religion. If Kohlberg is right, then all of these people are morally delayed, back in stage 1, back with the moral cavemen. In the absence of good reasons for our actions, we are morally immature, in need of a moral-reasoning class or the CliffsNotes to Kant's challenging prose. Kohlberg's diagnosis is at best incomplete, and, at worst, deeply flawed. Educators who have followed his diagnosis should take pause.

Some think that because these scenarios are artificial, removed from our everyday experiences, detached from the more common dilemmas that arise among friends and family as distinct from unknown others, and foisted upon us without opportunity for reflection, that they fail to provide insights into our moral psychology. They are silly, toy examples, designed for those in the ivory tower. As the moral philosopher Richard Hare stated the case, arguing against his own professional tribe, "the point is that one has no time to think what to do, and so one relies on one's immediate intuitive reactions; but these give no guide for what critical thinking would prescribe if there were time for it."[28]

These scenarios are indeed artificial. But it is *because* of their artificiality that they provide one vehicle—a scientific method—for understanding our moral intuitions.[29] This claim, although targeted at philosophical problems, forms the bedrock of many of the sciences, including psychology. For example, vision scientists use what are called ambiguous figures to explore how our attention and systems of belief interact with what we see and potentially can see. Consider the illustration below. What do you see?

Some of you most likely see a rabbit, others a duck, and yet others, both duck and rabbit in alternation. The information present on this

page isn't changing, but your classification of the image does. And if you show this to a three-year-old child, she will say that she sees either a rabbit or a duck, but she will not be able to flip-flop back and forth between these images. It is not until about four years of age that children can maintain two different beliefs in mind and spontaneously flip between them.

There are advantages to artificial and unfamiliar moral dilemmas. By using artificiality to strip the cases of familiarity, we are unlikely to develop judgments based on pure emotion, or some prior commitment to the case that was dictated by principles of law or religion. By making the individuals in these scenarios unfamiliar, we are more likely to guarantee impartiality, cordoning off self-serving biases that get in the way of achieving a universally valid moral theory. Moreover, by creating artificial dilemmas, we are free to change them, parameter by parameter. As the moral philosopher Frances Kamm states, "Philosophers using this method try to unearth the reasons for particular responses to a case and to construct more general principles from these data. They then evaluate these principles in three ways: Do they fit the intuitive responses? Are their basic concepts coherent and distinct from one another? Are the principles or basic concepts in them morally plausible and significant, or even rationally demanded? The attempt to determine whether the concepts and the principles are morally significant and even required by reason is necessary in order to understand why the principles derived from cases should be endorsed." I believe that as long as artificial examples are examined together with real-life cases, we will uncover important insights into the nature of our judgments.

Like all moral dilemmas, these two examples contain elements that map onto real-world phenomena: individuals who have the choice between two or more actions, where each action can be carried out, each action results in some morally relevant consequence, and once a decision has been made, there is at least one positive and one negative consequence with respect to human welfare. For example, following the devastation from Hurricane Katrina in the fall of 2005, a member of the Texas Army National Guard had this to say: "I would be looking at a family of two on one roof and maybe a family of six on another roof, and I would have to make a decision who to rescue."[30] Choosing between a rescue of few versus many is not restricted to the philosopher's armchair. And like other moral dilemmas, these also capture a conflict between competing duties or obligations, where a decision to take one path isn't immediately and transparently obvious. We can ask, for example, whether the two cases above depend on the absolute numbers (kill one hundred to save one thousand) or the degree to which the act is detached from the harm it causes (flipping a switch turns the trolley that hits a cow that lands on a hiker and kills him a day later). The conflict is maintained in both cases, the agent can choose, and what ought to be done isn't obvious.

Suppose that everyone reading these two cases delivers judgments that are automatic, consistent, and rapid, but based on poorly articulated or even incoherent explanations.[31] If there is a kind of consensus answer for each of these scenarios, then we need a theory to explain the consensus or universal view. Similarly, we must explain why humans have such intuitions but rarely come up with the underlying principles to account for them. We must also explain how we acquired these principles—in development and over the course of evolution—and how we use them in the service of making morally relevant decisions. Sometimes, we will find ourselves in conflict, simultaneously appreciating the force of our intuitions while recognizing that they lead to actions we should not accept.

My explanation for these disparate observations is that all humans are endowed with a *moral faculty*—a capacity that enables each individual to unconsciously and automatically evaluate a limitless variety of actions in terms of principles that dictate what is permissible, obligatory, or forbidden. The origin of this kind of explanation dates back to the economist Adam Smith, as well as David Hume who, although hanging his hat on the emotions, saw the expressive power of our moral faculty and the need to extract its principles:

"It may now be ask'd *in general*, concerning this pain or pleasure, what distinguishes moral good and evil, *From what principles is it derived, and whence does it arise in the human mind?* To this I reply, *first*, that 'tis absurd to imagine, that in every particular instance, these sentiments are produc'd by an *original* quality and *primary* constitution. For as the number of our duties is, in a manner, infinite, 'tis impossible that our original instincts should extend to each of them, and from our very first infancy impress on the human mind all that multitude of percepts, which are contain'd in the compleatest system of ethics. Such a method is not comfortable to the usual maxims, by which nature is conducted, where a few principles produce all that variety we observe in the universe, and everything is carry'd on in the easiest and most simple manner. 'Tis necessary, therefore, to abridge these primary impulses, and find some more general principles, upon which all our notions of morals are founded."

The most recent incarnation of the argument comes from the writings of the political philosopher John Rawls and the linguist Noam Chomsky who proposed that there may be deep similarities between language and morality, including especially our innate competences for these two domains of knowledge.[32] If the analogy is fitting, what should we expect to find when we look at the anatomy of our moral faculty? Is there a grammar and, if so, how can the moral grammarian uncover its structure? Let me explain why you should take these questions seriously by briefly showing how an analogous set of questions have opened the door to a startling set of discoveries in linguistics and its sister disciplines.

We know more than our actions reveal. This may sound like a platitude, but it actually captures one of Chomsky's primary intuitions about language, as well as other faculties of the mind. The mature speaker of a language—English, Korean, Swahili—knows more about language than his speech reveals. When I say that I know English, what I generally mean is that I can use it to express my ideas and to understand what others say to me in English. This is the commonsense notion of knowing. Chomsky's sense is different and refers to the unconscious principles that underlie language use and comprehension. It also refers to the unconscious principles that underlie certain aspects of mathematics, music, object perception, and, I suggest, morality.[33] When we speak, we don't think about the principles that order the words in a sentence, the fact that certain words fall into abstract categories such as nouns, pronouns, and verbs, and that there are restrictions on the number of phrases that can be packed into a sentence. If we did think about some of these aspects of language before speaking, we would either produce gibberish or it would take forever to have a fluid, interactive conversation. And when we listen to someone speaking, we don't consciously decompose their utterances into grammatical constituents, even though every speaker of a language can make instantaneous grammaticality judgments, decisions about whether a particular sentence is well formed or ill formed.

When Chomsky generated the sentence "Colorless green ideas sleep furiously," he intentionally produced a string of words that no one had ever produced before. He also produced a perfectly grammatical and yet meaningless sentence. The artificiality of Chomsky's sentence provides, however, one kind of insight into our language faculty: a distinction between the syntax of a sentence and its semantics or meaning. Most of us don't know what makes Chomsky's sentence, or any other sentence, grammatical. We may *express* some principle or rule that we learned in grammar school, but such expressed rules are rarely sufficient to explain the principles that actually underlie our judgments. It is these unconscious or *operative* principles that linguists discover, and that never appear in a textbook, that account for the patterns of linguistic variation and similarities. For example, every speaker of English knows that "Romeo loves Juliet" is a well-formed sentence, while "Him loves her" is not. Few speakers of English know why. Few native speakers of English would ever produce this last sentence, and this includes young toddlers just beginning to speak English. When it comes to language, therefore, what we express as our knowledge pales in relationship to the knowledge that is operative but unavailable to expression.

The language faculty maintains a repository of principles for growing a language, any language. When linguists refer to these principles as the speaker's *grammar,* they mean the rules or operations that allow any normally developing human to unconsciously generate and comprehend a limitless range of well-formed sentences in their native language. When linguists refer to *universal grammar,* they are referring to a theory about the set of all principles available to each child for acquiring any specific language. Before the child is born, she doesn't know which language she will meet; and she may even meet two if she is born in a bilingual family. But she doesn't need to know. What she does know, in an unconscious sense, is the set of principles for all the world's languages—dead ones, living ones, and those not yet conceived. The environment feeds her the particular sound patterns of the native language, thereby turning on the specific principles of only one language, or two if the parents are bilingual. The problem of language acquisition is therefore like setting switches. Each child starts out with all possible switches, but with no particular settings; the environment then sets them according to the child's native language.[34]

From these general problems, Chomsky and other generative grammarians suggested that we need an explicit characterization of the language faculty, what it is, how it develops within each individual, and how it evolved in our species, perhaps uniquely. I take each of these in turn.

What is it? To answer this question, we want to explain the kinds of processes of the mind/brain that are specific to language, as opposed to shared with other problem-oriented tasks, including navigation, social relationships, object recognition, and sound localization. In particular, we want to describe the principles that capture the mature individual's competence for language, as well as the machinery that enables these principles to function. And we want to characterize this system independently of the factors that impinge upon language production, or what the speaker chooses to say. For example, we use our ears when we listen to a person speaking and when we localize an ambulance's siren. But once the sound passes from our ears to the part of the brain involved in decoding what the sound is and what to do with it, the machinery changes, one system handling speech, the other nonspeech. Looking at the engineering of the brain, we see that speech perception recruits circuitry different from general sound localization. We also know that language isn't restricted to sound. We are equally capable of expressing our thoughts and emotions in sign language. Although spoken language taps our auditory sense, and sign language taps our visual sense, both recruit the same circuitry involved in ascribing meaning to a word and for stringing words together to create meaningful sentences. Both systems somehow tell us that we are in a language mode as opposed to a music mode. Some aspects of the language faculty are therefore unique to language, and some are shared with other systems.

Once the system detects that we are in a language mode, either planning to produce an utterance or listening to one, a system of rules is engaged, organizing meaningless sound sequences (phonemes) into meaningful words, phrases, and sentences, and enabling conversation either as internal monologue or external dialogue. When we speak about the language faculty, therefore, we are speaking about the normal, mature individual's *competence* with the principles that underlie their native language. What this individual chooses to say is a matter of her *performance,* which will be influenced by whether she is tired, happy, in a fight with her lover, or addressing an audience of five hundred at a political rally. Language competence refers to the principles that make sentence production and comprehension possible for every normally developing person. These principles *are* the language faculty. What we say, to whom, and how, is the province of linguistic performance, and includes many other players of the brain, and many factors external to the brain, including other people, institutions, weather, and distance to the target audience.

How does it develop? To answer this question, we want to explain the child's path to a mature state of language competence, a state that includes the capacity to create and understand a limitless range of meaningful sentences generated by other speakers of the same language. This boils down to a question of the child's initial state—of her unconscious knowledge of linguistic principles prior to exposure to a spoken or signed language—and the extent to which this state constrains not only what she learns and when, but what she can learn from listening or watching. Consider the fact that in spoken English people can use two different forms of the verb "is," as in "Frank is foolish" and "Frank's foolish." We can't, however, use the contracted form of "is" wherever we please. For example, although we can say "Frank is more foolish than Joe is," we can't say "Frank is more foolish than Joe's." How do we know this? No one taught us this rule. No one listed the exceptions. Nonetheless, neither adults nor young children ever use the contracted form in an inappropriate place. The explanation, based on considerable work in linguistics, is that the child's initial state includes a principle for verb contraction—a rule that says something like "'s" is too small a unit of sound to be alone; whenever you use the contracted form, follow it up with another word.[35] The environment—the sound pattern of English—triggers the principle, literally pulling it out of a hat of principles as if by magic. The child is born with this operative principle, even though she cannot express this knowledge.

How did it evolve? To answer this final question, we look to our history and recognize two distinctive parts: phylogeny and adaptation. A phylogenetic analysis provides a depiction of the evolutionary relationships between species, yielding twiggy branches of the tree of life. When we map out a portion of this tree, we obtain an understanding of which species are most closely related and how far back in time this relationship extends. We can also use our phylogenetic analysis to determine whether similar traits are homologous—shared between species due to common descent from an ancestor with this trait—or analogous—shared between species due to convergent evolution. When we witness homologies, we see historical or evolutionary continuities between species. When we witness analogies, we see historical or evolutionary discontinuities between species. We can therefore ask which components of our language faculty are shared with other species and which are unique? In cases where the components are shared, we can ask whether they evolved continuously from some common ancestor, or discontinuously as part of a common solution to a particular adaptive problem.

Consider the human child's capacity to learn words. Much of word-learning involves vocal imitation. The child hears her mother say "Do you want candy?" and the child says "Candy." "Candy" isn't encoded in the mind as a string of DNA. But the capacity to imitate sounds is one of the human child's innate gifts. Imitation is not specific to the language faculty, but without it, no child could acquire the words of its native language, reaching a stunning level of about fifty thousand for the average high school graduate. To explore whether vocal imitation is unique to humans, we look to other species. Although we share 98 percent of our genes with chimpanzees, chimpanzees show no evidence of vocal imitation. The same goes for all the other apes, and all the monkeys. What this pattern tells us is that humans evolved the capacity for vocal imitation some time after we broke off from our common ancestor with chimpanzees, 6–7 million years ago. Other species, more distantly related to us than any of the nonhuman primates, are capable of vocal imitation: passerine songbirds, parrots, hummingbirds, dolphins, and some whales. What this distribution tells us is that vocal imitation is not unique to humans. It also tells us, again, that vocal imitation in humans didn't evolve from the nonhuman primates. Rather, vocal imitation evolved independently in humans, some birds, and some marine mammals. This represents a relatively complete answer to the question *How did it evolve?*

To address questions of adaptation, we can look at the relationship between functional design and genetic success. To what extent does the capacity to vocally imitate provide selective advantages in terms of reproduction? Although vocal imitation may be used in many contexts—to acquire the lexicon, a local dialect, annoy parents by parroting their every word—what were the original selective pressures? What did vocal imitation originally evolve for and what is its current utility? Here, answers are much more tenuous. Though we can readily see the current advantages of imitation, especially in terms of maintaining local dialects, picking up traditions, and so forth, it is less clear why it evolved in some animal groups and not others.

To provide a complete description of the language faculty, addressing each of the three independent questions discussed requires different kinds of evidence. For example, linguists reveal the deep structure underlying sentence construction by using grammaticality judgments and by comparing different languages to reveal common abstract principles that cut across the obvious differences. Developmental psychologists chart the child's patterns of language acquisition, exploring whether the relevant linguistic input is sufficient to account for their output. Neuropsychologists look to patients with selective brain damage, using cases where particular aspects of language are damaged while others are spared, or where language remains intact and other cognitive faculties are impaired. Cognitive neuroscientists use neuroimaging techniques such as fMRI to understand which regions of the brain are recruited during language-processing, attempting to characterize the circuitry of the language organ. Evolutionary biologists explore which aspects of the language faculty are shared with other species, attempting to pinpoint which components might account for the vast difference in expressive power between our system of communication and theirs. Mathematical biologists use models to explore how different learning mechanisms might account for patterns of language acquisition, or to understand the limiting conditions for the evolution of a universal grammar. This interdisciplinary collaborative is beginning to unveil what it means to know a particular language, and to use it in the service of interacting with the world. We are in the midst of a comparable collaborative effort with respect to our moral faculty. We are now ready to appreciate and develop Rawls's insights, especially his linguistic analogy. I introduce the "Rawlsian creature," equipped with the machinery to deliver moral verdicts based on unconscious and inaccessible principles. This is a creature with moral instincts.

Rawlsian creature

A GRAMMAR OF ACTION

One way to develop the linguistic analogy is to raise the same questions about the moral faculty that Chomsky and other generative grammarians have raised for the language faculty. Here are Rawls's thoughts about this analogy:

> A useful comparison here is with the problem of describing the sense of grammaticalness that we have for the sentences of our native language. In this case, the aim is to characterize the ability to recognize well-formed sentences by formulating clearly expressed principles which make the same discriminations as the native speaker. This is a difficult undertaking which, although still unfinished, is known to require theoretical constructions that far outrun the ad hoc precepts of our explicit grammatical knowledge. A similar situation presumably holds in moral philosophy. There is no reason to assume that our sense of justice can be adequately characterized by familiar common sense precepts, or derived from the more obvious learning principles. A correct account of moral capacities will certainly involve principles and theoretical constructions which go beyond the norms and standards cited in every day life.[36]

In parallel with the linguist's use of *grammaticality* judgments to uncover some of the principles of language competence, students of moral behavior might begin by using *ethicality* judgments to uncover some of the principles underlying our judgments of morally permissible actions. Grammaticality judgments are delivered spontaneously, rapidly, and with little to no reflection. Ethicality judgments would be delivered similarly, but based on morally relevant actions. In the same way that grammaticality judgments emerge from a universal grammar of principles and parameters, the Rawlsian creature's ethicality judgments would emerge from a universal moral grammar, replete with shared principles and culturally switchable parameters. From this perspective, each culture expresses a specific moral grammar. The Rawlsian creature therefore places constraints on the range of possible variation, including the range of potential moral systems. A mature individual's moral grammar enables him to unconsciously generate and comprehend a limitless range of permissible and obligatory actions within the native culture, to recognize violations when they arise, and to generate intuitions about punishable violations. Once an individual acquires his specific moral grammar, other moral grammars may be as incomprehensible to him as Chinese is to a native English speaker.

To clarify the relationship between universality and cultural variation, consider the act of infanticide. For Americans, this is a barbaric act, characteristic of a group that requires a moral tutorial on child care. For the Eskimos, and several other cultures, infanticide is morally permissible, and justifiable on the grounds of limited resources and other aspects of parenting and survival. If two cultures see the world through completely different moral lenses, then our ethical values are only relative to the details of the local culture, and free to vary. There are no moral absolutes, no truths, no universals. From this perspective, the Eskimos would seem to be cold-hearted, uncaring parents. But this misses the point, and runs right by what is universal to all humans, including Americans and Eskimos: caring for children is a universal moral principle. In all cultures, everyone expects parents to care for their offspring. Within and across cultures, torturing infants as amusement or sport is forbidden. What varies across cultures are the conditions that allow for exceptions to the rule, including conditions of abandonment. The point here is simple: our moral faculty is equipped with a universal set of rules, with each culture setting up particular exceptions to these rules. We want to understand the universal aspects of our moral judgments as well as the variation, both what allows for it and how it is constrained.

A Humean creature would argue that the universality stems from our shared capacity not only to experience emotions, but to experience the same sort of emotions in certain contexts. The reason why everyone would find it morally abhorrent to watch or imagine an adult kicking a helpless infant is that everyone would experience disgust in this context. Our shared emotional code generates a shared moral code. Cultural variation emerges because individual cultures teach particular moral variants that, through education and other factors, fuse with emotions. Once fused, responses to moral transgressions are fast and unreflective, fueled

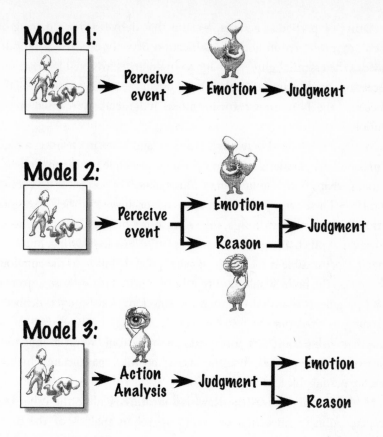

Model 1:

Perceive event → Emotion → Judgment

Model 2:

Perceive event → Emotion / Reason → Judgment

Model 3:

Action Analysis → Judgment → Emotion / Reason

by unconscious emotions. Rawlsian creatures respond in some of the same ways as Humeans, but rely on the causes and consequences of action as distinct from emotion.

We can take advantage of this coarse characterization of the Humean and Rawlsian creatures, as well as their more distant relatives, the Kantians, to set up three general models of our moral judgments. These models strip away the complexities of real-world cases to pinpoint some of the essential ingredients in the process of delivering a moral judgment.

Model 1 describes an archetypal Humean creature. Here, following the perception and presumed categorization of an action or event, there is an emotional response that immediately generates a moral judgment. We see one man with a knife, another man dead at his feet, and we classify this as murder, an action that generates a negative feeling, which generates a judgment of "forbidden." This is a deontological claim about

the nature of particular actions, a claim that derives from a pairing between any given action and a classification of right or wrong. Emotion
provides the essential glue. Damage to brain areas involved in emotional
processing would cause complete breakdown of moral judgments. This
is because the Humean creature's moral judgments are caused by the
emotions.

Model 2 is a hybrid between a Humean and Kantian creature, a blend
of unconscious emotions and some form of principled and conscious reasoning, perhaps based on utilitarian consequences or accessible deontological rules.[37] These two systems may converge or diverge in their assessment
of the situation; if they diverge, then some other mechanism must intrude,
resolve the conflict, and generate a judgment. Killing feels wrong, but sometimes it is permissible if it enables a greater good. Damage to the emotional
circuitry of the brain would lead to a breakdown of only those aspects of
moral judgments that depend upon the emotions, leaving cold, deliberate
reasoning to work out the rest. Conversely, damage to those parts of the
brain that enable conscious principled reasoning would result in judgments
that bypass consequences, focusing instead on rules that dictate which actions are permissible and which are not.

Model 3 characterizes the Rawlsian creature. Unlike the other two,
the perception of an action or event triggers an analysis of the causes
and consequences that, in turn, triggers a moral judgment—permissible,
obligatory, forbidden. Emotions, if they play a role, do so after the judgment. The man didn't intend to kill as a means to the greater good, but as
a foreseen consequence. His intent was to save five people in harm's way,
and killing one was the only solution. Analyses of the motivation or intentions underlying an action, together with analyses of intended and
foreseen consequences, provide the relevant material for our moral faculty. Emotions may only function to modulate what we actually do as distinct from what we comprehend or perceive as morally permissible. In
contrast to the other toy models, damage to the emotional circuitry has no
impact on the Rawlsian creature's moral judgments. Emotions are triggered by these judgments, not caused by them. Psychopaths may represent
a test case of this idea; they appear to deliver normal moral judgments, but
due to the lack of appropriate emotions, behave abnormally, with morally
inappropriate actions.

Only scientific evidence as opposed to philosophical intuition can determine which model is correct. What is important is to have all the options on the table, open for critical digestion. Until now, there has been no serious engagement with the Rawlsian creature (model 3). To engage, we need to achieve a level of detail that parallels current work in linguistics, extracting principles that can explain how we perceive actions and events in terms of their causes and moral consequences for self and other.[38]

The language faculty takes as input discrete elements that can be combined and recombined to create an infinite variety of meaningful expressions: phonemes ("distinctive features" in the lingo of linguistics) for individuals who can hear, signs for those who are deaf. When a phoneme is combined with another, it creates a syllable. When syllables are combined, they can create words. When words are combined, they can create phrases. And when phrases are combined, they can create *The Iliad*, *The Origin of Species*, or *Mad* magazine. Actions appear to live in a parallel hierarchical universe. Like phonemes, many actions lack meaning. When combined, actions are often meaningful. Like phonemes, when actions are combined, they do not blend; individual actions maintain their integrity. When actions are combined, they can represent an agent's goals, his means, and the consequences of his action or the omission of an action. When a series of subgoals are combined, they can create events, including the *Nutcracker* ballet, the World Series, or the American Civil War. This ability suggests that morality is based on a system of general principles or rules, and not a list of specific examples. We evaluate John's violent attack on Fred as a principle with abstract placeholders or variables for AGENT, ACTION, RECEIVER, CONSEQUENCE, MORAL EVALUATION. For example, the principle might read AGENT → HITS → RECEIVER → PAIN → IMPERMISSIBLE. Whether we also store information about John, the attack, and Fred in particular is another story.

By breaking down the principle into components, we achieve a second parallel with language: To attain its limitless range of expressive power, the principles of our moral faculty must take a finite set of elements and recombine them into new, meaningful expressions or principles. For language, we recombine words and higher-order combinations of these words (noun and verb phrases). For morality, we recombine actions, their causes and consequences.

Like Chomsky, Rawls suggested that we may have to invent an entirely new set of concepts and operations to describe the universal moral principles. Our more commonsense formulations of universal rules may fail to capture the mind's computations in the same way that grammar-school grammar fails to capture the principles that are part of our language faculty. For example, all of the following actions are universally forbidden: killing, causing pain, stealing, cheating, lying, breaking promises, and committing adultery.[39] Like other rules, these moral rules have exceptions. Thus, killing is generally forbidden in all cultures, but most if not all cultures recognize conditions in which killing is permitted or might be justifiable: war, self-defense, and intolerable pain due to illness. Some cultures even support conditions in which killing is obligatory: in several Arabic countries, if a husband finds his wife in flagrante delicto, the wife's relatives are expected to kill her, thereby erasing the family's shame. But what makes these rules universal? What aspects of these rules allow for cultural variation? Do the rules actually capture the relationship between the nature of the relevant actions (e.g., HARMING, HELPING), their causes (e.g., INTENDED, ACCIDENTAL), and consequences (e.g., INTENDED, FORESEEN)? Are there hidden relationships or principles, operating unconsciously, but discoverable with the tools of science? If, as Rawls intuited, the analogy between morality and language holds, then our commonsense accounts will be insufficient, requiring a more in-depth search beneath the surface. This search will uncover the set of principles that unconsciously guide our moral judgments of permissible, obligatory, and forbidden actions.

How does the moral faculty develop? To answer this question, we need an understanding of the principles guiding an adult's judgments. With these principles described, we can explore how they are acquired, asking such questions as: Does the child's environment provide her with enough information to construct a moral grammar, or does the child show competences that go beyond her exposure? Do we acquire our native moral norms with ease and without instruction, while painstakingly trying to remember the details of a new culture's mores? Is there a critical period for the acquisition of our moral norms?

At the most basic level, there must be some innate capacity that allows each child to build a specific moral grammar. No other species that we know of constructs elaborate moral systems. Something about human brains uniquely enables this construction, generation after generation. But we ultimately want to know what makes certain moral systems learnable, and what makes certain kinds of experiences morally relevant. Herein lies our greatest challenge, especially when contrasted with advances in linguistics. Unlike the army of linguists that have provided rich catalogs of what people in different languages say and comprehend, we lack a comparable catalog of people's actions and judgments in morally relevant contexts. In the absence of such information, coupled with a set of descriptive principles that can account for the mature state of knowledge, it is difficult to work out what the developmental problem really looks like. Questions of acquisition can only be sensibly raised when we understand the mature state.

How did the moral faculty evolve? Like language, we can address this question by breaking down the moral faculty into its component parts, and then exploring which components are shared with other animals and which are unique to our own species. With this breakdown in place, we can further ask whether the uniquely human components are unique to morality or shared with other systems of knowledge. We answer the uniquely human question by studying other animals, and we answer the uniquely moral question by studying other systems of knowledge, how they work and how they evolved.

One way to look at animal moral competence is to explore their expectations concerning rule followers and violators, whether they are sensitive to the distinction between an intentional and accidental action, whether they experience some of the morally relevant emotions and, if so, how these emotions play a role in their decisions. If an animal is incapable of making the intentional-accidental distinction, then it will treat all consequences as the same, never taking into account its origins. Seeing a chimpanzee accidentally fall from a tree and injure a group member is equivalent to seeing a chimpanzee leap out of a tree and injure a group member. Seeing an animal reach out and hand another a piece of food is indistinguishable from seeing an animal reach out for its own food while accidentally dropping a piece into another's lap. Finding parallels is as

important as finding differences, as both illuminate our evolutionary
path, especially what we inherited and what we invented.

Like all other domains of knowledge, our moral knowledge did not
evolve in a neural vacuum, isolated from other processes. Further, our
moral behavior depends upon other systems of the mind. What we are after is a description of those processes that are specific to morality as well
as those that are not specific but play an essential supporting role. For example, we would not be able to evaluate the moral significance of an action if every event perceived or imagined flitted in and out of memory
without pausing for evaluation. Based on this observation, it would be incorrect to conclude that memory is a specific component of our moral
anatomy. Our memories are used for many aspects of our lives, including
learning how to play tennis, recalling our first rock concert, and generating expectations about a planned vacation to the Caribbean. Some of
these memories reference particular aspects of our personal lives (autobiographical information about our first dentist appointment), some allow us
to remember earlier experiences (episodic recall for the smell of our
mother's apple pie), some are kept in long-term storage (e.g., travel routes
home), and others are short-lived (telephone number from an operator).
Of course, memories are also used to recall morally forbidden actions, to
feel bad about them, and to assess how we might change in order to better
our own moral standing. Our memory systems are therefore part of the
support team for moral judgments, but they are not specific to the moral
faculty.

Like memory, our conceptual representations of others' beliefs, desires, and goals also figure into both moral and nonmoral processes. Consider the distinction between an intentional and an accidental action. This
distinction is part of our folk or commonsense notion of others' minds,
including such mental states as *belief*, *desire*, and *intention*. We infer
these invisible properties of the mind based on indirect measures, such as
where someone is looking or reaching, or where someone has been. Many
moral distinctions depend on the capacity to distinguish between intentional and accidental actions, even though this difference is not specific to
the moral domain. What is the difference between "Joe intentionally hit
John" and "Joe accidentally hit John"? Linguistically, three out of the four
ingredients are identical. The same individuals are involved in both

scenes, and the action of hitting is the same as well. If we can uncover the cause of Joe's action, then we ascribe responsibility to Joe if he hit John intentionally. In contrast, if Joe hit John accidentally, then although the consequences for John may be the same, we might not want to hold Joe responsible.

How we judge the moral relevance of someone's actions may also influence how we attribute cause. This shows the interaction between the more general folk psychology and more specific moral psychology. Consider the following scenario:

> The vice president of a company went to the chairman of the board and said, "We are thinking of starting a new program. It will help us increase profits, and it will also harm the environment." The chairman of the board answered, "I don't care at all about harming the environment. I just want to make as much profit as I can. Let's start the new program." They started the new program. Sure enough, the environment was harmed.

> *How much blame does the chairman deserve for what he did?*
> *Answer on a scale of 1 (considerable blame) to 7 (no blame):* ___

> *Did the chairman intentionally harm the environment?*
> *Yes___ No ___*

When subjects answer this question, they typically say that the chairman deserves blame because he intentionally harmed the environment. In contrast, when they read a virtually identical scenario in which the word "help" replaces the word "harm," and "praise" replaces "blame," they typically say that the CEO deserved little praise and did not act to intentionally help the environment. At the heart of these scenarios is whether a side effect of an action is perceived as intentional or not. In these cases, there is an asymmetry. When the side effect of the action is a negative outcome, people are more willing to say that the agent caused the harm. This is not the case when the outcome is positive or helpful. Recent studies with children show that such effects are present as early as at three years of age, suggesting that we are endowed with a capacity that is more likely to perceive actions as intentional when they are morally bad than when they are morally good.[40]

Let me give one final illustration to hammer home the point about processes that are necessary for our moral judgments and behavior but are not unique to morality. Our emotions and motivational drives are coupled to our moral judgments and actions, but are not specific to morality. We are fearful and happy about many things that carry no moral significance, including fear of snakes and happiness about our work or an ice cream cone. Paralleling what I said about the intentional-accidental distinction, a small shift in context can also change the significance of these core emotions. To an observer, there is something wrong (morally relevant) about a man who enjoys the ice cream cone that he has just taken away from a child who is now crying; the man should feel bad, not happy, for having done something impermissible. If these emotions are triggered by morally relevant actions, then they must be part of the moral faculty. This is the classic view that dates back at least to Hume, and has been carried forward into the present by the social psychologist Jonathan Haidt, who proposes that we are equipped with four families of moral emotions: (1) *other-condemning*: contempt, anger, and disgust; (2) *self-conscious*: shame, embarrassment, and guilt; (3) *other-suffering*: compassion; (4) *other-praising*: gratitude and elevation. These moral emotions run the show. They provide us with our intuitions about what is right or wrong, and what we should or shouldn't do.

It is impossible to deny that we experience guilt, compassion, and gratitude, and that these emotions materialize in our minds and bodies in the context of moral behavior, planned or imagined. These experiences, however, leave open two questions: What triggers these emotions and when do they arise in the course of moral evaluation? For an emotion to emerge, something has to trigger it. Some system in the brain must recognize a planned or completed action, and evaluate it in terms of its consequences. When an emotion emerges in a context that we describe as morally relevant, the evaluative system has identified an action that often relates to human welfare, either one's own or someone else's. The system that perceives action, breaking the apparently seamless flow of events into pieces with particular causes and consequences, must precede the emotions. For example, when we feel bad about an action that causes harm, loss, or distress to another, usually someone familiar, we call this feeling "guilt."[41] We usually judge the action as wrong or impermissible. I don't deny the feeling. But I do challenge the causal role attributed to our feelings. Guilt might cause our judgments or might follow from our judgments. The same may be true of the other moral emotions. I will have much more to say about this later on. For now, the only relevant point is that our emotions are not part of the dedicated and specialized components of the moral faculty.

Our moral faculty enables each normally developing child to acquire any of the extant systems of morality. Below is a rough sketch of the Rawlsian creature's moral anatomy—in essence, its design specs. This characterization follows directly from the linguistic analogy, taking it at face value. It is a road map for the rest of the book.

ANATOMY OF THE RAWLSIAN CREATURE'S MORAL FACULTY

1. The moral faculty consists of a set of principles that guide our moral judgments but do not strictly determine how we act. The principles constitute the universal moral grammar, a signature of the species.
2. Each principle generates an automatic and rapid judgment concerning whether an act or event is morally permissible, obligatory, or forbidden.
3. The principles are inaccessible to conscious awareness.
4. The principles operate on experiences that are independent of their sensory origins, including imagined and perceived visual scenes, auditory events, and all forms of language—spoken, signed, and written.
5. The principles of the universal moral grammar are innate.
6. Acquiring the native moral system is fast and effortless, requiring little to no instruction. Experience with the native morality sets a series of parameters, giving birth to a specific moral system.

7. The moral faculty constrains the range of both possible and stable ethical systems.

8. Only the principles of our universal moral grammar are uniquely human and unique to the moral faculty.

9. To function properly, the moral faculty must interface with other capacities of the mind (e.g., language, vision, memory, attention, emotion, beliefs), some unique to humans and some shared with other species.

10. Because the moral faculty relies on specialized brain systems, damage to these systems can lead to selective deficits in moral judgments. Damage to areas involved in supporting the moral faculty (e.g., emotions, memory) can lead to deficits in moral action—of what individuals actually do, as distinct from what they think someone else should or would do.

Features 1–4 are largely descriptions of the mature state, what normal adults store in the form of unconscious and inaccessible moral knowledge. Features 5–7 are largely developmental characteristics that define the problem of acquiring a system of moral knowledge, including signatures of the species and cultural influences. Features 8–10 target evolutionary issues, including the uniqueness of our moral faculty and its evolved circuitry. Overall, this anatomical description provides a framework for characterizing our moral faculty.

I HAVE PROVIDED a rough sketch of how we should think about our moral psychology in light of what we know about language. Based on the characterization of the language faculty that Chomsky initiated, and that generations of linguists have developed and criticized, we are now ready to follow suit and explore the nature of our moral faculty. The analogy to language will be strategically useful, forcing us to address novel questions about the nature of moral knowledge. The analogy will also reveal deep parallels between these domains as well as new insights into their differences. Differences are to be expected, given their apparent functions: morality depends upon an impartial judge, language doesn't; morality regulates social interactions, while language contributes to this enterprise but also provides a vehicle for our thoughts. By moving deeper into the principles underlying both systems, we will uncover what is shared, what is unique, and how each domain of knowledge evolves and naturally develops within each child.

I have also intentionally left one issue wide-open, interpretable in at least two ways: Are the Kantian, Humean, and Rawlsian creatures all a part of our moral psychology or is the Rawlsian creature running solo, with the Kantian and Humean jumping in when we decide to act on our moral convictions? I won't adjudicate between these possibilities here, because I have yet to provide the relevant evidence. The answer is forthcoming.

PART I

Universal Declarations

2

JUSTICE FOR ALL

---

*Military justice is to justice what military music
is to music.*

—GROUCHO MARX[1]

R ICHARD DAWKINS'S SMASH-HIT BOOK *The Selfish Gene* provided a beautifully written introduction to sociobiology, an approach to human and animal behavior that placed the gene at
the center of the stage, responsible for the evolution of altruism,
violence, parenting, deception, and sexual conflict. Many have interpreted
Dawkins's book as a subversive piece of science writing, designed to eliminate free will, justify abhorrent human actions by appealing to our biology,
and put us face-to-face with those nasty little self-serving strands of DNA.
Dawkins, and other evolutionary biologists, have forcefully denied such
charges in a volley of comments that is all too reminiscent of the great 1860
Oxford debate between Bishop Wilberforce and Thomas Henry Huxley.[2]
Having read Darwin's *Origin of Species* and seen its potential for under-
mining religious stricture and its moral regulations, Wilberforce launched
an attack, armed with sarcasm and rhetorical flourishes. As the story goes,
he concluded his diatribe, then turned to Huxley and asked whether it was
through his grandmother or grandfather that he should claim descent from
an ape? Huxley responded: "[A] man has no reason to be ashamed of having an ape for his grandfather. If there were an ancestor whom I should

feel shame in recalling it would rather be a man—a man of restless and versatile intellect—who, not content with an equivocal success in his own sphere of activity, plunges into scientific questions with which he has no real acquaintance, only to obscure them by an aimless rhetoric, and distract the attention of his hearers from the real point at issue by eloquent digressions and skilled appeals to religious prejudice."

Although Dawkins may be Huxley's reincarnation and the champion of the selfish-gene approach—a topic that I will revisit later—few have noticed the following line out of the very same book: "Kin selection and . . . reciprocal altruism . . . are plausible as far as they go but I find that they do not begin to square up to the formidable challenge of explaining cultural evolution and the immense differences between human cultures around the world. . . . I think we have got to start again and go right back to first principles. For an understanding of the evolution of modern man we must begin by throwing out the gene as the sole basis of our ideas on evolution."[3] Unpacking this comment, Dawkins rightly suggests that there is more to human nature than pure, unadulterated self-interest. What, however, is responsible for the more beneficent aspects of human behavior? How much comes from a mind that gets an emotional high from helping others? How much comes from the rules and regulations that each society imposes to curtail our egoistic tendencies? How might we work toward a just society, one based on principles of fairness or some other metric that benefits those in need or who work hardest? Wouldn't we be wise to listen to Mother Nature even if, in the end, we tell her thanks, but no thanks?

What has allowed us to live in large groups of unrelated individuals that often come and go is an evolved faculty of the mind that generates universal and unconscious judgments concerning justice and harm. Over historical time, we have invented legal policies and religious doctrine that sometimes enforce these intuitions, but often conflict with them. The third American president, Thomas Jefferson, renown for his vision of justice, eloquently stated the case for our moral faculty, while pointing to the tension between intuition and conscious reasoning:

> He who made us would have been a pitiful bungler, if he had made the rules of our moral conduct a matter of science. For one man of science, there are thousands who are not. What would

have become of them? Man was destined for society. His morality, therefore, was to be formed to this object. He was endowed with a sense of right and wrong merely relative to this. This sense is as much a part of his nature, as the sense of hearing, seeing, feeling; it is the true foundation of morality . . . The moral sense, or conscience, is as much a part of man as his leg or arm. It is given to all human beings in a stronger or weaker degree, as force of members is given them in a greater or less degree. It may be strengthened by exercise, as may any particular limb of the body. This sense is submitted indeed in some degree to the guidance of reason; but it is a small stock which is required for this: even a less one than what we call Common sense. State a moral case to a ploughman and a professor. The former will decide it as well, and often better than the latter, because he has not been led astray by artificial rules.[4]

Jefferson's comment captures several themes that will resurface here and in the following chapters: intuitive judgments versus consciously reasoned policy, innate capacities and acquired values, the common man's intuitions versus the educated man's reasoning. With Jefferson's thoughts in mind, I now turn to an exploration of how universal judgments of fairness constrain the range of cross-cultural variation, and the extent to which people are aware of the principles driving their moral judgments.

To set the stage, let us return to the sports car and charity dilemmas discussed in chapter 1. Both cases ask, when is it morally obligatory to help someone else—to act unselfishly?

1. Sports car. A man is driving his new sports car when he sees a child on the side of the road with a bloody leg. The child asks the car driver to take her to the nearby hospital. The owner contemplates this request while also considering the $200 cost of repairing the car's leather interior.

   Is it obligatory for the man to take this child to the hospital?

2. Charity. A man gets a letter from UNICEF's child health care division, requesting a contribution of $50 to save the lives of

twenty-five children by providing oral rehydration salts to eliminate dehydrating diarrhea.

Is it obligatory for the man to send money for these twenty-five children?

Although our selfish genes and our selfish psychology push us to answer "no" to both cases, some other part of our psychology pulls us to answer "yes" to case 1 and, for a much smaller fraction of people, to case 2 as well. To understand the nature of this psychological push-pull, I will break down each case into some similarities and differences. The goal here is to dissect each dilemma into relevant dimensions, including the identity of each agent and recipient, the relationship between agent and recipient, the set of actions and consequences along with their unfolding over time, and the consequences of action or inaction for both agent and recipient. The breakdown and analysis are by no means exhaustive, but rather an appetizer.

With each scenario filtered into a set of relevant dimensions, we can begin to see why we are pushed in one direction for case 1 and another for case 2. I have highlighted in gray the case-discriminating dimensions. It is striking how the cost-benefit analysis tilts in the opposite direction from what one might expect. Case 1 costs more, relative to case 2. Case 1 costs the agent time and $150 more and saves only one life—actually, only one leg—relative to twenty-five lives. It is hard to imagine a legal principle or religious doctrine that would adjudicate in favor of an action that leads to one saved leg over twenty-five saved lives. In addition, there is an asymmetry in the cause of the recipient's problem. In case 1, either the child is to blame for walking in the road or the cause of her injury is ambiguous. In case 2, the children are not to blame; they are the victims of a harsh environment and extreme poverty. With this asymmetry in play, and assuming that an individual's responsibility affects in some way our moral judgments, we should be more willing to give to charity than we are to stop and help the injured child. What, then, might explain the reversal in our intuitions' polarity?

The agent and recipient are unfamiliar, unrelated, and from a different social group. But perhaps we unconsciously code these cases differently,

CASE-RELEVANT DIMENSIONS	CASE 1-SPORTS CAR	CASE 2-CHARITY
Agent	One human	One human
Recipient	One child	Twenty-five children
Action—general	Helping/saving	Helping/saving
Action—specific	Physical aid	Financial aid
Negative consequence of action for agent	Spends $200, invests time in helping child to hospital	Spends $50, no time investment
Positive consequence for recipient[s] of action	Child's leg is saved	Twenty-five children survive
Urgency of action vis-à-vis consequence for recipient	High	Medium
Relationship between agent and recipient	Unfamiliar, non-kin	Unfamiliar, non-kin
Time lag between action and consequence	Short	Long
Is consequence a direct or indirect response to action?	Direct	Direct
What is consequence of inaction?	One child loses a leg, and possibly more	Twenty-five children die
Is action personal or impersonal?	Personal	Impersonal
Alternatives to agent's helping action	Unlikely, but another driver might approach	Likely, including other donors, governments, agencies
Probability that action directly causes consequence	High	Low from agent's perspective, high from agency's perspective
Is recipient causally responsible for personal situation and need for help?	Either "yes" or "ambiguous"	No

assuming that agent and recipient are from the same group in case 1 but most definitely not in case 2. Many may assume that case 2 involves a white American male and black African girls. Although this assumption, and the ethnic-racial discrimination it entails, may play some role, we could rewrite these cases to equate the relationships between agent and recipient. If the car driver had been a fifty-year-old white American male, and the child had been a black African teenager from the Sudan, this case would still have more pull than the charity case with respect to helping. In fact, even if the mail reader was a black African man from the Sudan but presently living in the United States, we would nonetheless perceive less of an altruistic pull.

One additional dimension, flagged in the last chapter, stands out: case 1 entails an up-close-and-personal act, while case 2 entails a distant and impersonal act. Action at a distance generates a weaker altruistic pull, because we lack the evolved psychology. Helping individuals that are out of arm's reach, sometimes out of sight, is a newly developed pattern of action and interaction.

We can put these details together to generate a principle that robustly pushes us to help in case 1, but causes uncertainty or negation in case 2:

> *If we can directly prevent, with a high degree of*
> *certainty, something bad without sacrificing anything of*
> *comparable significance, we are obliged to do it.*[5]

This is a start. It represents a principle that would make it through the Kantian five-point method. It isn't perfect. How certain is certain? How do we work out the cost-benefit analysis that accompanies the notion of *sacrificing of comparable significance*? By framing this principle as starkly as I have, we gain some purchase on the kinds of parameters involved in mediating our judgments of fairness.

VEILED IGNORANCE

Moral philosophers have long been interested in our sense of justice, of what counts as fair, and how we might derive the relevant principles. In the twentieth century, John Rawls stands out as one of the most important contributors to this problem. When Rawls published *A Theory of Justice*, its ideas were so profound and so important to disciplines outside of philosophy that the book became a classic, translated into dozens of languages, read by politicians, lawyers, economists, anthropologists, biologists, psychologists, and thousands outside of academia. It has been quoted in political discourse, in discussions of sporting matches, and even in the popular American television show *The West Wing*. Oddly, many current discussions of the evolution of morality, and fairness in particular, either ignore Rawls or misinterpret him.

What was Rawls after? Having served in a war and thought about inequalities, he spent a lifetime trying to understand our institutions, their modus vivendi, their policies of justice. Central to his thinking was an identity relationship, the principle of *justice as fairness*. What this identity relationship implies is that fairness isn't a component of justice or a form of justice or related to justice. Fairness *is* justice. Like the British philosophers of the Enlightenment, especially David Hume, Rawls believed in a moral sense, a sense of justice that was designed on the basis of principles that "determine a proper balance between competing claims to the advantages of social life." He also believed, paralleling Hume, that we can understand the nature of our moral sense by using the tools of science. Unlike Hume, however, Rawls placed little emphasis on the emotions. Rather, unconscious principles drive our moral judgments. This perspective, highlighted in the last chapter, has four desirable features.[6]

The first feature is strategic. We should explore the principle of justice as fairness, and the moral faculty more generally, in the same way that linguists following in the tradition of the generative grammarians, most notably Chomsky, have explored the language faculty. As I briefly sketched in chapter 1, Rawls stepped into this position by recognizing the many parallels between these systems of knowledge, as well as the potential validity of the approach taken in linguistics. For example, like language, moral systems are limitless in their scope of expression and interpretation. From a finite and often limited set of experiences, we project our intuitions to novel cases. Children take in a limited set of linguistic experiences, but output a broader range of linguistically appropriate utterances. What comes out is much richer than what went in.

Moral input and output appear similarly asymmetric. Mrs. Smith gives her son Fred a bag of candy and says that he should share it with his friend Billy. Fred gives Billy one piece and keeps the rest for himself. Billy says, "That's not fair." Mrs. Smith agrees. Fred is unlikely to conclude that sharing with Billy requires an equitable distribution of the candies, whereas sharing with others permits an unequal distribution. Fred is most likely to conclude that when it comes to sharing, everyone gets a fair shake. Fred has an intuitive understanding of the principle of fairness, and his local culture hands him the parameter space, the range of distributions or values that count as fair. This example doesn't show that the process of generalizing from a single instance to a more general principle is specific to morality. It is possible that the generalization process is the same for language, mathematics, object categorization, and morality. Other evidence is necessary.[7]

If children are born with a set of moral principles, then this foundation helps solve the acquisition problem. The poverty of the experience is no longer a problem for the child. From a few examples handed down through the local culture, she can derive the proper principles. The poverty of the stimulus argument, made famous by Chomsky's thinking on language, can be translated into a simple four-step method that sets the scientific process in motion:

1. Identify a particular piece of knowledge in mature individuals.
2. Identify what kind of input is necessary or indispensable for the learner to acquire this piece of knowledge.
3. Demonstrate that this piece of knowledge is not present or available from the environment.
4. Show that the knowledge is nonetheless available and present in the child, at the earliest possible age, prior to any relevant input.

During language acquisition, children produce constructions that are to be found nowhere in their experience. If the child is generating appropriate constructions, then she unconsciously has the knowledge. In the context of justice, here is how Rawls put these ideas together: "We acquire a skill in judging things to be just and unjust, and in supporting these judgments with reasons. Moreover, we ordinarily have some desire to act in accord with these pronouncements and expect a similar desire on the part of others. Clearly this moral capacity is extraordinarily complex. To see this it suffices to note the potentially infinite number and variety of judgments that we are prepared to make. The fact that we often do not know what to say, and sometimes find our minds unsettled, does not detract from the complexity of the capacity we have." This final sentence leads to a second feature of Rawls's perspective.

Rawls suggested that we may often pronounce a judgment about what is fair or unfair, permissible or impermissible, without knowing why—without being able to justify our actions or give an explanation that is consistent with our behavior. The fact that we may act without knowing why raises a question: What does it mean to know, and, in particular, to know about the principles of fairness? Rawls's suggestion, building on the linguistic analogy, was that many of our morally relevant judgments emerge rapidly, often without reflection, in the absence of heated emotion, and typically, without access to a clear justification or explanation. Moreover, these judgments tend to be robust, as evidenced by the vehemence with which individuals stick to their intuitions in the face of reasonable alternative judgments.

The simple point I am making is this: there are different ways of *knowing*. For each particular domain of knowledge, we may find that what we do or perceive is based on operative principles that we can't express; our ability to express such principles may only emerge when we are formally trained in the relevant discipline: linguistics, music, vision, acoustics, mathematics, economics. When people give explanations for their moral behavior, they may have little or nothing to do with the underlying principles. Their sense of conscious reasoning from specific principles is illusory. And even when someone becomes aware of an underlying principle, it is not obvious that this kind of understanding will alter their judgments in day-to-day interactions. Having conscious access to some of the principles underlying our moral perception may have as little impact on our moral behavior as knowing the principles of language has on our speaking.

The third characteristic of Rawls's position is an attempt to unite the unconscious but operative principles with those that are expressed when we reflect on our actions, those that have occurred or are about to happen.

Rawls suggested that we resolve moral conflict by means of *considered judgment*. Considered judgments are made rapidly, automatically, without reflection, with full confidence, in the absence of heated passion, and without explicitly self-motivated interests. They are also made without any awareness of specific moral principles or rules. When we judge an action as fair, we do so without simultaneously thinking "Poor sucker! He doesn't realize that I'm making out like a bandit. I've just violated the principle of retributive justice, getting away with a much shorter conviction relative to the crime." As Rawls explains, "Considered judgments are simply those rendered under conditions favorable to the exercise of the sense of justice, and therefore in circumstances where the more common excuses and explanations for making a mistake do not obtain." Once such judgments are in place, however, they are open to revision, refinement, and possibly outright rejection. These actual judgments are open to different constraints, including whether the person is in a bad mood, can recall the details of the argument, is distracted, or is willing to devote the time needed to work through the dilemma. What is important, therefore, is to distinguish between the principles that guide judgments under ideal conditions with those that underlie judgment in the face of an actual moral dilemma, happening in the here and now and requiring an immediate response.

The final characteristic of Rawls's approach comes from a thought experiment designed to reveal the unconscious principles of justice embedded in our moral faculty. In what he described as the *original position*, a group of people assemble to discuss the core principles of justice as fairness. Their mission is to evaluate such issues as how to dispense compensation for individual actions, how natural advantages or disadvantages should weigh in with respect to the final outcome, and how such principles impact upon the structure of institutions and the individuals that comprise them. This proposal follows in the tradition of other contractarian philosophers, dating back at least as far as Thomas Hobbes in the sixteenth century. For Rawls, the contract was an idealization and a method for figuring out principles of justice that are impartial, immune to self-serving biases: "The conception of the original position is not intended to explain human conduct except insofar as it tries to account for our moral sentiments and helps to explain our having a sense of justice."

This approach mirrors Chomsky's on language: "Any interesting generative grammar will be dealing, for the most part, with mental processes that are far beyond the level of actual or potential consciousness; furthermore, it is quite apparent that a speaker's reports about his behavior and his competence may be in error. Thus, a generative grammar attempts to specify what the speaker actually knows, not what he may report about his knowledge."[8] For morality, we must be prepared to find that what an individual knows has little to nothing to do with what he or she reports or chooses to do.

The problem, recognized by generations of scholars, is how to develop a method that extracts principles of justice in the face of the countervailing pressures of self-interest. As Adam Smith noted over two hundred years ago, "Every individual necessarily labours to render the annual revenue of the society as great as he can. He generally, indeed, neither intends to promote the publick interest, nor knows how much he is promoting it. . . . He intends only his own gain, and he is in this, as in many other cases, led by an invisible hand to promote an end which was no part of his intention." More playfully, he wrote, "It is not from the benevolence of the butcher that we expect our dinner, but from his regard to his own self-interest."

Rawls believed that we could access the core principles of our sense of justice by setting up constraints on how the discussion proceeds: When individuals debate the principles of justice, they must operate under a *veil of ignorance*. The veil covers up knowledge of their own or anyone else's personal characteristics such as age, wealth, religious beliefs, health, or ethnicity. These characteristics, so Rawls argued, get in the way of clearheaded thinking about justice. They are self-serving biases, morally irrelevant features that distort our capacity to articulate the principles of justice from an impartial perspective. The veil forces impartiality, tapping into the fact that every participant should be globally selfish—wishing that everyone obtain the best possible deal, the most resources, and the best opportunities. Oddly, this move has an eerily familiar resemblance to Marxist politics, in which individual significance dissolves into group significance, regardless of effort or talent.

Whom do we recruit to sit behind the veil and hash out the principles of justice? For Rawls, the participants should be capable of rational judgments, what he called "reasonable men." These people, independent of wealth, so-cial status, race, gender, nationality, or religion, must have the following characteristics: (1) a mature intellectual and emotional core; (2) a typical level of education given their age; (3) the ability to deliver reasonable judg-ments, most of the time; (4) a sufficient level of sympathy toward others to consider their feelings, interests, and potential suffering. Individuals with these characteristics should generate spontaneous judgments, "apparently determined by the situation itself, and by the felt reaction to [them] as a re-sult of direct inspection; . . . [they are] not determined by the conscious ap-plication of some rule, criterion, or theory. One model of a spontaneous judgment might be the kind that we make when we see unexpectedly some object of great natural beauty. We have no preconceived ideas about it, but, upon seeing it, we spontaneously exclaim that it is beautiful. A spontaneous judgment will be similar in this respect, but will be made concerning virtu-ous actions, various goods of life, and the like."[9]

Rawls believed that under the veil of ignorance, everyone would agree to two central principles, principles of justice that are part of our moral fac-ulty: (1) all members of society have equal right or access to basic liberties, and (2) the distribution of social and economic goods should be set to ben-efit the least advantaged members of society; this second principle allows for unequal distribution, but only if those at the bottom profit. Rawls further argued that since these principles are part of our moral faculty, not only will they percolate up from under the veil of ignorance, but they are justifiable and should be adopted by all reasonable members of society.

Though there has been much discussion of Rawls's principles, for now I want to make two small points. First, we must distinguish between the processes that are responsible for implanting these principles in our heads and the processes that lead us to accept or reject them. Even if these principles are part of our innate endowment, we need not accept them. If we reject them, deciding that other principles are more consistent with our sense of justice, we must be prepared for conflict and instability. Sec-ond, even if Rawls's specific principles of justice fail, his methodological proposal focuses our attention on the appropriate dimensions of the prob-lem, including issues of self-interest, impartiality, operative principles, and spontaneous judgments.[10]

Rawls's use of the linguistic analogy is important because it raises the possibility that some aspects of our perception of fairness may rely upon principles that operate outside of our awareness. His use of the original position is important in setting out certain methodological criterion for tapping principles of fairness. His discussion of justice as fairness was an important ideal. When it comes to humans, in real situations where fairness is at stake, how do we fare?

JUST PARAMETERS

Every morning, at around seven-thirty a.m., Richard Grasso was handed a set of papers. Moments later, he stood up from his desk and rang a bell. This ritual, approximating the opening moments of a boxing match, was part of his job as chairman and CEO of the New York Stock Exchange. On September 17, 2003, he resigned from his job. Why would someone in the process of renegotiating a contract of approximately $140 million in compensation step down? Following a communal display of disgust, Grasso decided that he should move on.

On the one hand, the public's response makes no sense. If you look at Grasso's annual income in the year 2001, it was slightly higher than that of the late-night television host David Letterman and the basketball star Shaquille O'Neal but less than that of the equally huge basketball star Michael Jordan and the pop singing icon Britney Spears. We may be envious of these celebrities, but I haven't yet heard a war cry of "unfair!" On the other hand, the public's response makes complete sense when looked at from the perspective of fairness as the distribution of resources based on effort. Does Grasso work harder than and contribute more to society than a plumber, artist, teacher, or architect, who may work sixty hours per week? Does Grasso have a skill that is more in demand or demanding than the CEOs of other Wall Street firms? No and no. Grasso was taking more than his fair share. Our species was rightly pissed off.

The Grasso case emerged in the face of other corporate debacles, including Enron and WorldCom. Grasso didn't cheat or deceive. In contrast, the CEO of Enron, Kenneth Lay, kept his personnel in the dark as he sequestered millions of dollars for personal gain. In each of these cases, however, the public smelled a rat.

Thus far, I have discussed fairness without saying much about its meaning, the principles that drive an exchange between individuals, whether they are universally shared, free to vary according to cultural quirks, or some combination of universal and culturally variable factors. Adopting the analogy to language, one would expect a universally held principle of fairness that varies cross-culturally as a function of parametric variation; experience with the native environment triggers the culture's specific signature of fairness and fair exchange. Once parameters are set, judgments of fairness may seem as incomprehensible across cultures as judgments of grammaticality for word order. But what is the relationship between a principle and parameter?

The linguist Mark Baker[11] suggests cooking as a simple metaphor for thinking about principles and parameters. Principles are like cooking recipes in at least two ways: certain ingredients—called "parameters"—are necessary, while others are optional, and once an ingredient has been added, it interacts in important ways with other ingredients. When we make a cake, we have to do some things in a particular order, such as turning the oven on before we are ready to put the batter in, setting the temperature to the right level to avoid burning, mixing the batter thoroughly, and putting the batter in a mold that will allow for expansion. Some of the ingredients we put into the cake are necessary if we want to bake a particular kind of cake. A sweet cake requires an ingredient of sugar or a sugar substitute. It also requires eggs and baking soda to rise. Other ingredients, such as chocolate, are entirely optional, as is the amount of the ingredients. Looking at cooking in this way shifts the continuous process of ingredients unfolding and mixing over time to a discrete process whereby particular decisions are made at each step in the recipe. This is the way parameters work in language, and, I suggest, in morality as well.

There is another way for us to think about the recipe metaphor. A recipe creates a product: a bouillabaisse, fish curry, or soufflé. It should then be possible to reverse the problem and, to some extent at least, figure out the ingredients and the process of assembly. A good chef would come close, tasting and smelling the key ingredients, figuring out how they blend. But even the best chef would have difficulty pinpointing the proportions of each ingredient and the order in which they were put together. Reverse engineering would yield a recipe, but there would be differences with the original. Although having the recipe undoubtedly yields greater homogeneity in the end product, variation will arise. If ten different people follow the same recipe, odds are high that each will produce a different cake. One might be lighter, another sweeter, yet another drier than the others. Differences in the output might be driven by the quality of each cook's ingredients, their oven, attention, cooking experience, and patience. These are all matters of performance, and may say nothing at all about what they know, consciously or unconsciously.

To appreciate the importance of parametric differences across languages, and anchor the analogy with morality, consider the *word order* parameter. Languages differ in how they put words together to form a question. No living language, however, takes the order of words used in a statement of fact and then reverses their order to create a question, as in "The White House is where the president lives" to "Lives president where the White House is?" This is an absolute constraint, with no exceptions. There is also the distinction between words classified as nouns and not verbs. All languages make this distinction. Other universals provide options, but only a limited range. For example, most of the languages of the world use either a subject-verb-object or a subject-object-verb order. If languages had the complete freedom to change, we would expect to find as many different patterns of word order as we find different kinds of cereal or clothing styles.

There must be constraints on language change, because in development, the language we hear sets up the operative principles and parameters. Once an American-born child experiences her native language and creates sentences by placing the subject before the verb, and the verb before the object, it becomes difficult for her to order words in any other way. Whereas every child is born with the capacity to acquire a range of possible word orders, once the native language is acquired, alternative forms are hard to put in place.

If the principles-and-parameters approach is right, at least in some form, then we can characterize linguistic diversity by identifying how each of the core parameters is set. Flip a few parameters this way and that, and you get English. Flip the same ones a different way, add a few more parameters, and you get Chinese. If this approach is correct, then we can not only explain the diversity of all *possible* languages, we can also identify a set of *impossible* languages, those that are not learnable based on the constraints set up by the innate principles that constitute our universal grammar. This is comparable to saying that we can't echolocate like a bat because we don't have the necessary machinery. There are constraints on what is learnable. Some of the constraints are internal to the system— language for humans, echolocation for bats—and some are external, including limitations imposed by memory, the capacity of the vocal tract to produce sound, and of the hearing system to decode it.

There is disagreement among linguists about the significance of parameters in thinking about the design of language. Independently of how this shakes out, my own sense is that the notion of a parameter is useful for thinking about morality and fairness more specifically. In the same way that our universal grammar provides a toolkit for building a specific grammar, in which certain principles and parameters hold and others do not, our universal moral grammar provides a different toolkit, enabling us to implement particular principles and parameters but not others. This framework raises new questions, such as: For a given ethical system, which ingredients are optional and which are mandatory? When particular ingredients are implemented, how does this impact upon future change?

Let's return to the public outcry against Richard Grasso, what we mean by "fair," and the principles that underlie our judgments of an equitable exchange. Surveying the conceptual terrain, the linguist George Lakoff produced a ten-tiered taxonomy of fairness (see table).[12]

Lakoff's taxonomy forces us to recognize that claims of fairness are vacuous in the absence of more precise specification, clarifications of the form "I acted fairly because I allocated resources *according to need.*" It makes no sense to ask the generic question "Was it fair?" as our judgment is dictated by the particular details of the exchange or distribution. To deliver a judgment of fairness, we must assess the relevant parametric variation, whether it targets distribution, opportunity, responsibility, power, or some other commodity. Lakoff's taxonomy also raises questions about the relationship between the different forms of fairness—whether a society that implements one form necessarily implements or excludes some or

TYPE OF FAIRNESS	SPECIFIC EXAMPLE OF FAIRNESS
Equality of distribution	Every person gets one meal
Equality of opportunity	Everyone is eligible to apply for the job
Procedural distribution	Benefits are established by the rules of the game
Rights-based fairness	You get what you are entitled to (e.g., property)
Needs-based fairness	Those who need more, get more
Scalar distribution	Those who work harder, get more
Contractual distribution	You give based on what you promised
Equal distribution of responsibility	Effort is equitably shared
Scalar distribution of responsibilities	Those who can do more have greater responsibility
Equal distribution of power	Everyone can vote

all of the others. For example, are societies that commit to equality of distribution unlikely to also commit to a scalar distribution based on work effort within the same context? Resolving these problems not only will impact upon academic discussions of how humans trade off self-interest for more altruistic dispositions, but will also shed light on policy, and how we think about human rights more generally.[13]

GAMES FOR ADULTS

Survivor is the daddy of reality TV. It was created to provide entertainment, and perhaps a soupçon of insight into how our ancestors lived as hunter-gatherers on the savanna, struggling in games of cooperation and competition. In its first incarnation, sixteen "castaways" descended onto a gorgeous Polynesian island and divided into two tribes. They participated in a Darwinian game of survival, where the fittest would take home a million-dollar prize. As in nature, competition occurred both within and between tribes; some of the competition within tribes was built out of co-operation, trust, and commitment, as two or more individuals formed alliances to overturn a third. It was as if these people, ripped from their day jobs as truck drivers, corporate trainers, teachers, and retired navy personnel, had studied a page out of Jane Goodall's chimpanzee diaries or out of a book about hunter-gatherer tribes. After thirty-nine grueling days, Richard Hatch emerged as the winner. Hatch had this to say about his victory: "I really feel that I earned where I am. The first hour on the island, I stepped into my strategy and thought, 'I'm going to focus on how to establish an alliance with four people early on.' . . . I didn't want to just hurt people's feelings or do this and toss that one out. I wanted this to be planned and I wanted it to be based on what I needed to do to win the game. I don't feel I was diabolical. There were ethics in this game . . . I don't regret anything I've done or said to them, and I wouldn't change a thing."

Hatch's comments capture several key elements of the show, and also serve to explain why *Survivor* and its entertainment offspring have garnered more than 50 million viewers a week in the United States. We like to watch other people struggle with temptation and conflict, and we like to make pronouncements about whether particular actions are permissible or not. At the heart of Hatch's comments are issues of personal responsibility, fairness, loyalty, desire, and social strategizing. Hatch says that he earned his reward and didn't feel guilty about his strategic plotting. For him, this was a game with rules, and he played fair and square. He didn't intend to hurt anyone directly, even though he acknowledges that he might have along the way; his intent was to win, not to create emotional chaos for others. When tempted by a more advantageous alliance, he took the opportunity. When offered the opportunity to defect, he controlled himself when it counted most.

Hatch's victory encapsulates the balancing act between temptation and control that is characteristic of our species, and many that preceded us. It also raises questions about the principles guiding our judgments about particular actions, especially what we consider fair and ethical. Only the tools of science can shed light on this problem.

The intuition driving most economic games of cooperation is that the human mind has been designed to maximize payoffs—money, food, mates, babies. If the payoffs to defection are higher than the payoffs to cooperation, individuals defect. This intuition was captured in the suspense movie *The Score*, starring Ed Norton and Robert De Niro. The plot centers around the planning and implementation of a heist to steal a sixteenth-century French scepter under heavy surveillance. By masquerading as a mentally handicapped janitor, Norton has made all the necessary contacts with the guards, and is now intimately familiar with all the entries and exits. Norton, however, needs De Niro's skills as a thief and safecracker. Knowing this, De Niro demands a sixty-forty deal. Norton initially rejects this offer as unfair, motivated by the conviction that he alone developed the plans for the heist. De Niro throws his weight around and makes it a sixty-forty offer or no deal. Norton agrees. As the heist unfolds, it becomes clear that Norton has changed his mind: he plans to escape with the loot, leaving De Niro high and dry. De Niro, however, anticipates Norton's devious motives, outwits him, and reverses the situation, leaving Norton without a penny. This plot illustrates a simple truism about cooperation: Once individuals realize that there are opportunities for differential payoffs, each player rapidly works out the differences, often unconsciously and with lightning speed, moving toward the strategy that maximizes individual returns. And in many games of cooperation, defection is the best individual strategy.

To examine how people allocate resources, economists create simple games designed to capture some corner of reality. Traditionally, economists start from the assumption that people are self-interested and will do what they can to maximize their payoffs.[14] Two of the best studied games are the dictator and ultimatum, classically played between a pair of individuals, with all information about a player's history, identity, and reputation withheld. Thus, like Rawls's imaginary scenario for constructing principles of fairness, these games are also played behind a veil of ignorance. Unlike Rawls's scenario, however, these games eliminate the opportunity for negotiation. Each game is set up in such a way that one individual plays the role of proposer while the other plays the role of responder. Both players know the rules of the game as well as the starting pot of money. In the dictator game, the proposer starts off with $10 and has the option of giving the responder some portion of the $10 or none at all. Once the proposer announces his offer, the game ends; the responder has no opportunity for negotiation. The ultimatum game starts out the same way, with the proposer announcing an offer of some portion of the $10 or nothing at all. Next, the responder either accepts or rejects the proposer's offer. If the responder accepts the offer, he keeps the offered amount and the proposer keeps what is left. If the responder rejects the offer, both players walk away empty-handed.

If, as economists generally assume, proposers are rational money-maximizers—*Homo economicus*—then in both games they should make the smallest possible offer. This logic should be transparent to anyone thinking through the responder's position: responders have no option in the dictator game; but in the ultimatum game, they should accept any amount offered. However, when this game is played in industrial societies, such as the United States, Britain, and Japan, with varying amounts of money in the initial pot, and with adult players of different socioeconomic backgrounds, the results from both games indicate something quite different. In the dictator game, there is no cost to offering the lowest amount, because the responder must take whatever is offered. Many proposers follow this logic and offer zilch. Others, however, play by an apparently irrational rule and offer half of the initial pot.

The ultimatum game is different because the responder can reject an offer—perhaps a form of retaliation or spite—leaving both players with no money. But why would a responder do that? A rational money-maximizer should offer a small amount, say $1 out of an initial pot of $10. Responders should accept the offer, since something is better than nothing. But here, too, proposers offer in the range of $5, and responders reject offers under $2. In both games, therefore, players appear to play irrationally. No one told them to play fair and split the pie in the dictator game. No one told them to be foolish and reject $2 in the ultimatum game. Given that the players in these games are not brain-damaged patients with a deficit in decision-making, why are they so irrational? What happened to our rational, self-interested, money-hungry player? What happened to *Homo economicus*?

The standard explanation for these results is that although we may have evolved as *Homo economicus*, we are also born with a deep sense of fairness, concerned with the well-being of others even when our actions take away from personal gain. As the mathematical biologists Martin Nowak and Karl Sigmund put it, "The fiction of a rational 'Homo economicus' relentlessly optimizing material utility is giving way to 'bounded rational' decision-makers governed by instincts and emotions,"[15] a view that defines the anatomy of *Homo reciprocans*. In the ultimatum game, proposers are given money without paying any costs—a freebee. Responders who reject offers may be foolish in terms of turning down some cost-free cash, but their rejections tell an important tale. In the eyes of the recipient, and in the mental calculator that is charged by emotions, some offers are downright unfair. After all, an experimenter simply handed the proposer some cash. What is fair is something approximating half of the initial pot. At least that is the story so far, based on one game, played among young adults in industrial countries.

Cooperative interactions are often repeated, sometimes with the same set of people, sometimes with a new pairing each round. Let's say that you have just played one round of the dictator game with person A, who offers nothing, and one round with person B, who offers $5 out of the $10 starting pot. Now it's your turn to play one round of the dictator game with A and one with B. Will you make equal offers to A and B or different offers? If you are like most people playing repeated games with knowledge of what others offered in earlier rounds, you will give $0 to A and around $5 to B. People play repeated games with such strategies because they use reputation to guide cooperation. Mathematical models of this problem reveal that fairness evolves as a stable solution to the ultimatum game if proposers have access to information about a receiver's past behavior.[16] When it comes to group level activity, reputation fuels cooperation and provides a shield against defection.[17]

Many of the games that experimental economists play involve anonymous players who come and go following a single-shot interaction. In some sense, this might better capture our early evolutionary history as nomadic hunter-gatherers who perchance encountered another group and perhaps had opportunities to exchange goods or information. As we shifted from a hunter-gatherer lifestyle to a more sedentary existence, however, two features of our environment changed, which would forever alter the problem of cooperation: group size increased, and we started to rely on shared resources, locally available through the development of agriculture and farming. With an increase in the number of people, opportunities to interact and cooperate increased, especially with genetically unrelated individuals. This placed increasing pressure on our capacity to remember who did what to whom and when. Couple an increase in group size with a sedentary lifestyle, and you have a new problem of resource use and sharing. Instead of moving from place to place while hunting and gathering, our shift to agriculture created a dependency on what we produce and use nearby. Although hunter-gatherers cooperate and share, they typically do not confront what the social scientist Garrett Hardin has called the "tragedy of the commons."[18] The basic idea here is that there is some public resource that everyone is free to use, and thus potentially overuse. Given the problem of overuse, how can a group prevent selfish overuse by one or more individuals? If everyone uses the resource in private, then there is no way to catch cheaters, individuals tempted to take a bit more than what the group has decided is a fair allotment.

In the fourteenth century, British villages repeatedly fell victim to the logic of the commons. Each village was associated with a common pasture for their cattle and sheep. The pasture represented a shared resource. But since household wealth increased with the number of animals grazing on the pasture, the temptation to acquire more emerged. More animals meant more use of the pasture. More use of the pasture, less pasture. Less pasture, more competition. More competition, more strife. More strife, less village cohesion. Eventually, village after village dissolved. Private ownership emerged as a counterstrategy of control over the temptation to cheat. Each household had its own pasture to maintain. But soon, this strategy failed as well, as individual greed led to new ways of acquiring more land. Soon, there were big and small landowners, and soon after that, rich households with land and poor households with nothing at all. This increasing economic division spread throughout the world, often without concern for the next generation. The travel writer Dayton Duncan captures this laissez-faire attitude in describing the voices of Texas cattlemen, who, after overstocking the land in the post–Civil War period, had this to say at a town meeting: "Resolved that none of us know, or care to know, anything about grasses, native or otherwise, outside of the fact that for the present, there are lots of them, the best on record, and we are after getting the most of them while they last."[19]

One way to maintain cooperative use of the land is to make resource use public knowledge. An individual's image or reputation can thus play a critical role in cooperation.[20] Experimental economists have run several laboratory experiments with college students, showing that both reputation and punishment have positive effects on cooperation.[21] For example, in a set of experiments carried out by the economist Ernst Fehr, individuals played a public-goods game, where each player could pay to punish noncooperators. Although individuals initially attempted to reap the benefits of the commons without incurring a cost, stable contributions to the public good emerged once individuals realized that others in the group could pay to punish the slackers. That individuals will pay to punish cheaters shows that moral indignation can fuel actions that are of immediate personal cost but of ultimate personal benefit as public goods accrue.

One interpretation of punishment in these games might be that it is selfishly motivated. If I punish others, my status goes up relative to their status. A suite of results make this interpretation implausible. Consider once again the original ultimatum game. If responders wanted to reduce another's status, they should reject anything less than a fifty-fifty split. They don't. In an ultimatum game where the proposer has $10 but can offer only $2 or $10, offers of $2 should be rejected. They are not. Punishment certainly does reduce another's status, but sometimes people punish to make clear who is in and who is out, even when it costs them personally.

The results of these games provide evidence of a profound property of human nature: the only way to guarantee stable, cooperative societies is by ensuring open inspection of reputation and providing opportunities for punishing cheaters. There will always be weak individuals who are unable to control the temptation to defect and take more than they should. By spotlighting such individuals, and providing mechanisms for punishing them, we may safeguard the public goods that are the right of all human beings—a principle of Rawls's justice as fairness.

A growing group of anthropologists and economists have taken the evidence discussed thus far as a signature of a uniquely human cognitive adaptation. Whereas we inherited a largely selfish nature from our ancestors, we also evolved a uniquely human psychology that predisposes us toward a different form of altruistic behavior: *strong reciprocity*.[22] Reciprocal altruism, as originally proposed by the evolutionary biologist Robert Trivers, is based on selfishness: I scratch your back with the expectation that you will scratch mine later. I am only scratching your back because I know that I will need you to scratch mine in the future. My action is purely selfish. Strong reciprocity is not. As defined by the leading proponents of this position, strong reciprocity is a "predisposition to cooperate with others and to punish those who violate the norms of cooperation, at personal cost, even when it is implausible to expect that these costs will be repaid either by others or at a later date." Although strong reciprocity is not selfish, it is strategic: only cooperate with those you can trust and nail those who are untrustworthy because they have cheated. The consequence of punishment may of course be beneficial to the punisher: cheaters can revert to being good citizens. But the punisher's intent is not to convert. It is to make cheaters pay by excluding them from the circle of cooperators. It is to make explicit the difference between in-group and out-group. Fehr's behavioral experiments support this explanation, because individuals pay to punish even if they will never interact again with the cheaters. Their punishment cannot, therefore, be designed to bring them back into the fold, to convert them from sinners to virtuous cooperators. Further, those who punish most are also those who contribute most in public-goods games, which suggests that they have the most at stake, and have the greatest interest in maintaining the circle of cooperators; as expected, cheaters both contribute and punish least.

All of the studies discussed in this section lead to the conclusion that we have evolved the capacity to punish those who cheat and selectively focus our cooperative efforts on those who are trustworthy. But this is a highly Western, developed, industrial-nation account of history. What about the East? What about the small-scale hunter-gatherer societies in Africa, Australia, and South America?

EVEN THE BONGO BONGO

Social anthropologists are fond of pointing out cultural exceptions to apparently universal patterns of human behavior: but not in the Bongo Bongo! Experimental economists have largely restricted their tests to a single-subject population: university students. To what extent do these subjects—the equivalent of laboratory mice for psychologists—provide results that generalize to other humans, allowing for more broad, sweeping conclusions about our species?

An extraordinary collaborative project between anthropologists, psychologists, and economists focusing on small-scale, nonindustrial societies begins to address the question of whether equity distribution is a universal, part of our species-specific mind-set. In the same way that all humans share a universal grammar but might speak Chinese, English, or French, it appears that all humans share a universal sense of distribution fairness, with cross-cultural differences coupled to local quirks of exchange, justice, power, and resource regulation. The idea here, returning to our analogy with language, is that fairness is a universal principle with the potential for parametric variation and constraints. Cultures set the parameters based on particular details of their social organization and ecology, and these settings constrain what are optional forms of exchange and distribution.

Taking advantage of fieldwork conducted on each of the globe's continents, the anthropologist Joseph Henrich and his colleagues ran a variety of bargaining games in fifteen small-scale societies, including foragers, slash-and-burn horticulturalists, nomadic herders, and sedentary small-scale agriculturists.[23] Subjects played each game anonymously and for potential payoffs equivalent to one or two days of salary. In contrast to results from ultimatum games played among college students in industrial societies, the most common offer in this sample ranged from 15 to 50 percent. Responders either never rejected low offers or rejected them as often as 50 to 80 percent of the time. This variation suggests, contrary to the standard economic models, that there is no clear-set point for what constitutes a fair offer in this game. Rather, there is significant cross-cultural variation that maps onto social norms within each culture. For example, among the horticulturist Au and Gnau of Papua New Guinea, individuals offer in the range of industrial societies (40 percent), but reject at much higher rates (50–100 percent). The explanation for this high rejection rate appears to be that among the Au and Gnau, gift-giving plays a central role. Accepting gifts, even if unsolicited, sets up a commitment for future reciprocation. If the gift is large, then the receiver is in a subordinate position and is expected to return an equally large gift. Thus, receiving an unsolicited gift may make these people anxious. Anxiety may cause them to reject even quite fair offers. This game is a nightmare for them. In contrast, the foraging Ache of Paraguay accept low offers and typically offer more than 40 percent. This pattern of generosity coincides nicely with the generally cooperative tendencies of the Ache, where hunters invariably share their catch with the rest of the camp. The slash-and-burn horticulturist Machiguenga of Peru made the lowest offers (15 percent) of all the peoples sampled, and rarely rejected these low offers (10 percent). This pattern also fits well with their cultural practices, involving little cooperation, exchange, and sharing beyond the family unit. Overall, each society expresses some sense of fairness, but societies vary with respect to their perception of inequity and their willingness to punish by means of rejecting offers; some societies reject offers when they are deemed unfair, while others never reject, regardless of the amount offered.

These simple economic games suggest that fairness is a universal principle with parameters set, presumably in early development, by the local culture. For example, in the ultimatum game, there are parameters concerning the responsible agent, the original source of the resources, the dependency on others for acquiring resources, and the option of rejecting an offer. Once set, the psychological signature of the culture constrains what counts as a fair and permissible transaction—for what counts as an inequity in terms of resource distribution. If a culture sets the agent parameter to responsible (as opposed to a setting of no responsibility), and the rejection parameter to nonoptional (as opposed to optional), then players might perceive the ultimatum game as akin to the dictator game, where there is no opportunity to reject an offer. This pattern looks like that of the Machiguenga of Peru, as well as the Hadza hunter-gatherers of Tanzania.

It is interesting to note that no culture in the sample presented by Henrich and colleagues made offers of less than 15 percent or more than 50 percent in the ultimatum game. If the parameter view is correct, and the strong analogy to language holds, then no culture will ever reject offers under 15 percent, and no culture will ever offer more than 50 percent. If they do, such patterns will exist for an eye blink of human history. At least this is one prediction that emerges from these studies and that field anthropologists can test.

Industrial societies do not fully capture our species' psychological signature. In the same way that laboratory mice do not capture the riches of the world's fauna, university students do not capture the riches of human nature. Economic theory that takes *Homo economicus* as its target subject is doomed to failure. Economic theory that looks to the opposite extreme, assuming that we are simple you-scratch-my-back-and-I'll-scratch-yours *Homo reciprocans*, doesn't work, either. Individuals are not acting on the basis of pure self-interest. Nor are they acting on the basis of a universally agreed-upon set point for fairness. Rather, the patterns of exchange are heavily influenced by local cultural practices. Economic theory can only work if it recognizes a universal principle of fairness, while also acknowledging that cultures will tweak the parametric variation in order to constrain what counts as a permissible exchange. Our moral faculty outputs principled intuitions about distributive justice. But there will undoubtedly be differences between our intuitive judgments about a just transaction and our behavior in the same situation. Many may have the intuition that a particular action is unfair, but when it comes to making a decision, ignore the intuition and feed self-interest. We should not, however, assume that because people act unfairly or give incoherent justifications for their actions, that the idea of a moral faculty is mere fiction. What we know and how we choose to act will often occupy separate universes.

NEANDERTHAL WELFARE

In 1993, the alternative rock group Crash Test Dummies wrote a hit tune about our ancestral past as cavemen, with the refrain: "See in the shapes of my body leftover parts from apes and monkeys." But they could have time-traveled a bit more, as the economist Ken Binmore did when he suggested that the intuitions underlying Rawls's social contract model can be seen in the architecture of our mind, leftover circuitry from the cavemen.[24]

Hunter-gatherers provide the best window into the original social contract, because there is no philosopher-king pensively developing and dictating rules. Hunter-gatherers are largely egalitarian. When a major decision is called for, it is a decision that is achieved collaboratively. In fact, as the anthropologist Christopher Boehm has pointed out, when modern humans evolved, they reversed the typical dominance hierarchies of our primate relatives, and made it impossible for any particular individual to rise in status. As Binmore notes, however, this egalitarian system requires a careful alignment of different factors: "A hunting and gathering way of life in itself does not guarantee a decisively egalitarian political orientation; nomadism and absence of food storage also seem to be needed. Nomadism in itself does not guarantee egalitarianism either, for after domestication of animals some pastoral nomads were egalitarian but others became hierarchical. Nor does becoming sedentary and storing food spell the end of an egalitarian ethos and political way of life. Neighbors of the Kwakiutl such as the Tolowa and Coastal Yurok also lived in year-round villages with food storage, but they kept their leaders weak and were politically egalitarian."

In our hunter-gatherer past, the most important playground for our moral principles would have been food sharing. Due to the unpredictability of hunting large game—as opposed to the relative predictability of the gathering part—a self-reinforcing norm for fairness evolved, relying heavily on sharing with group members, fueled by social pressures to maintain the status quo. Based on these patterns, Binmore proposes that we are born with a principle of fairness that targets equity distribution, but the particular content of this principle varies depending upon the local ecology and random quirks of a particular culture.

General evidence in favor of Binmore's proposal comes from studies of experimental economics in which bargaining games are set up to simulate certain aspects of hunter-gatherer life, including small groups of familiar individuals where reputation and punishment play a role in guiding behavior. Thus, in games where players know each other, can chat prior to making an offer, have information about other players' reputations as generous or stingy, and can punish those who cheat, cooperation emerges triumphant over freeloading cheaters. More specific evidence that these laboratory games do simulate real hunter-gatherer life comes from the anthropologist Kim Hill and his long-term study of the Ache of Paraguay.[25] As I previously mentioned, the Ache are a highly cooperative people and when they play the ultimatum game, proposers offer close to 50 percent of the initial amount and responders rarely reject. To what extent does this contrived game map onto their natural patterns of behavior, especially in the context where it matters most: foraging for unpredictable and highly valuable game? Although individuals hunt alone, both men and women spend close to 10 percent of foraging time in personally costly cooperative foraging; on some days, individuals may devote over 50 percent of their time to cooperative foraging.

Where, then, does this leave us in terms of justice as fairness? Are any of the signatures of our caveman past within us? Are we aware of the principles? Would we support Rawls's difference principle, attending to the worst off as a standard for resource distribution? Most experimental economists don't ask their subjects to justify offers or responses in bargaining games. They think that justifications are inherently messy, and open to the vagaries of life, including current testing conditions, complicated emotions, and fuzzy recall. Some economists, however, and some psychologists interested in economic principles, have not only examined justifications in traditional bargaining games, but also run experiments to determine whether Rawls's abstract ideas have any success in the real world.[26]

Recall that Rawls's method of exploration involves extracting principles of fairness by assuming broad self-interest among the participants, equal ability, and the need to cooperate to achieve particular ends. From this foundation, we achieve impartiality by requiring individuals to work out the principles behind a veil of ignorance, in the absence of heated passions. Rawls believed that this two-pronged approach would guarantee a moral victory, a just society. He summarized his conviction in the last sentence of *A Theory of Justice*: "Purity of heart, if one could attain it, would be to see clearly and to act with grace and self-command from this point of view."

The political scientists Norman Frohlich and Joe Oppenheimer brought American college students into a laboratory setting, provided them with the rules outlined by Rawls's contractarian approach, and allowed them to discuss and settle on a set of principles focused on distributive justice. Just as Rawls predicted, subjects readily settled on a principle of fairness. But the winning principle was not quite as Rawls predicted. No group selected the difference principle, where distribution is anchored by the worst off. Instead, groups settled on a principle that maximized the overall resources of the group while preventing the worst off from dropping below some preestablished level of income. This principle provides a safety net for those who are disadvantaged, for whatever reason, while allowing for extra benefits to flow toward those who contribute more to society. Returning to the taxonomy sketched by Lakoff, the winning principle combines the parameters of equality distribution with scalar distribution.

One problem with these experiments is that they don't allow for the kind of long-term reflective equilibrium or contemplation that Rawls envisioned. What would happen if those subjects involved in the original veiled negotiation then went out into the world, carried out their day jobs, and then reported back months later on how the parameterized principle worked out? Would subjects stick to the principle once they could see whether it delivered on justice as fairness? Returning to the linguistic analogy, to what extent does our competence to judge what would be a just society line up with our performance, carrying forward into action the principles that we think are fair? Even though our emotions may play no role in judging the relevant principles, are they engaged when we are at work, collecting and distributing scarce resources?

Frohlich and Oppenheimer designed an experiment to bring our judgments of fairness into a situation of implementation. There were three groups. Two groups read through a set of potential principles for redistributing income and then collectively selected one principle of distribution to govern a taxation scheme; one group required a unanimous vote, the other a majority. Since subjects selected a principle without knowing about the task, they were effectively operating under a veil of ignorance, incapable of predicting either their own productivity or status in the local economy. For the third group, an experimenter imposed a distribution principle that maximized the mean income of all group members, while not allowing anyone to go below a certain minimum. Each individual in each group then went to "work," correcting spelling mistakes that were fabricated from the convoluted writings of the American sociologist Talcott Parsons.

The pay scale was the same for each group, and was based on productivity. Following a round of work, the experimenter calculated each individual's posttax earnings, indicated the yearly salary, and then redistributed the earnings based on the principle in play. Each group played three rounds of this work-pay-redistribution process and, after each round, evaluated their attitudes toward the principle and its productivity.

Individuals playing in the open-choice groups selected a principle that maximized average income while maintaining an eye on the minimum level of income, or floor. Thus, all three groups started by applying the same principle, even though only two groups freely chose this principle. In discussion, most groups mentioned the veil of ignorance as a constraint, discussed their goal as maximizing personal income (selfishness), and saw work ethic as an important component, something worthy of extra compensation. Attitudes toward these principles were high, and showed little change over the course of the experiment. However, when subjects had the freedom to choose, and vote unanimously, their satisfaction and confidence in the principle were significantly higher than when the same principle was imposed on them. The average-income-and-floor principle emerged as the clear winner. As a principle, it was stable after multiple iterations of the work-pay-redistribution cycle, but functioned to instill confidence in people, both those at the top and those on the floor. Contrary to many current political analyses, an income-distribution principle that allows for inequalities while taking care of those who are most in need does not reduce incentives to work hard, nor does it create a sink of free riders, individuals who suck the welfare system dry. Those who received from other players, and who actively participated in deciding the best principle, almost doubled their efforts in order to contribute to the overall income. In contrast, those working under the same regime, but with the principle imposed, cheated and decreased their efforts, because they perceived redistribution through taxes as their right.

We can derive a simple equation from these experiments, a rule of thumb that adds a bit of bedrock to the construction of a cooperative society:

Freedom of choice + discussion of principles = justice
as fairness

In Rawls's wording, liberty, considered judgments under a veil of ignorance, and contemplation to reach reflective equilibrium, are essential ingredients for not only discovering but implementing the principles of a just society. As many modern, large-scale corporations have learned, giving every worker, independent of their contribution to productivity, a feeling that they are valuable and have a voice, increases productivity, satisfaction, and cooperation. *I Love Lucy* aficionados will recall the marvelous episode in which Lucy and Ethel are working in a fudge factory. They fall behind in handling, and have no opportunity to stop the process except to eat the fudge as it whizzes along the assembly line. Worker dissatisfaction and frustration at its finest.

Frohlich and Oppenheimer concluded their study by stating that "Concern for the poor and weak, a desire to recognize entitlements, and sensitivity to the need for incentives to maintain productivity all enter into subjects' deliberations regarding a fair rule." Subjects in this case were college students. Follow-up studies in Poland, Japan, Australia, and the Philippines yielded almost identical results, causing Rawls to remark: "If the results hold up it may be that the difference principle cuts across the grain of human nature."[27] Of course, it is not quite Rawls's difference principle, and it may not cut across human nature. As we learned from the bargaining studies in small-scale societies, individuals in industrial nations capture only a small corner of the variation when it comes to judgments of fairness. What the Frohlich and Oppenheimer experiments also skip is Rawls's intuition that the principles underlying justice as fairness may be like the principles of grammar, scripts engraved in the mind, operative without our awareness. In all of the studies run thus far, an experimenter hands subjects a list of potential principles or imposes one on them. Rawls's social-contract model imagines individuals operating under a veil of ignorance, discussing potential routes into securing a just society. What spontaneously percolates up are two principles, one for basic liberties and the other for tolerating inequalities only if they allow the worst off in society to rise in status. Few have argued with the first principle. The political right, however, has attacked the difference principle on the grounds that it is offensive to take away and redistribute what someone has rightly earned. The political left has attacked the same principle for its toleration of inequities. What emerges from Frohlich and Oppenheimer's work is that people are not bothered by inequities so long as the least well off can live a satisfactory life. And no one finds it offensive to have their earnings redistributed so long as the average income in their group is better off. These results leave open the role of more intuitive processes, including unconscious principles and emotions in guiding both people's judgments and behavior in the context of distributing or accessing resources.

Some of the deepest insights into our intuitive sense of fairness comes from the work of the psychologist and Nobel laureate in economics Daniel Kahneman and his lifelong collaborator, psychologist Amos Tversky.[28] Based on a framework called "prospect theory," they argue that people approach problems involving moral conflict with rules of thumb or heuristics that operate over losses and gains. Our sensitivities to pleasure and pain provide signatures of the interaction between mind, brain, and behavior. When we consider the value or utility of a resource, we do so in reference to our current state and the extent to which obtaining the resource will significantly change this reference state. Like most other organisms, humans are averse to loss, including lost opportunities as well as decreases in current resources. Problems involving delayed gratification are hard, because of conflict between waiting for a large gain and missing out on an opportunity for an immediate but smaller gain. Taking the immediate but small gain is a loss of opportunity for a larger one. But waiting also depends on the individual's sense of current needs. If you are starving, waiting may not be an option. Fairness can therefore be assessed in terms of gains and losses relative to the individual's subjective experience of how good or bad things are right now. What is important about this perspective, and counter to much work in economics and political science, is that it mandates an understanding of current subjective experience in order to predict the utility of changing this state.

To illustrate the approach and bring the conversation back to fairness, consider an example from another recent Nobel laureate in economics, the political scientist Thomas Schelling. Would you vote in favor of a tax policy that provides a larger exemption for rich families with children than poor families with children? Most people reject this policy because of a simple heuristic or intuitive rule of thumb: determination of gains and losses should be blind to current wealth. Although this intuition works in the specific case, it fails for more general cases and illustrates our vulnerability to wordplay. The standard tax table references a family without children. In theory, it could reference a family with two children. Under this system, families without children would pay extra. Would you vote in favor of a tax policy that imposes equal surcharges on poor and wealthy families without children? Most people perceive this policy as unfair. The framing of these questions flips our intuitions around. It leaves us with a sense that taxation is arbitrary nonsense. The solution to this problem is to set up the after-tax income for every level of pretax income for families with and without children. Looking to final income fails to acknowledge gains and losses relative to a reference point.

A simple way to illustrate the anchoring effects of a reference point in our perception of fairness comes from cases where an institution or company changes the sale price of a particular commodity, contingent upon some change in the environment. For each policy below, indicate whether you—the consumer—think it is fair or unfair:

- A hardware store raises the price of snow shovels during a spring blizzard.

    FAIR ☐   UNFAIR ☐

- A landlord raises the rent of a sitting tenant after learning that the tenant has found a good job in the area and will not want to move.

    FAIR ☐   UNFAIR ☐

- A car dealer raises the price of a popular model above list price when a shortage develops.
  FAIR ☐ UNFAIR ☐

- An employer who is doing poorly cuts wages by 5 percent, to avoid or diminish losses.
  FAIR ☐ UNFAIR ☐

- A landlord who rents apartments in two identical buildings charges higher rent for one of them, because a more costly foundation was required for that building.
  FAIR ☐ UNFAIR ☐

Most people—without reflection—consider the first three cases as UNFAIR and the last two as FAIR. As Kahneman's work shows, the key idea that explains all of these cases is that of a reference transaction—what I would characterize as another parameter of the principle of fairness. As a parameter—rarely expressed in people's justifications—it designates a profit and set of reference terms for the individual or group in power with respect to what is offered. The individual(s) in power can change the terms, but actions that abuse this power are judged as unfair. What determines the content of the rule will vary by culture and by commodity, but the principle is general, abstract, and apparently applied with little to no conscious reflection. In parallel with the unconscious nature of linguistic judgments, it appears that calculations of losses, gains, and judgments of fairness emerge from our unconscious setting of the reference transaction parameter.

The reference transaction is only one way in which our judgments about gains and losses, or pleasure and pain, are guided by unconscious principles and parameters that are part of the Rawlsian creature's design. In a beautiful line of experiments, Kahneman has demonstrated that subjects' judgments during a task do not match their retrospective evaluations of the experience at the time. In these studies, we witness how our subjective sense of pleasure and pain interface with our judgments of particular actions. In one study, people held one hand in painfully cold water

(14°C) for sixty seconds and then dried off. Either before or after this event, they put the other hand in the same 14°C water for sixty seconds, and then experienced a gradual warming of the water to 15°C; the latter is still painfully cold. When subsequently asked which experience they would repeat, most selected the second. This is surprising, if one's view of hedonic experience is that more pain is worse than less pain. It is expected if, following Kahneman's lead, you consider the *peak* and *end* experience. When the water warms up, this feels relatively good and that constitutes the final experience. The peak experience is the same in both conditions. We therefore make our judgments based on peak and end experiences, blind to overall duration.

Our unconscious calculation of peak-end experience should guide other judgments, including our evaluation of resource distribution. For example, if the overall distribution of income in case A peaks at $100 and ends with $80, while case B yields a steady state of $50, case A will be perceived as preferable even if the total amount of money received is the same in both cases. This experiment hasn't been run, but there is evidence that people perceive a gradually increasing income as better than a steady or decreasing income even if each yield the same overall amount of income. If the predicted pattern of experience is confirmed, it might lead to some odd dynamics in terms of the perception of fairness. Someone given income distribution B in year two of a job might perceive this allocation as unfair if in the previous year he received income distribution A—a bizarre judgment if fairness operates over total income as opposed to its distribution over time.

As Kahneman explains, "Duration neglect remains a cognitive error . . . built deep into the structure of our tastes and is probably impossible to prevent." Although we may consciously wish to increase the duration of pleasure and decrease the duration of pain, deep within our mind is a system that is running on autopilot, out of reach from our most strongly held convictions that things should be otherwise. Kahneman sums this point up nicely: ". . . even in the treacherous domain of subjective experience and judgment there is often a fact of the matter, which is not always accessible even to sophisticated intuition. Intuition alone would not have led us to drop the elegant representation of consequences in terms of final states, in favor of the more cumbersome and seemingly arbitrary language of gains and losses. Intuition alone would not persuade us of the pitfalls of an evaluative memory that each of us has trusted for a lifetime. And the intuitions evoked by carefully crafted thought experiments will not reliably yield correct predictions of the responses to cases seen in between-subjects designs. In short, I have tried to convince you that it could occasionally be useful to supplement philosophical intuition by the sometimes non-intuitive results of empirical psychological research." When it comes to making judgments, be they about temperature, a fair exchange, or the conflict between harming and letting another die, we may not have access to the underlying principles. Rawls was at least partially right. And Chomsky's more general insight about unconscious and inaccessible principles was dead on! Kahneman's views and work fit well with these ideas.

Should fairness always be the guiding principle when it comes to exchange and distribution? Is fairness always the best or most appropriate principle with respect to individual welfare? Is maximizing average income with a floor constraint something that our ancestors would have selected, or that hunter-gatherers today would select if they played any of Frohlich and Oppenheimer's games? Could an individual's well-being ever drop under a fairness norm? There are currently no answers concerning the universality of Rawlsian principles of justice.[29] And no one has yet engaged the hunter-gatherers of Africa, South America, and Alaska in a veiled social-contract game. Where answers have emerged, however, is in understanding the implications of fairness principles for individual well-being or, more generally, welfare economics. And here is where our social norms and sense of fairness are often at odds with individual welfare and especially the protective legal policies that have been developed by our institutions.

In a masterful treatment of welfarist policies, the economists and legal scholars Louis Kaplow and Steven Shavell suggest that adherence to fairness often leads to unfairness with respect to distribution, and, consequently, may raise rather than deter cases of defection at the hand of temptation.[30] The core insight is that most notions of fairness focus on normative principles—loosely, rules that dictate what one ought to do—independently of their consequences for an individual's well-being. Fairness-based principles are, by definition, therefore, unfair, because they take into account factors other than individual welfare, such as Rawls's

focus on the worst-off. As Kaplow and Shavell state, "a principle of fairness may favor a legal rule that prevents sellers from disadvantaging buyers in some way, even though the rule will result in buyers being hurt even
more, taking into account that they will pay for the protection through
higher prices. One would then have to ask whether such a result is really
fairer to buyers." This perspective leads to the strong claim that all laws
should ignore issues of fairness and focus instead on welfare economics.
Lest they be misunderstood, Kaplow and Shavell do not deny that our
minds generate strong intuitions about equitable distribution of resources
and opportunities. What they do deny is that our institutions should be
slaves to these intuitions. What we need to explore, therefore, is when our
intuitions lead us astray with respect to individual welfare. And herein lies
the classic tension between descriptive and normative principles, or between what is and what ought to be. Although we must be cautious not to
equate these different levels of analysis, it is also a mistake to reject outright what our intuitive moral psychology brings to the table with respect
to issues of justice.

PUNISH OR PERISH

In the computer-animated movie *AntZ*, Z is a worker ant who is troubled
by his lot in life, his designated role, his lack of options, the trauma imposed upon him as the middle child in a family of five million. In a contemplative moment, he expresses his angst: "There's got to be something
better out there . . . I wasn't cut out to be a worker . . . I'm supposed to
do everything for the colony. What about my needs? What about me?" Z
strikes out, leaves his colony for greener pastures—Insectopia. This is a
powerful expression of free will and an explicit violation of ant social
norms, for which he is ostracized and then hunted down. Z ultimately
returns to the colony, saves it from the destructive intentions of its new
warlord General Mandible, and consummates his passion for his true
love, Princess Bala, once again violating the mandates of the ant caste system. As the camera pans back on the ant hole, Z has the last word: "There
you have it: your basic boy-meets-girl, boy-likes-girl, boy-changes-the-
underlying-social-order story."

Social norms are rules and standards that limit behavior in the absence of formal laws.[31] Many of our own norms evolved from the simpler systems that exist in animals to regulate territoriality and within group-dominance relations, and that subsequently ballooned into other domains, especially patterns of food sharing. Some have their origins in matters of health, others as signals of group allegiance. Norms—especially the mechanisms that keep them running—represent a launching point for exploring how our species developed new systems for maintaining justice as fairness.

One outcome of a social norm is that it functions as a group marker, a signature of shared beliefs concerning such problems as parental care (Is sex-specific infanticide permissible?), food consumption (What is taboo?), sexual behavior (Are multiple partners tolerated or admonished?), and sharing (Is gift exchange expected or optional?). When we violate a norm, we feel guilty. When we see someone else violating a norm, we may feel angry, envious, or outraged. These emotional responses can lead to change, as individuals say mea culpa for their own sins and lash out against others by means of gossip, ostracism, and violent attacks.

Social norms have two interesting psychological properties. On the one hand, they are complicated rules that dictate which actions are permissible. Although we may be able to articulate the underlying rules, they operate automatically and often unconsciously. When we follow a norm, we are like subjects seeing a visual illusion, like ants adhering to the mandates of colony life and caste organization. Although we can imagine reasons why norms shouldn't be followed, and, more specifically, why we as individuals shouldn't follow them in every circumstance, we are more often than not immune to this counterevidence. Norms wouldn't be norms if we could tamper with them, constantly questioning why they exist and why we have to follow them. Their effectiveness lies in their unconscious operation, and their power to create conformity. On the other hand, although social norms often exert an unconscious hand of control, we do sometimes violate them. When we do, or observe someone else in violation, our brains respond with a cascade of emotions, designed both to register the violation and to redress the imbalance caused. When we break a promise, we feel guilty. Guilt may cause us to reinstate the relationship, repairing the damage done. When we see someone else break a promise, we feel angry, perhaps envious if they have made out with resources that we desire. When a brother and sister have intercourse and are caught, their incestuous consummation represents a violation of a social norm. It can trigger shame in the siblings, sometimes suicide, and often moralistic outrage in both genetically related family members and interested but unrelated third parties. Again, these processes operate outside of our conscious systems of control. An emotion's effectiveness relies upon two design features: automaticity and shielding from the meddling influences of our conscious, reflective, and contemplative thoughts about what ought to be. The Humean creature has stepped up to the plate.

Norms have another interesting property: They emerge spontaneously in different societies and, ultimately, come face-to-face with a government or more formal legal body that may adopt a different take on what is fair, just, sensible, or stable. The legal scholar Eric Posner nicely summarizes this historical process:[32]

> In a world with no law and rudimentary government, order of some sort would exist. So much is clear from anthropological studies. The order would appear as routine compliance with social norms and collective infliction of sanctions on those who violate them, including stigmatization of the deviant and ostracism of the incorrigible. People would make symbolic commitments to the community in order to avoid suspicions about their loyalty. Also, people would cooperate frequently. They would keep and rely on promises, refrain from injuring their neighbors, contribute effort to public-spirited projects, make gifts to the poor, render assistance to those in danger, and join marches and rallies. But it is also the case that people would sometimes breach promises and cause injury. They would discriminate against people who, through no fault of their own, have become walking symbols of practices that a group rejects. They would have disputes, sometimes violent disputes. Feuds would arise and might never end. The community might split into factions. The order, with all its benefits, would come at a cost. Robust in times of peace, it would reveal its precariousness at moments of crisis.

Now superimpose a powerful and benevolent government with the ability to make and enforce laws. Could the government selectively inter-

vene among the continuing nonlegal forms of order, choosing to transform those that were undesirable while maintaining those that were good?
Could it tinker with the incentives along the edges, using taxes, subsidies,
and sanctions to eliminate, say, the feuds and the acts of discrimination,
without interfering with neighborly kindliness and trust?

As I discussed earlier, all societies have a normative sense of fairness.
What varies between cultures are the range of tolerable responses to situations that elicit judgments of fairness. In essence, each culture sets the
boundary conditions, by tweaking a set of parameters, for a fair transaction. No one articulates these boundaries. Everyone, however, learns
where they are. When formal laws intervene, it is typically because the operative principles underlying a social norm cause harm to individuals. To
reiterate Kaplow and Shavell's point, although most people have strong
intuitions about fairness, and in all societies problems of exchange rely on
unstated norms of fairness, the outcome of these judgments is often not
what is best from the perspective of welfare. Formal laws can therefore
play a trump card, overriding our intuitions about fairness. The difficulty
lies in the fact that the law is riding against the current. This conflict can
create anxiety, and, possibly, escalate the temptation to cheat, because
the law is perceived as unfair. Ultimately, therefore, legal policy must not
only establish why particular principles are justified, but what happens to
those who violate them. Punishment is one answer. Here again, important
new developments in the field of experimental economics and theoretical
biology highlight the importance of punishment in the evolution of cooperation, and, in particular, in the unique forms of human civilization
that have emerged and persisted over thousands of years in the face of
temptations to break them apart. They also set the stage for thinking
about how our formal laws must attend to and interface with our intuitions about punishment.

Let us return to the idea of strong reciprocity, a potential solution to
the problem of cooperation in large groups that may represent a uniquely
human cognitive adaptation. Strong reciprocity arises when members of a
social group benefit from adherence to the local norms and are willing to
punish violators even when the act of punishment is costly and there is no
opportunity to see the person again. When individuals reject offers in the
ultimatum game, they incur a personal cost in order to impose an extra

cost on the proposer. The interpretation of these findings is that responders consider low offers unfair. Thus, they would rather have nothing than accept a low offer that perpetuates the belief that an asymmetric exchange is just fine. If you play an ultimatum game with two responders, and introduce the notion of competition, the dynamics flip. If both responders accept the offer, they each have a 50 percent chance of taking the proposed amount. If one accepts and the other rejects, the accepting responder takes the proposed amount while the rejecter leaves empty-handed. In this game, the effectiveness of punishment—rejecting the offer—is functionally taken away, and thus both responders typically accept the offer, and accept lower offers than in the classic ultimatum game. Proposers, recognizing the quandary faced by each responder, make lower offers than in the noncompetitive ultimatum game. Proposers know that they can get away with a more egregiously asymmetric offer. This result allows us to make two points: Many of us look for opportunities to feed self-interest, and, in the absence of punishment, this temptation might very well win out. In any negotiation, it therefore pays to look out for opportunities to punish and be punished. Your mind runs the calculation automatically and unconsciously.

Ernst Fehr designed an experiment to look at when people take advantage of opportunities to punish, even when it is personally costly. The game involved three subjects: an allocator, recipient, and third-party observer. The allocator starts with one hundred monetary tokens, the recipient with no tokens, and the observer with fifty tokens; at the end of a game, subjects turn in their tokens for money. The game is played once, and all players remain anonymous. The allocator can pass any proportion of his or her tokens to the recipient, and the recipient has no other option but to accept what has been offered; this phase is like the dictator game. The experimenter then tells the observer about the allocator's offer. With this information, the observer can use some of his or her tokens to punish the allocator; for each token invested in punishing, the allocator's pot decreases by three tokens.

Results show that observers invest in punishing for all offers less than fifty tokens, the amount invested in punishing increases as the offer decreases, and recipients fully expect and believe that observers will invest in punishing as a function of the size of the allocator's offer. Since these are genetically unrelated players, and this is a one-shot game, the observer has little to gain from investing in punishment, except the emotional high that may come from screwing someone who is not playing fair.[33] The observer not only gets no monetary returns from punishing allocators but actually walks away with less money than someone playing a turn-the-other-cheek strategy. These results also suggest that the fine imposed is proportional to the inequity: as the inequity increases, so, too, does the punishment. What is unclear in this game is why the allocator doesn't give more to the recipient, knowing that the observer can punish. The fact that many allocators provide nothing, or amounts substantially less than fifty tokens, suggests that unlike recipients, they don't believe the observers will punish. When these games are repeated, however, allocators provide increasingly large amounts, suggesting that they may have to learn about the possible costs of a low-ball offer.

People look out for punishment in bargaining games and use the psychological leverage that punishment can impose to vary their strategies. Theoretical models and simulations of the evolution of cooperative human societies provide further support for these bargaining games. Once group size exceeds that of a typical hunter-gatherer group—about 150—punishment is necessary, in one form or another, to preserve stable cooperation.[34] To what extent are the experimental and theoretical results representative of the cooperative dynamics of societies currently living without formal laws of enforcement? Countless ethnographies have explicitly made the point that punishment through shame, ostracism, scapegoating, and outright violence is essential for maintaining egalitarianism and adherence to social norms in small-scale societies. For example, scapegoating among the Navajo takes the form of triggering fear and shame in those who contemplate violating a norm, whereas among the Eskimos, scapegoating creates guilt and fear that a bearer of a supernatural power, such as a witch, will intervene.[35] There are, however, only a handful of studies exploring how often these strategies are called upon and how effective they are.

The anthropologist Polly Wiessner analyzed a couple hundred conversations among the Ju/'hoansi bushmen of Botswana to explore whether the patterns of punishment are as effective and strategic as the strong-reciprocity thesis predicts. Conversations provide a reasonable window into bushmen punishment, as they do not use witchcraft or socially designed duels, and rarely engage in violence. In foraging societies, such as the bushmen, punishment carries potential costs, including loss of a potential ally, forager, or child caretaker, severing of ties, incitement of violence, and damage to personal reputation when the apparent charges are too severe. In conversation, individuals were about eight times more critical of others than they were praiseful. In an egalitarian society such as the bushmen, praising another creates an immediate inequity. Acknowledging explicitly that someone was kind enough to share spotlights that individual as capable of sharing. Acknowledging the compliment is equivalent to boasting, an act that is open to a new round of punishment.

Among the Ju/'hoansi, men mostly criticized other men, primarily about politics, land use, and antisocial behavior. Women, in contrast, criticized both men and women, explicitly mentioned jealousy as a motivating force, and focused on nickel-and-dime stinginess over special possessions (beads, clothing) as well as the failure to share, maintain kinship obligations, and deflect sexually inappropriate behavior. Both men and women maintain obligations within couples by means of individual enforcement, only infrequently raising conflict to the level of a group decision; this effectively reduces the costs of punishment and the odds of defection. This pattern is consistent with the experimental economics literature, implying that those individuals who are directly injured by norm violations are also most willing to incur the costs of punishment. Both sexes tended to focus on men as the primary target of punishment. This sex difference emerged because men were more likely than women to boast about their hunting abilities; due to their dominant role in food sharing, men were more likely to be criticized for inequities. Although the Ju/'hoansi, like other hunter-gatherer societies, are highly egalitarian, there are subtle differences in perceived status that emerge in individuals' subjective impressions. Some men and women are seen as strong, others as weak. When the strong are punished, through mockery, pantomime, or criticism, they usually resort to self-mockery, which helps their reputation and maintains the egalitarian nature of the society. Regardless of a violation's form, a striking feature of Ju/'hoansi punishment is that it rarely involves violence. Out of the entire sample of norm violations, only 2 percent resulted in violence.

Wiessner's work on the bushmen converges with more descriptive accounts of other hunter-gatherer societies. On the whole, individuals cooperate by means of enforcing obligations. Punishment is an important force, designed to redress the imbalance that may arise when someone boasts, gossips, bickers, or reneges. But when the machinery underlying punishment is carefully scrutinized, there is little evidence that bushmen pay to punish. As Wiessner summarizes, "With a few notable exceptions the goal of Ju/'hoansi [punishment] is to bring the transgressors back in line through skillful punishing without losing familiar and valuable group members. Although punishment is frequent and potentially costly, there is little evidence that these costs were borne by altruistic behavior. Rather the Ju/'hoansi have developed an array of cultural mechanisms that allow conformity to norms . . . at little cost."

Added on to Wiessner's work on the bushmen is Marlowe's study of the Hadza hunter-gatherers.[36] When individuals played Fehr's third-party game, allocators allocated little, and observers passively observed, keeping their own resources and rarely incurring the cost of punishing even the most selfish allocators. Punishment was an important force in our hunter-gatherer past, but not in the form of the altruistic punishment.

It is unclear whether strong reciprocity was the hunter-gatherer solution to the problem of free riders in small social groups, or whether it is a recent cognitive invention. Regardless, the experimental games reveal that humans living in industrial societies have principled intuitions about what is or is not punishable, and use various forms of punishment to maintain social norms. What happens then when formal legal policy runs head-to-head with existing social norms?

THE CLASH

During the spring of 2003, Jo Hamlett, mayor of Mount Sterling, Iowa, population 40, proposed a new ordinance: anyone caught lying would be punished by a fine. The story made headlines. Why? As stated, it hardly sounds like news. After all, a criminal accused of murder or robbing a bank is sworn to tell the truth and nothing but the truth. Lying in court is a crime and is punishable. But Mayor Hamlett wasn't after people who might be lying about crimes against human life or property. Rather, he wanted to put an end to all the tall tales. How tall? Very: a hunter took down twelve deer with a bow and arrow; a boy saw a bullet zip by his head and then traced its origin back to the man who took aim; a man trapped a rat with a three-pound tail. Though many of the townsfolk were skeptical of the mayor's true intentions—did he want to put an end to lying or pick up some extra cash to pave the roads?—the latest vote left the city council with a 2-2 split. Most residents of Mount Sterling, however, thought the ordinance was ludicrous, or, as one person remarked, "It's like banning sex from the whorehouse."

The case of Mount Sterling will never make it to the Supreme Court. It will soon be forgotten. But it provides a small window into a larger problem: the clash between human intuitions about punishment and the legal view from on high about how to manage society, how to handle those who fail to cooperate and thus threaten human welfare. Punishment may not represent the silver bullet in every situation, but when our capacity to punish evolved, it provided a new system of control on the temptation to cheat. When we made the conditions for punishment explicit, either in the form of verbal pronouncements at town meetings or written into the bylaws, we changed the social landscape.

Imagine a society in which, by mandate of the town council, all acknowledged altruists have a green *A* tattooed on their forehead. In the words of Richard Dawkins, this would be the cultural equivalent of a *green beard*, an emblematic signal designed to offload the costs of searching for a proper partner. Those lacking a green *A* may or may not be cooperators, but those with an *A* would come with a higher approval rating, perhaps even a money-back guarantee. Thinking this through, it doesn't take long to see that back-room tattoo shops would emerge, providing cheaters with an opportunity to pay for a counterfeit *A*. Soon the system would decay, only to be countered down the line by another labeling mechanism or a system of recognizing counterfeit tattoos.

Now imagine an alternative scenario. Instead of brandishing the altruists, brandish the cheaters with a scarlet *C*. Anyone caught reneging on a promise, failing to reciprocate, or free-riding the system without helping to punish, is emblazoned with a *C*. Would this function as a deterrent? Would it stabilize cooperation and preserve justice? Displaying a *C* would make most people feel shame, perhaps guilt for their crime. Some might even beg for forgiveness, repent, plead to pay for their sins, and have the tattoo erased. This system was, of course, the centerpiece of Nathaniel Hawthorne's profound though fictional writings on punishment through shame, a form of retribution that has been revived in the last decade, bringing back visions to many legal scholars of public floggings, burnings, and decapitations.[37]

Scarlet-letter punishments—as they are now called—potentially solve two problems. They provide safety for the community by flagging its criminals and they deter future offenses by instilling shame, guilt, or fear. In recent times, these punishments have taken on a variety of forms, including bumper stickers for individuals convicted of drunk driving, full-body placards for individuals caught shoplifting, and registration of sex offenders with the department of public safety. There are at least three reasons why this form of punishment has been on the rise, primarily in the United States: skepticism that prison time and fines function as deterrents, concern that overcrowding in the prisons requires a cheaper solution, and intuitive beliefs that shaming is just desert for the public's sense of moral outrage—shaming is a form of revenge. Some critics of scarlet-letter punishments argue that they are barbaric, a form of evil-meets-evil. Others argue that their effectiveness as a deterrent is linked to the size and intimacy of a society. Whereas shaming was effective in Hawthorne's colonial America, it is unlikely to be effective today, as few people live in the same community from birth to death, and fishbowl communities are few and far between. Historically, scarlet-letter punishments disappeared for precisely this reason, swiftly replaced by prisons and monetary fines. These cycles show that to understand the reasoning machinery behind our judgments about punishment, we must consider the socially evolved intuitions of the mind together with current environmental pressures. Or, to quote the Nobel laureate in economics, Herbert Simon, our rational judgments are "shaped by a scissors whose two blades are the structure of the task environments and the computational capabilities of the actor."[38] Finally, some argue that scarlet-letter punishments constitute a violation of privacy. Whether this is, in fact, true is debatable, given the belief by some that rights and due process are the privilege of noncriminals. As the legal scholar James Whitman has argued, the harshness of attitude toward justice and treatment of criminals by the legal system of the United States is certainly not representative, and provides a striking contrast with continental Europe where criminals are treated more humanely.[39]

Scarlet-letter punishments, like most other legal approaches to criminal acts, stem from principles of fairness, of *lex talionis* (the law of retaliation), of biblical commandments handed down in Exodus:

> *And if any mischief follow, then thou shalt give life for life,*
> *Eye for eye, tooth for tooth, hand for hand, foot for foot,*
> *Burning for burning, wound for wound, stripe for stripe.*

The Bible's stance on violations is carried through to the earliest code of law, or the Code of Hammurabi. This legal document is entrenched in a system of retaliation and revenge, leading some scholars to claim that the original function of law was to deal with inappropriate or unacceptable behavior by retaliating in kind. Unlike the biblical proportionality equation, however, the Hammurabi Code appears extreme and disproportional: stealing domestic animals from another resulted in a tenfold fine if from an ordinary man, a thirtyfold fine if from a god or a member of court, and death if the thief had no income to pay the fine; if a judge tried a case in error, he paid a fine twelve times that originally set and lost his job as a judge; anyone taking a slave of the court was put to death; anyone making a hole in a house in order to steal would be killed and buried in front of the hole.

A true proportionality or equity scheme seems barbaric and dated, something that a civilized society would never consider. Surprisingly, perhaps, such deliberations are apparent today in the supposedly modern and sophisticated legal system of the United States. In Memphis, Tennessee, a judge ordered that it was permissible for a victim to enter a burglar's house, unannounced, and take something of comparable value. A Florida judge ordered a young teenager to wear an eye patch because he had thrown a brick and blinded the victim in one eye. A man in Baltimore caught selling fraudulent insurance policies to horse trainers was ordered to clean out the stalls of a mounted police station. I should say, quoting Dave Barry's signature phrase, "I am not making this up." Really.

The fact that such cases are real, having reemerged after a hiatus of quiescence, shows that aspects of our intuitive psychology of punishment continually resurface and clash with our existing legal systems, raising the important question of whether legal systems of punishment are inherently unstable. When the legal system intervenes and imposes a penalty on a crime, its effectiveness depends in part on whether the public puts faith in the law and its analytic treatment of the offense. When the public lacks faith, bedrock can turn to sand. As the philosopher Alvin Goldman points out, "When punishment does not at least approximate giving satisfaction to the victims of crime and to those in the community who wish to demonstrate their moral outrage, these individuals will take it upon themselves to extract punishment instead of, or in addition to, that officially imposed. This would be likely to lead to an escalation of private vendettas, substituting reigns of private terror for law and relative tranquility."[40] And, of course, the history of crime and punishment is dotted with such episodes.

Is there a way around the instability? I doubt it. But paying attention to human intuition, rather than ignoring it, is a good start. We must recognize the seductive power of seeing punishment in light of principles of fairness, and to design legal systems that indicate the pitfalls of this intuition, case by case.[41] Legal systems, in turn, must recognize that if they go against people's tastes for punishment, they may create more problems, as individuals seek revenge and take the law into their own hands. Obviously, the law is not foolproof. It will sometimes punish the innocent and fail to punish the guilty. It also can't act as the arbiter for all matters unscrupulous. As Clarence Darrow, the famous defense attorney in the Scopes Monkey Trial, remarked, "The law does not pretend to punish everything that is dishonest. That would seriously interfere with business."

Many aspects of law enforcement are based on the retributivist perspective: When someone does something wrong, they deserve to be punished. Some retributivists see punishment as obligatory, others as just permissible. The motivation underlying punishment is explicitly not a deterrent. Rather it is guided by the belief that the appropriate response to a wrongdoing is punishment. Once punishment has been established, the next question is, how much? The retributivists' answer is that the magnitude of the punishment should match the severity of the crime. As the evolutionary psychologists Martin Daly and Margo Wilson claim, this proportionality perspective may be a core part of human nature: "Everyone's notion of 'justice' seems to entail penalty scaled to the gravity of the offense."[42] This view certainly lines up with the results from experimental economics. But as a general principle of punishment, it fails. What would be a proportional punishment for a kidnapper, rapist, child molester, or serial killer? The only sense in which this perspective is at all coherent is as a rough yardstick. As Kaplow and Shavell suggest, "the notion that punishment should be proportional to the gravity of an offense does make some sense as a crude proxy principle if the purpose of the legal system is to promote individuals' well-being. All else being equal, greater harms tend to warrant greater sanctions. The reason is that the social value of any deterrence [or any incapacitation or rehabilitation] achieved through punishment is greater when the harm is greater; hence, when harm is greater, it will usually make sense for society to incur greater costs of punishment in order to prevent harm."[43] Recall, however, that retributivists don't place any value on deterrence, so this can't save the proportionality view, or what might be considered fair punishment. Even the Bible recognized this problem, providing an alternative to an eye-for-an-eye: "If a man shall steal an ox or a sheep, and kill it, or sell it, he shall restore five oxen for one ox, four sheep for a sheep."[44] This perspective is much closer to Hammurabi's Code, and suggests that those developing the earliest laws appreciated the potential deterrence function of punishment.

If our intuitions about punishment are based in part on the principles and parameters that guide fairness, then we must evaluate whether it makes sense to follow these intuitions in modern society. We can safely assume that these intuitions evolved prior to or during our life as hunter-gatherers, an assumption that I will further examine in parts II and III. In such small-scale societies, fairness was most likely an effective proxy for judging punishable acts. What worked then may not work today. For example, consider the taxes we pay to support law enforcement. All taxpayers are presumably supportive, because more effective enforcement translates into greater safety from criminals. But is the cost we pay to enforce the law fair? And is the level we pay appropriate, given the success of such enforcement in deterring those who are tempted to break the law? If we assume that law enforcement is an imperfect system, then we must acknowledge three negative consequences: some criminals will slip through the system without penalty, some innocent people will be punished by imprisonment or fines, and we will pay for law enforcement that is unfair in terms of failing to punish some criminals and inappropriately punishing the innocent. The retributivist doesn't take these negative consequences into account, and thus doesn't engage these aspects of fairness.

To ignore the potential deterrence function of punishment, in any form, is to ignore one of its most significant educational functions. If the punishment is high enough, it will deter. Just imagine what would happen if a million-dollar fine was imposed on petty thieves or people who fail to put money in a parking meter? Surely the rates would go down. But herein lies the clash between the effectiveness of punishment as a mechanism of control and the sense of fair punishment. A million-dollar fine is not fair, given that these crimes are not severe. And yet, this kind of fine would definitely deter and perhaps eliminate such crimes. As the legal scholar David Dolinko points out in reference to the imperfection of law enforcement, "We would also benefit from admitting frankly, as deterrence theorists, that punishing criminals is a dirty business but the lesser of two evils and thus a sad necessity, not a noble and uplifting enterprise that attests to the richness and depth of our moral character. Indeed, I think one could argue that it is the deterrence theorists, with their utilitarian outlook, who truly 'respect' the criminal by acknowledging that inflicting pain on him is, in itself, *bad*, and not to be done unless it can be outweighed by its good consequences."[45]

It is our nature, perhaps, to judge situations based on notions of fairness. The fact that we have such intuitions, and often can't justify them by appealing to carefully articulated principles, in no way mandates that we should be slaves to them. Ignoring them is equally misguided. Legal policy will often clash with our intuitions. To maximize the effectiveness of punishment, we must recognize the psychological expectations that people hold, often unconsciously. Revenge, fairness, deterrence, and education are all part of the equation, built into our moral faculty over evolutionary time.

3

GRAMMARS OF VIOLENCE

Revenge is a kind of wild justice, which the more man's nature runs to, the more ought law to weed it out.
—Francis Bacon[1]

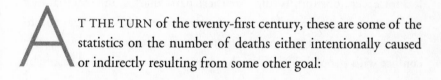T THE TURN of the twenty-first century, these are some of the statistics on the number of deaths either intentionally caused or indirectly resulting from some other goal:

- The death toll for the American-initiated war in Iraq was over two thousand in a period of one year.
- Internationally, approximately one million cases of suicide per year, or one death every forty seconds.
- The number of firearm murders ranges from a low of about fifty to seventy-five people per year in countries such as Japan, Australia, and the UK, to highs of ten thousand to forty thousand in Brazil, South Africa, Colombia, and the USA.
- Approximately five thousand deaths per year, mostly of women, from actual or perceived marital infidelities—so-called *honor killings*—with a heavy concentration in Arabic countries such as Pakistan, Jordan, and the Gaza and West Bank areas of Palestine.

There are also equally remarkable numbers for acts of nonfatal violence including rape, spousal abuse, and aggression toward children. If
there are principles underlying the impermissibility of violence, we might
question whether anyone is listening. Given that the capacity to kill or
harm more generally is universal, to what extent do individual cultures
tighten or release the reigns? For example, the aboriginal Semai of the
Malaysian rain forests have the lowest reported levels of violence in the
world, with norms that forbid physical and verbal aggression. In contrast, the Yanomamo of South America—often referred to as "the fierce
people"—not only use violence to settle disputes, but expect members of
their tribe to show off with dramatic acts of aggression. Violence among
the Yanomamo is a virtue, rewarded in the ultimate evolutionary currency:
more matings, more genes.

In this chapter, I push harder on the linguistic analogy by asking whether there are universal principles underlying our judgments
of permissible harm, including actions that lead to death. Although
most cultures may have a general rule that blocks killing, all cultures allow for exceptions to this rule. Are there particular parameters that each
society sets, triggering culture-specific signatures? How do legal and religious systems intervene on the patterns of violence and how do they
conflict with people's intuitions about permissible killing and justifiable
punishment?

PERMISSIBLE KILLING

Thou shalt not kill. As one of the Ten Commandments, the rule is clear,
accepted by Jews, Catholics, and Protestants, and leads to the moral
judgment that killing is forbidden. In Hebrew, the word *ratsach* was
typically translated as "kill," but sometimes more appropriately translated or interpreted as "murder." The shift to murder was designed to
cover cases such as the legitimate killing of certain animals for food.
Even within the narrower context of murder, Hebrew scripture specified many exceptions to this commandment. It was deemed permissible to murder an engaged woman seduced by another man, individuals

worshipping a god other than Yahweh, strangers entering the temple, and women practicing magic. Historically, many Christian groups interpreted this commandment with respect to an in-group versus out-group distinction: Killing an outsider is fair play. Beginning around the fifteenth century, and continuing for several hundred years, Roman Catholics and Protestants unearthed and killed thousands of heretics and worshippers of Satan, deemed dangerous to the sanctity of the church and its community. The Bible also permitted certain forms of killing within group, including abortions and infanticide. Psalm 137:9 doesn't mince words: "Happy shall he be, that taketh and dasheth thy little ones against the stones."

When individuals or groups in power voice a disclaimer on the permissibility of killing, hatred captures a new outlet, the temptation to eliminate outsiders grows, and violence can turn to addiction. As a précis to exploring what triggers violence and what controls it, we need to understand the psychology of violence, and, especially, our intuitions concerning when it is permissible to harm another individual. Characterizing these intuitions will help uncover the logic beneath our judgments of permissible acts of violence, the extent of cultural variation, and the challenges that societies face in attempting to control a perhaps all-too-natural consequence of anger, envy, jealousy, and revenge.

Consider a classic moral dilemma originally proposed by the philosopher Phillipa Foot,[2] and briefly mentioned in chapter 1: the trolley problem. Foot's goal was to gain some purchase on the distinction between killing and letting die. This distinction lies at the core of many biomedical and bioethical decisions, especially euthanasia and abortion. More generally, the goal of this family of examples was to assess how our moral judgments are mediated by overall goodness (*virtue ethics*), good versus bad consequences (*utilitarianism* or *consequentialism*), categories of right or wrong actions (*deontological principles* or *nonconsequentialism*), and the relevance of the psychology of acts versus omissions as well as foreseen versus intended consequences. For these cases to have some teeth with respect to the issues at hand, it is necessary to assume that everything reported is true, with no added information or assumptions.

1. **Bystander Denise**. Denise is a passenger on an out-of-control trolley. The conductor has fainted and the trolley is headed toward five people walking on the track; the banks are so steep that they will not be able to get off the track in time. The track has a side track leading off to the left, and Denise can turn the trolley onto it. There is, however, one person on the left-hand track. Denise can turn the trolley, killing the one; or she can refrain from flipping the switch, letting the five die.

Denise

Is it morally permissible for Denise to flip the switch, turning the trolley onto the side track?

My initial intuition—one that appeared automatically, with no reflection—is that it is permissible for Denise to flip the switch; most philosophers share this intuition. Now let's take apart the intuition and the features of this particular scenario. Denise has control over the trolley's path. Because the track forks, Denise has the option of allowing the trolley to maintain its course by doing nothing, or flipping the switch and changing the trolley's course. Regardless of Denise's behavior, at least one person will die. Doing nothing results in five people dying, whereas flipping the switch results in one person dying. If Denise allows the trolley to stay its course, then she has omitted the action of flipping the switch, thereby allowing the five to die. Flipping the switch counts as an intended action that results in one person dying. Although the act of flipping the switch is intended, Denise's goal is not to kill the one person. Rather, her goal is to save the five. To see that this is the case, imagine an empty side track. Here, flipping the switch results in a cost-free rescue of the five and presents no moral dilemma at all. The goal is to save the five. Failing to flip the switch seems forbidden. Rescue seems obligatory since there are no costs. Since Denise doesn't know the hikers, and since all of them are equally irresponsible for walking in an area that puts them at risk, the problem boils down to a relatively straightforward calculation: Is killing five worse than killing one when everything else is equal? On utilitarian grounds, where maximizing the good provides the only relevant yardstick, the answer is unequivocally "yes." But this scenario is far more complicated. To illustrate how, while also revealing a flaw in the utilitarian's perspective, consider a second case.

2. **Bystander Frank**. Frank is on a footbridge over the trolley tracks. He knows trolleys and can see that the one approaching the bridge is out of control, with its conductor passed out. On the track under the bridge there are five people; the banks are so steep that they will not be able to get off the track in time. Frank knows that the only way to stop an out-of-control trolley is to drop a very heavy weight into its path. But the only available, sufficiently heavy weight is a large person also watching the trolley from the footbridge. Frank can shove the large person onto the track in the path of the trolley, resulting in death; or he can refrain from doing this, letting the five die.

Is it morally permissible for Frank to push the large person onto the tracks?

My intuition, and that of philosophers discussing this case, is that it is not permissible for Frank to push the large person. Why not? The outcome or consequence is the same as in the first case. An actor or agent's behavior results in one person dying and five others surviving. On utilitarian grounds, it should be permissible to push the large person to save the five. But for some reason, the utilitarian calculus fails here. What are some of the salient differences between these two cases? In both cases, omitting the action (not flipping the switch, not pushing the person) results in five dead people. Inaction is presumably permissible in both cases. It is hard to imagine a law that would forbid inaction or make it punishable unless the utilitarian calculus is a trump card that always wins; no country or culture that I know of invokes this utilitarian card as a trump. In case 1, Denise's action leads to a redirected threat: she takes responsibility and causes the trolley to kill the lone hiker; as noted above, her goal is not to kill the lone hiker but it is an anticipated or foreseen consequence. In case 2, Frank's action leads to the direct harm of an innocent person. In both cases, there is nothing about the hikers that gives them an upper hand in terms of the right to live. In case 2, however, the large person is not a participant in the events that are about to unfold. This large person is an innocent bystander, and should have the right to stay uninvolved.

The distinction between a bystander and a participant is not, however, that simple. We may share the intuition that it is wrong to push the large person onto the tracks, whereas it is permissible for us to flip a switch that causes the trolley to kill one person, but what distinguishes these cases? The large person is safe if Frank refrains from pushing. But similarly, the lone hiker is also safe if Denise allows the trolley to stay its course. All of the hikers are irresponsible—they shouldn't be walking on the tracks. The conductor has responsibility for his trolley, whereas neither Denise, Frank, nor the large person have any responsibility. We might think that if the side track had been empty, it would be not only permissible for Denise to flip the switch, but obligatory: The act is cost-free and therefore any witness to the scene must act and flip the switch. In contrast, if Frank is alone—no large person nearby—he has no other option than to watch the trolley kill the five hikers. For some, the key issue lies in the Kantian imperative that it is impermissible to use a person merely as a means to an end. Frank, but not Denise, is faced with an opportunity to use a person as a means to prevent the trolley from running over the five hikers. That is forbidden. Others think that the key difference between these cases is that Denise involves impersonal harm whereas Frank involves personal harm. Denise simply flips a switch, an act that in and of itself has no emotional pull. Frank, in contrast, has to physically push a person, an act that is rich in emotional pull. And yet others think that we can distinguish these cases by appealing to the principle of double effect: Harming another individual is permissible if it is the foreseen consequence of an act that will lead to a greater good; in contrast, it is impermissible to harm someone else as an intended means to a greater good. For Denise, killing one person is a by-product of taking an action that saves five; saving five is the greater good. For Frank, the large person is harmed as an intended means to saving the five hikers. The principle is violated because the harm is not a foreseen consequence but an intended means.

It is not possible to pinpoint the principle driving our different judgments about Denise and Frank. There are too many differences between these cases: impersonal versus personal harm, redirecting threat versus introducing a new threat, intended harm versus foreseen harm, one track versus two tracks, and innocent bystander versus irresponsible hiker. And it is easy to conjure up small changes to these cases to show that at least some of the distinctions are irrelevant. For example, would it make any difference if, instead of pushing the person, you hit a button (impersonal, no contact) that immediately launched him onto the track, killing him, stopping the trolley, and saving the five? I doubt it. We need new cases that reduce the differences. We need to follow the path of any good scientific experiment and reduce the number of possible factors causing differences between groups. If group A likes roller-coaster rides and group B doesn't, it is impossible to pinpoint the reason for this difference if group A consists of older women and group B consists of younger men. Here, both age and gender can explain the group differences. We need to test groups of older women against groups of older men, groups of older women against groups of younger women, and groups of older men against younger men. Now consider two new trolley cases.

3. **Bystander Ned**. Ned is taking his daily walk near the trolley tracks when he notices that the approaching trolley is out of control. Ned sees what has happened: The conductor has passed out and the trolley is headed toward five people walking on the track; the banks are so steep that the five hikers will not be able to get off the track in time. Fortunately, Ned is standing

next to a switch that he can throw, which will temporarily turn the trolley onto a side track. There is a heavy object on the side track. If the trolley hits the object, the object will slow it down, thereby giving the hikers time to escape. The heavy object is, however, a large person standing on the side track. Ned can throw the switch, preventing the trolley from killing the hikers, but killing the large person. Or he can refrain from doing this, letting the five hikers die.

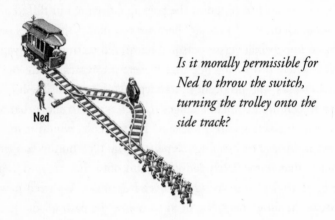

Is it morally permissible for Ned to throw the switch, turning the trolley onto the side track?

4. **Bystander Oscar**. Oscar is taking his daily walk near the trolley tracks when he notices that the approaching trolley is out of control. Oscar sees what has happened: The conductor has passed out and the trolley is headed toward five people walking on the track; the banks are so steep that the five hikers will not be able to get off the track in time. Fortunately, Oscar is standing next to a switch that he can throw, which will temporarily turn the trolley onto a side track. There is a heavy object on the side track. If the trolley hits the object, the object will slow it down, thereby giving the hikers time to escape. There is, however, a person standing on the side track in front of the heavy object. Oscar can throw the switch, preventing the trolley from killing the hikers, but killing the person in front of the weight. Or he can refrain from doing this, letting the five hikers die.

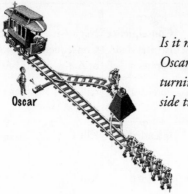

*Is it morally permissible for
Oscar to flip the switch,
turning the trolley onto the
side track?*

Oscar

My intuition is that it is not permissible for Ned to flip the switch, but
it is permissible for Oscar to flip the switch. If Ned flips the switch, he is
committing intentional harm. The only way to save the five hikers is by
turning the trolley onto the side track and using the large person as a
means to stop the trolley. If Oscar flips the switch, he is causing harm, but
as a foreseen side effect. For Oscar, the goal is to use the heavy weight as a
means of stopping the trolley. The fact that a person is standing in front
of the weight is unfortunate, but it is not Oscar's intent to kill this person.
And paralleling Ned, the only way for Oscar to save the five hikers is to
flip the switch and turn the trolley onto the looped track. If the weight is
absent, the trolley runs over the person on the loop and then runs over the
five hikers ahead. Like Frank and the large person on the footbridge, the
weight—not the person—provides the means to the greater good of saving five hikers.

Some may think that my intuitions are off. After all, Ned's case looks
almost exactly like Denise's. Both Ned and Denise are redirecting a threat.
But that is also true of Oscar. Both Ned and Denise are performing an
impersonal act: flipping a switch. So is Oscar. Both Ned and Denise have
the intent of saving five hikers. So does Oscar. Others may think Ned's
action is as permissible as Denise's and Oscar's. But if that is your intuition, then explain why Ned can flip the switch but Frank can't push the
large person? Both Frank and Ned are using the large person as a means
to an end. If the Kantian imperative holds, then it is impermissible for
both Frank and Ned to act. The personal-impersonal distinction doesn't
work here. What does appear to work is the principle of double effect.

Case	Act	Emotional Quality of Act	Negative Consequence Is Intended/ Foreseen	Negative Consequence of Act	Positive Consequence of Act
1. Denise	Flip switch	Neutral/ impersonal	Foreseen	Kill 1	Save 5
2. Frank	Push man	Negative/ personal	Intended	Kill 1	Save 5
3. Ned	Flip switch	Neutral/ impersonal	Intended	Kill 1	Save 5
4. Oscar	Flip switch	Neutral/ impersonal	Foreseen	Kill 1	Save 5

This is easy to see if we break the principle down into some of its essential components and then compare each case as illustrated above. What emerges from these cases is a key insight: It is impermissible to cause an intended harm if that harm is used as a means to a greater good. In contrast, it is permissible to cause harm if that harm is only a foreseen consequence of intending to cause a greater good.

This is by no means a complete analysis. Philosophers discussing these cases have gone further, illustrating where the principle of double effect breaks down, and the kinds of parametric extensions that are necessary to account for our judgments.[3] The main point here is that seemingly small changes to a moral dilemma can radically alter our intuitions.

Everything I have said about Denise, Frank, Ned, and Oscar comes from my intuitions and the intuitions that many if not most philosophers share. It also comes from analyses of principles, which, in turn, come from staring long and hard at these cases and others like them. The aim is to find principles that can account for the nature of our moral intuitions, as opposed to the varieties of utilitarian consequences. It is a tradition that is consistent with my characterization of the Rawlsian creature. If your interest is in human nature, however, you should find this diagnosis unsatisfying on two counts. First, my judgments and justifications, together with those articulated by moral philosophers, may only emerge after careful immersion in such dilemmas. Further, all those who have written on this topic are highly educated, brought up in a Western culture, and over thirty. What we need, if we are interested in the nature of human moral judgment, is a sense of what other people say when confronted with these kinds of dilemmas. These "other people" must include a sample of the world's diversity, including the young and old, educated and not, male and female, devoutly religious and explicitly atheistic, and inhabitants of large and small groups embedded in sprawling cities, rural countrysides, tropical jungles, and expansive savannas. Second, although the analysis that I provided above begins to uncover some of the relevant principles and parameters underlying our judgments, it is unclear whether anyone reading these cases for the first time consciously reflects on the problem and then delivers a judgment based on reasoning from these principles. It is possible that our judgments reflect intuition, percolating up from unconscious and inaccessible principles of action. And as Rawls intuited, it is also possible that our commonsense descriptions of these principles are wrong, requiring far more abstract concepts to capture what lurks beneath the surface of our perception.

JUDGMENT DAY

When moral philosophers sit down and examine the nature of moral judgments, they take one of two different paths. They either attempt to deduce, using reason and logic, what individuals ought to do based on the facts at hand, or let intuition play its course, followed up by a search to work out the nature of the intuition, what it means in the service of developing the prescriptive principles. Scientific evidence plays no role in either approach. When philosophers, such as Judith Thomson and Frances Kamm, argue that it is permissible for Denise to flip the switch but not for Frank to push the large person, they argue on the basis of intuition and clear-headed, emotion-free, impartial reflection. They also argue that such clarity of thought comes from training in philosophy and that even smart but untrained students from some of our most prestigious universities will give faulty intuitions and incoherent justifications. It is not about being smart, but about learning to reason and achieve clear intuitions. For many, perhaps most moral philosophers today, it wouldn't matter if millions of people had different intuitions from their own. The philosophers would stick to their intuitions because of the conviction that intuitions need tuning up. But this is precisely where philosophy and science part company, and where profound questions about the nature of intuition arise. Is there a difference between the intuitions of a philosopher who has thought long and hard about different moral dilemmas and your average Joe or Jane? When the questions on the table concern applied issues such as euthanasia, abortion, suicide, and other contexts in which harm is deemed permissible, philosophical insight is important, but so, too, are the views of people who vote, whether by electronic ballot or by discussion with elder clansmen. When philosophers speak about intuitions concerning killing and letting die, they should care about what others think, even if the intuitions that emerge are from the voice of untutored reason. And if they are willing to engage this voice, then they have passed the proverbial buck to scientists who can uncover how people judge different moral dilemmas, what causes cross-cultural variation or universality, what biases dominate and perhaps even obscure reason, and how the mind computes whether an act is permissible or forbidden. It is time to take philosophical insight to the streets.

The psychologist Lewis Petrinovich was the first to explore how people untrained in philosophy judge the classic trolley problems, along with a lifeboat dilemma, in which one individual's life is sacrificed in order to save five others.[4] Each subject read a scenario and then rated—on a scale—whether they strongly agreed or disagreed with a particular action. If, as Petrinovich argued, an evolved psychology underpins our moral judgments, then the pattern of responses should be consistent across cultures, and sensitive to evolutionarily relevant dimensions that impact upon survival and reproduction, such as kinship, ethnic group, and species: When given a choice, we should save kin over non-kin, humans over other animals, and individuals within our ethnic group over those outside.

For American (southern California) and Taiwanese college students, numbers win when no other distinguishing features bias one action over the other. It is therefore permissible to flip a switch to kill one if the intent is to save five and if none of the six are known to the agent. When Petrinovich's scenarios revealed information about identity, then subjects saved kin over non-kin, friends over strangers, humans over nonhumans, and politically safe or neutral individuals over politically abhorrent monsters. In both cultures, subjects judged that it was permissible to save an unknown person over an endangered gorilla, and to sacrifice several people with politically abhorrent beliefs (e.g., Nazis) over one person with politically and emotionally neutral beliefs. These results held across both groups, even though the Taiwanese students tended to follow Eastern religions that favor inaction over action. Like Americans, Taiwanese students judged that it was permissible for a bystander like Denise to flip the switch and kill one to save five, but impermissible to push a heavy man in front of the trolley, killing him but saving five. As Petrinovich concludes, individual moral judgments reflect evolved, universal decision-making processes that increase genetic fitness.[5]

Petrinovich's work shows that the sciences can contribute in interesting ways to our understanding of moral judgments. With respect to uncovering the design of the Rawlsian creature, however, we haven't advanced far enough. Given that subjects selected among different options for each scenario, choosing the extent to which they agreed or disagreed with different solutions to the dilemma, it is unclear whether the responses represent intuitions based on unconscious analyses of action, flashes of emotion, or conscious reasoning. Further, because Petrinovich didn't ask subjects for their justifications, we don't know if their decisions were based on intuition or conscious reasoning from explicit moral principles; the fact that kinship, familiarity, and political affiliation influence moral judgments doesn't license the conclusion that people use these parameters to consciously reason about moral dilemmas.

The philosopher and legal scholar John Mikhail designed a series of trolley problems aimed at uncovering a fragment of the Rawlsian creature's moral grammar, especially the rules guiding judgments of permissible harm.[6] Mikhail's test differed from Petrinovich's in several ways: all of the individuals in the dilemmas were anonymous, which created impartiality and forced subjects to judge each case based on other dimensions, especially the relationship between intention, action, and consequence; following each scenario, subjects judged whether an action was morally permissible (no rating scale) and then justified their answer; subjects included males and females, a group of eight-to-twelve-year-old children, and several nonnative Americans living in the United States, especially an immigrant Chinese population. Mikhail was explicitly interested in exploring principles that appear in common law, especially their bearing on people's judgments, and their accessibility to conscious reflection. Here's how he defined two of the target principles, the second being familiar:

1. The principle of *prohibition of intentional battery* forbids unpermitted, unprivileged bodily contact that involves physical harm.
2. The principle of *double effect* is a traditional moral and legal principle . . . according to which otherwise prohibited acts may be justified if the harm they cause is not intentional and the act's foreseeable and intended good effects outweigh its foreseeable bad effects.

Like Petrinovich, Mikhail found no evidence that gender, age, or national affiliation influenced the pattern of permissibility judgments. He also found no evidence that straightforward deontological, utilitarian, or conditional rules accounted for differences in permissibility judgments across moral dilemmas. For example, if subjects thought that killing was always wrong (deontological), that actions maximizing the overall good were always preferred (utilitarian), or that morally neutral acts such as flipping a switch were always permissible (conditional—if an act is of type X, do it), then subjects should have judged Denise's, Frank's, Ned's, and Oscar's actions as either perfectly permissible or forbidden—in all four cases, killing is involved and action maximizes the overall good. Mirroring the intuitions I spelled out earlier, however, most subjects said that Denise can flip the switch (killing one, but saving five), Frank can't push the large person (killing him, but saving five), Ned can't flip the switch (killing one as a means of saving five), and Oscar can flip the switch (killing one as a foreseen side effect of saving five). In addition, a casual look at people's justifications led Mikhail to suggest that people are largely incoherent when it comes to explaining their judgments. For example, some subjects conveyed surprise that they had provided different answers for two scenarios resulting in the same outcome—five survive and one dies. Others said that they went with their *gut* response, an *instinct*, or an *intuition*. When it comes to certain moral dilemmas, then, it appears that people have confidence in their judgments, but are clueless with respect to their justifications. This pattern of responses begins to throw light on the nature of our moral faculty and its operative principles.

For Mikhail, the Ned-Oscar contrast provided the key insight, ruling out the Humean creature because both cases entail an impersonal act with the same consequences. What explains the different judgments is the principle of double effect, a principle that is operative but not expressed when people are asked to justify their responses. As Rawlsian creatures, we are equipped with a mental barometer that distinguishes between killing as a means and killing as an unintended but foreseen side effect. Killing is wrong if it is intended as a means to some end. Killing is permissible if it is unintended but a foreseen by-product of a greater good.

Mikhail concludes from his findings that "we can explain how people manage to make the moral judgments they do by means of rational principles—specifically, the intentional battery prohibition and the principle of double effect." When Mikhail says "by means of," he doesn't mean that subjects consciously reflect upon these principles. Rather, these principles operate unconsciously, but directly influence judgment. And when Mikhail says "rational principles," he simply means that there is a logic that relates intentions to actions and actions to consequences. Thus, the intuitive knowledge underlying our moral judgments is like the intuitive knowledge of language, physics, psychology, biology, and music, topics that I return to in parts II and III. We know that two solid objects can't occupy the same space at the same time and that a solid object will fall straight down if unsupported and unobstructed on the way down. But we know these factoids in the absence of a course in physics and without being aware of them. Mikhail's claim, and the key idea driving my argument for the moral faculty, is that much of our knowledge of morality is similarly intuitive, based on unconscious and inaccessible principles for guiding judgments of permissibility. Given subjects' apparent inability to provide reasons for their moral judgments, these findings fly in the face of the Kantian creatures' anatomy, and of Kohlberg's criteria for moral advancement.

Mikhail's results also conflict with the views of Carol Gilligan, a developmental psychologist who criticized Kohlberg for focusing on boys as his target sample, and for restricting issues of morality to justice as opposed to the more caring and nurturing aspects of our moral behavior. Developing girls, Gilligan argued, show a different path to moral maturity than developing boys. When girls confront moral dilemmas, they are more concerned with issues of caring. Boys, in contrast, are more concerned with issues of justice. Girls want to ensure that their relationships are functional and sufficiently nurtured, boys care about what is fair, even if it tarnishes their relationships. In neither Mikhail's nor Petrinovich's sample was there evidence of a gender difference in the pattern of judgments. When it comes to judging the moral permissibility of harm and rescue, boys and girls, as well as their more mature incarnation as men and women, look like clones. Gender differences may play a role in performance, and the justifications that the sexes give. But when it comes to our evolved moral faculty—our moral competence—it looks like we speak in one voice: the voice of our species.

Mikhail's work takes important steps beyond the original studies by Petrinovich. But it also leaves open many questions about the universality of human moral judgment and our capacity to access the underlying principles. Mikhail tested American students as well as American children, non-American but Western adults beyond college age, and a sample of recent Chinese immigrants. This test population does not, however, warrant the conclusion that "people"—presumed to represent *Homo sapiens*—make comparable judgments about moral permissibility. It does not warrant the conclusion that our species is endowed with a universal moral grammar. Some cultures endorse infanticide and spousal murders as punishment for promiscuity, suggesting a significant cultural spin on the intentional-battery prohibition. This variation could represent differences in what people actually do as opposed to how they perceive these actions, a difference in people's performance as opposed to their moral competence. But these differences may also reflect variation in competence, thereby challenging the idea of a universal moral grammar, or minimally, forcing an emendation to the nature of its underlying principles.

Mikhail quotes several justifications that imply a psychological cleavage between what people say is morally permissible and what they offer as a justification or explanation of their judgment. The mind adjudicates when it comes to moral dilemmas, but guards its operative principles, leaving individuals to express principles that provide either weak or incoherent support for their judgment. Mikhail's results trigger several questions: Does knowing the deep principles of our moral faculty change our moral judgments? If I am told that the principle of double effect provides the right justification for judging Oscar's action as permissible, but not Ned's, will I take this lesson home with me and forever watch out for cases where someone violates this principle? Am I right in thinking that someone who uses another as a means to the greater good is always guilty of a moral transgression? If a mother smothers her crying baby in order to save a hundred refugees from being detected by a violent group of guerrillas, she has violated the principle of double effect, but would we judge her act as forbidden? If your answer is "no," is it because of the 1:100 ratio, the fact that the greater good so clearly outweighs the negative consequence of her action? Why is it obvious that numbers win? Isn't the life of the child as valuable as the life of any of the hundred? Is it because a mother has a choice when it comes to her child's life?

To begin answering some of the unresolved issues raised by Petrinovich's and Mikhail's work, my students Fiery Cushman and Liane Young and I created the Moral Sense Test (moral.wjh.harvard.edu). Within the first year of opening the site, and with only an English version of the test, we had collected data from sixty thousand subjects, covering 120 countries. The sample included children as young as seven and adults as old as seventy; males and females; individuals with no education, primary school, secondary school, college, Ph.D.s, MDs, and JDs; atheists, Catholics, Protestants, Jews, Buddhists, Hindus, Muslims, and Sikhs; and 120 ethnicities. Like Petrinovich and Mikhail, we also used trolley problems and other moral dilemmas involving questions of harm, rescue, and the distribution of beneficial resources such as medicine. We also manipulated the intentions, actions, and consequences of each scenario, holding constant the numbers of individuals and their anonymity, blocking any form of partiality. We explored a wider range of dilemmas varying in content (not only trolleys, but stampeding elephants, burning houses, rescue boats, dispensation of limited drugs), wording (killing, saving, running over), the time allotted to answer (as much time as needed or speeded, requiring an immediate response following the question), and the identity of agents and targets (unknown bystander, test subject as agent or target/victim). For each dilemma, we asked whether an action was permissible, obligatory, or forbidden. We also asked subjects to justify their judgments, and then analyzed their explanations in terms of coherent and incoherent responses.

Consider once again Denise, Frank, Ned, and Oscar. Based on a sample of several thousand subjects taking the test, and responding to one of these cases as their first moral dilemma, about 90 percent said that it was permissible for Denise to flip the switch, whereas only about 10 percent said it was permissible for Frank to push the large person. Although these were all English speakers with access to the Internet, the judgments were consistent across subjects with widely different ages, ethnicities, backgrounds in religion, general education, and specific knowledge of moral philosophy. However, when justifying why Denise's action is permissible but Frank's is not, 70 percent of subjects looked like children at stage 1 of Kohlberg's stepladder—clueless. Insufficient answers included appeals to God, emotions, hunches, gut feelings, deontological rules (killing is wrong), utilitarian consequences (maximize the greatest good), and my favorite, "Shit happens!" Of those we judged as sufficient, about half mentioned some aspect of the principle of double effect (it isn't permissible to intentionally use the heavy man to save the five), while the others focused on the distinction between personal and impersonal harm. The fact that most people have no idea why they draw a distinction between these cases, reinforces the point that people tend to make moral judgments without being aware of the underlying principles. On the other hand, perhaps those who do invoke a proper justification are older and wiser, generally well educated, and specifically trained in moral philosophy and the law. Nothing in our sample of subjects suggested such a division. Older, educated, religious Asian women were as likely as younger, uneducated, atheist European and American men to provide insufficient justifications. Many felt as Groucho Marx did, stating, "Those are my principles. If you don't like them, I have others."

Ned and Oscar provide a different picture. About 50 percent of the subjects thought that Ned could flip the switch, whereas approximately 75 percent thought that Oscar could flip the switch. Although there is no clear cutoff for what counts as a majority, people seem split on Ned but are more convinced that Oscar should throw the switch. With a larger and more diverse sample of subjects, it is also the case that Mikhail's conclusion holds up to some extent: More people see Oscar's action as permissible than Ned's. And the only difference between these two cases is that the negative consequence of Ned's action serves as a means to the positive consequence, whereas Oscar's action leads to a foreseen negative consequence. But do people have access to this distinction, to something like the principle of double of effect? Although these cases are virtual clones, leaving only one morally relevant distinction, only about 10 percent of subjects justified their answer with this distinction. Again, there was nothing special about these people. For reasons that we don't yet understand, a small minority of people have access to the relevant principles, but they are not all tutored in moral philosophy.

Are we now licensed to conclude that judgments concerning certain forms of harm are universal? Not yet. Though the number of subjects taking our Web-based test was substantial, and the analyses of cultural and demographic variables revealed either small or nonsignificant effects, there is one glaring problem: All of these subjects were Internet-savvy. All surfed the Web to some extent, whether it was to read the news, buy something from Amazon, or download music. Moreover, everyone taking the test could at least read English, and a majority were native English speakers from the United States, Canada, Australia, and England. To remedy this problem and extend the cross-cultural reach, my students and I translated the Web site into five other languages—Hebrew, Arabic, Indonesian, Chinese, and Spanish—and have begun testing the same small-scale societies that anthropologists have tested on the economic-bargaining games. Though it is too early to say for sure, the general pattern is the same: Permissible harm is sensitive to parametric variation, and judgments are not guided by consciously accessible principles.

Independent of how the cross-cultural work turns out, the linguistic analogy generates clear predictions. We do not expect universality across the board. Rather, we expect something much more like linguistic variation: systematic differences between cultures, based on parametric settings. Thus, in parallel with the cross-cultural studies of the ultimatum game, we expect differences between cultures with respect to how they set the parameters associated with principles for harming and helping others. For example, as in the case of bargaining and judgments of inequity, do cultures take into account parameters such as the agent's responsibility to act, the utilitarian outcome, and whether the consequences of an action are intended or foreseen? I believe that some aspects of the computation are universal, part of our moral instinct. I doubt very much that people will differ in their ability to extract the causes and consequences of an action. Everyone will perceive, unconsciously, the importance between intended and foreseen consequences, intended and accidental actions, actions and omissions, and introducing a threat as opposed to redirecting one. The central issue in thinking about cross-cultural variation is to figure out how different societies build from these universal factors to generate differences in moral judgments.

When we read several of these dilemmas in sequence, over a short period of time, I have the distinct feeling that my thinking bleeds across dilemmas. How I respond to one dilemma appears to influence how I respond to others, either because of my shifting emotional state or because I am trying to maintain some semblance of logical consistency across dilemmas. My father's responses to some of these dilemmas represent a perfect illustration, especially given his training as a hyperrational, logical physicist. I first asked him to judge the Denise case. He quickly fired back that it was permissible for her to flip the switch, saving five but killing one. I then delivered the Frank case. Here, too, he quickly judged that it was permissible for Frank to act, pushing the large person onto the tracks. When I asked why he judged both cases in the same way—why they were morally equivalent—he replied, "It's obvious. The cases are the same. They reduce to numbers. Both are permissible because both actions result in five people surviving and one dying. And saving five is always better than saving one." I then gave him a version of the organ-donor case mentioned in chapter 1. In brief, a doctor can allow five people, each needing an organ transplant, to die, or he can take the life of an innocent person who just walked into the hospital, cutting out his organs to save the five. Like the 98 percent of our internet subjects who judged this act as impermissible, so did my father. What happened next was lovely to watch. Realizing that his earlier justification no longer carried any weight, his logic unraveled, forcing him to revise his judgment of Frank. And just as he was about to undo his judgment about Denise, he stopped and held to his prior judgment. I then asked why only Denise's action would be permissible. Not having an answer, he said that the cases were artificial. I am not recounting this story to make fun of my father. He has a brilliant mind. But like all other subjects, he doesn't have access to the principles underlying his judgments, even when he thinks he does.

To better understand how experience in thinking about moral dilemmas influences moral judgments, we looked more closely at Ned and Oscar. Some subjects judged only Ned or Oscar, some judged both in one session, and some judged one during the first session and the other during a second session administered one month later. Reiterating the information above, when a subject responds to only one of these cases, half say that Ned's action was permissible while three-quarters say that Oscar's action was permissible. If, however, you get both, either in the same session or in two different sessions, you fall into one of two groups: you say that either both actions are permissible or both are impermissible. This finding makes three points: Our experience with these dilemmas influences our judgments, the impact on judgment does not translate into our justifications and ability to access the underlying principles, and there appear to be people who for unknown reasons are more likely to judge certain situations as permissible or impermissible.

In this section, I have focused on the trolley problems to illustrate how a science of morality can capitalize on the linguistic analogy to begin uncovering some of the principles and parameters underlying our moral judgments. It is a sketch of what our moral grammar might look like. It is a description of principles that can account for some aspects of moral knowledge—of our competence in judging moral dilemmas.

MACHO CULTURES

How we perceive another's actions may say little about our own actions, about what we would or potentially could do in situations where violence is an option. What we need is a better sense of what real people do in real situations in real cultures.

Paralleling the study of language, one path to discovering whether our moral faculty consists of universal principles and parameters that allow for cultural variation is to tap into the anthropological literature with its rich descriptions of what people across the globe do when confronted with selfish and beneficent options. Although there are many universal patterns of violence in our species' history[7]—for example, men are responsible for a disproportionately large number of homicides, and of these, most are young men between fifteen and thirty years of age—there are significant differences within and between cultures that must be explained. In fact, the best predictor of violence is the number of unmarried young men! And if you work through the logic of this point, you will soon realize that societies where polygamy is supported are most vulnerable to such violence, because some men grab the lion's share of spouses, leaving others with none.[8]

When confronted with cross-cultural variation, most scholars in the humanities and social sciences throw up their arms, suggesting that the patterns of violence observed are simply subject to the local vagaries of a culture and its climate. Like a bottle adrift at sea, subject to the ocean's currents and patterns of wind, cultures of violence float, arresting for a period of time before changing again, unpredictably. In contrast, some psychologists and anthropologists argue that they can explain the variation by looking deep into our evolved psychologies. Cultural variation is only possible because of specialized psychological mechanisms that enable particular forms of learning. Once again, recall how linguists in the Chomskyan tradition have characterized the language faculty: a universal toolkit for building a specific set of languages and not others. Our evolved capacity for building language enables us to build a wide variety of languages, and also limits that range, making some languages unlearnable or, if learnable, unstable. We can conceive of an evolutionary approach to violence in the same way: Our biology imposes constraints on the pattern of violence, allowing for some options but not others; which options are available and selected depends upon prior history and current conditions. As the evolutionary psychologists Margo Wilson and Martin Daly suggest, "dangerous competitive violence reflects the activation of a risk-prone mindset that is modulated by present and past cues of one's social and material success, and by some sort of mental model of the current local utility of competitive success both in general and in view of one's personal situation. Thus, sources of variability in addition to sex might include the potentially violent individual's age, material and social status, marital status, and parental status; local population parameters such as the sex ratio, prevalence of polygamy, and cohort sizes; and ecological factors that affect resource flow stability and expected life span."[9] In terms of a general theory of morality, we want to exclude possible biases that may cause us to see as just situations that favor the in-group over the out-group; I use the notion of "exclude" here as a practical desideratum, acknowledging upfront that we inherited from our primate cousins (and probably before) a highly partial and biased mind-set, one designed initially to favor kin. It is this in-group bias that must be overcome if we are to advance an impartial moral theory.[10] In terms of understanding how our moral faculty interfaces with other aspects of mind and society, we want to understand how biases—conscious and unconscious—develop, survive, and break down.

Here's a simple test to read your aggressive temperament. If someone bumped into you and shouted "Asshole!" as they walked away, would you be mildly annoyed or really ticked off? If you're an American, feel really ticked off, and are tempted by the thought of pummeling the person who bumped into you, chances are you grew up in the South. If you think I am being presumptuous and prejudiced, you are only partially right. In the eyes of many, the South has always seemed a contradiction, trapped between a caricature of Miss Manners and the machismo of a Marlboro-smoking, trigger-happy guy. But careful studies by the social psychologists Richard Nisbett and Dov Cohen suggest that the macho side dominates, leading to a culture of honor.[11]

Here's how Nisbett and Cohen explain the cultural differences. Cultures of honor, such as the American South, are characterized by a common point of origin. They develop in situations where individuals have to take the law into their own hands because there is no formal law in place to guard against competitors who can steal valuable resources. A psychology of violence emerges. You can steal domestic herding animals (cattle, goats, horses, sheep), but you can't steal farming crops—you, of course, could steal some potatoes or carrots, but not enough to make a serious dent in the owner's resources. Cultures of honor therefore tend to develop among herding peoples, not farmers with crops. Many of these cultures have emerged over history and across the continents, including such herding peoples as the Zuni Indians of North America, the Andalusians of southern Spain, Kabyle of Algeria, Sarakatsani of Greece, and Bedouins of the Middle East.

The settlement of the South and North of the United States high-lights the relationship between resources, violence, and social norms. Scottish and Irish herders developed the South, whereas Dutch and German farmers developed the North. During the period of settlement, there were either no laws or poorly enforced ones. Consequently, as a means of protecting their property and their livelihood, the herders of the South developed their own means of protection: the rule of retaliation, or *lex talonis*. What started out as a macho response to animal property subsequently spread to other parts of life, including marital infidelities. If a man caught someone sleeping with his wife—caught in flagrante delicto—it was not only appropriate but expected of him to defend his honor by killing the offenders. The Southerners absorbed this attitude—this culture of honor—and carried it through to many facets of life.

Nisbett and Cohen used these historical data as a starting point for their studies, and in particular, the bumping experiments used to introduce this topic. Although the origin of the North-South distinction is several hundred years old, many perceive the South as holding on to a culture of honor. This perception is reinforced by an analysis of city names.[12] Cities in the South are twice as likely as in the North to include words such as "gun," "kill," and "war" (Gun Point, Florida; War, West Virginia), while cities in the North are more than twice as likely as those in the South to have words such as "joy" and "peace" (Mount Joy, Pennsylvania; Peace Dale, Rhode Island). Although city names may represent remnants of the past, these patterns continue into the present, as evidenced by the greater number of businesses in the South with names such as Warrior Electronics, Gunsmoke Kennels, and Shotguns BarBQ.

If the machismo of the past has staying power into the present, then Southerners should respond more aggressively than Northerners in situations where their potential honor is at stake, even if they are away from their home turf. Nisbett and Cohen tested this possibility by running the bump experiments with undergraduate men at the University of Michigan; with the exception of their home of origin—South or North—all subjects had similar socioeconomic and ethnic (white non-Hispanic, non-Jewish) profiles. In one variant of the experiment, a member of the experimental team—technically known as a "stooge"—approached a subject in a narrow hallway, bumped the subject, and either walked on or shouted "Asshole!"

before walking away. When the stooge bumped and insulted the Southerners, they reported greater anger, showed a massive stress response as indicated by an increase in the hormone cortisol, as well as an increase in testosterone, one indicator of aggressive intent. Northerners found the bump and verbal insult more amusing and showed no noticeable change in cortisol or testosterone. In a second experiment, one stooge bumped and insulted the subject, and soon thereafter, a second stooge—a six-foot-three-inch, 250-pound male—approached. Not only did Southerners experience greater anger than Northerners, but they were unwilling to move when the hulk approached. Having been insulted once, they had no intention of being insulted again. They were fighting for the status of king of the hallway.

When Northerners are insulted, they can ignore it, inhibiting the impulse to strike back either verbally ("Yeah, well, you're an asshole too!") or physically. Southerners have a different physiological set point. The Southern system of control is weaker than the Northern, at least at this point in history. These results show that culture can push around our aggressive tendencies, specifically the threshold for triggering our impulses to fight. All humans have the capacity for aggression. Each human has a different boiling point. Humans in some cultures have more similar boiling points than humans in other cultures. In the South, not only are people more likely to respond aggressively to insult, but they expect others to respond violently to insult. If a Northerner sees someone walk away from an insult, that is the proper thing to do. If a Southerner sees someone walk away from an insult, he's a wimp.

What is surprising about these observations is that the culture of honor mind-set can commandeer the psychology for so long after its emergence. Moreover, and as Nisbett and Cohen point out, this psychology can bias attitudes in a variety of closely related public and political arenas:

> A variety of laws, institutions, and social policies requiring the participation of many people in a shared meaning system is consistent with the culture-of-honor characterization of the South. These include opposition to gun control; a preference for laws allowing for violence in protection of self, home, and property; a preference for

strong national defense; a preference for the institutional use of violence in socializing children; and a willingness to carry out capital
punishment and other forms of state violence for preventing crime
and maintaining social order. In addition, individuals acting in their
institutional roles . . . are more forgiving of honor-related violence
and are more inclined to see such violence as justified by the provocation of another.

None of the culture of honor patterns in the United States can be explained by South-North differences in temperature, poverty, or a history
of slavery. For example, although temperatures in the South are higher
than those in the North, the highest homicide rates within the South are
in the cool regions, and the regional difference only appears among whites
and not blacks. Similarly, although the South is generally poorer than the
rest of the country, homicide rates in the South are higher than in the
North when cities or towns of comparable income are compared, and
cities of equivalent income within the South can show large differences in
homicide rates. Finally, although the South's use of slavery outpaced that
of the North, the lowest homicide rates in the South fall within regions
where slavery was more common.

Nisbett and Cohen's analyses show how the psychology supporting a
particular social norm can resist change even when the original trigger or
catalyst has long since disappeared. Southerners no longer need to defend
their herds, but their psychology is immune to the changing landscape.
In the case of honor cultures, the possibility that a defector will be
tempted by taking resources away from a competitor has generated a reflexive response to threat that takes the form of violence. In some sense,
the systems of control are compromised as the pressure to defend honor
dominates. Cultures of honor also showcase the economic notion of
discounting—of giving in to the immediate temptation of killing a competitor who has threatened one's resources as opposed to waiting for a
nonviolent, alternative solution. Although impulsivity and impatience
are typically seen as maladaptive responses, it is more likely that there has
been selection on humans to respond to opportunities for immediate
gains, especially when the prospects for the future are uncertain or grim.
Sometimes it may pay to knock off a competitor rather than wait for

some committee of elders to deliver their sagacious decision. Let me develop this idea a bit more.

Procuring food often entails some kind of delay. For the hunters in hunter-gatherer societies, there is an investment of time into finding and ultimately bringing down prey. A hunter may confront numerous prey while out on a hunt, but must decide whether to forgo one opportunity for something better. Often, the decision will be between a small prey item now versus the possibility of a larger prey item later. For herding cultures, there is a decision between killing an individual when it is young, small, and not yet reproductive, or waiting until it is older, when it can reproduce, and, in some cases, provide milk. Patience has virtues, and also requires control in the face of temptation. Evidence I will discuss in chapter 5 shows that there are innate differences in our capacity to delay gratification, with individual differences remaining constant over our lives: impulsive kids become impulsive adults, while patient kids become patient adults. Children who are impulsive, and take the smaller reward immediately as opposed to waiting for the larger reward, are more likely to end up as juvenile delinquents, alcoholics, gamblers, students with poor grades, and adults with unstable social and marital relationships. They discount the future, and the temptation for immediate gratification rules them. One unfortunate by-product of this discounting function is that such individuals may see violence as the easiest, quickest, and most effective short-term solution to a problem of resource inequity.

Violence and low-odds risk-taking may look like maladaptive strategies, but in an environment where the prognosis for future success is poor, these may constitute the most adaptive strategies.[13] For example, Daly and Wilson have shown that in Chicago, male life expectancy, which varies from fifty-five to seventy-seven years, depending upon the socioeconomic status of the neighborhood, is the best single predictor of homicide rates. Excluding death by homicide, if you live in a neighborhood where men rarely make it to sixty, murder rates are as high as one hundred out of every one hundred thousand; this rate drops to about one out of every one hundred thousand in neighborhoods where men live past seventy. In addition to life expectancy, differences in income among households—what is known as the Gini index—also contribute to both micro- and macroscopic differences in the levels of violence. Across the United States and Canada, and within the neighborhoods of Chicago, violence rates are highest in areas where inequities are highest. When envy rears its ugly head, the systems of control collapse, giving way to violent temptations.

Immediate opportunities to gain resources trigger actions that will maximize our status. At this level, we are universal opportunists. At a different level, our moral faculty generates judgments about equity, and justice more generally, pulling back the reigns of self-interest. Each culture then imposes its own set of signature constraints on when individuals can engage these opportunistic and strategic mental programs. This represents the signature of parametric variation.

Authority, dominance hierarchies, and obedience play an additional role in regulating our violence. Obedience to authority is a fundamental aspect of human nature, a characteristic that we see early in life, as children are exposed to their parents' rules. Home court rules. In the early 1960s, the social psychologist Stanley Milgram[14] conducted one of the classic studies of authority, bringing to light how systems outside of our moral faculty can impose significant constraints on what we do. They tell a harrowing tale of human nature, of the power of authority and the blind slavishness of obedience. They also speak directly to a prototypical problem of moral conflict—how one decides between two competing actions when one conflicts with our conscience and intuitions about what is morally right, and the other conflicts with the requests of an authority figure. Ultimately, one action loses and the other wins.

Milgram's studies involved three individuals: two subjects and an experimenter. One subject was naïve, sampled from a pool of people living in New Haven, Connecticut. The other subject was a stooge, a trained, middle-aged, male actor. Although rigged beforehand, the experimenter appeared to randomly assign each subject to one role, either teacher or learner; the naïve subject always drew the teacher role, the actor the learner role. Prior to the test, the experimenter brought the teacher and learner into a room and informed them of the experiment, pointing out that its main goal was to explore how punishment influences a person's capacity to learn. In the first phase of the experiment, the learner read and memorized a list of paired words, such as blue-box, nice-day, and wild-duck. In the next phase, the experimenter strapped the learner into a chair and attached electrodes leading to the shock machine; in the experiment, there was, of course, no shock, but the actor responded as if there was. The experimenter then informed the teacher and learner that shock would be used to assess its effectiveness in learning. The experimenter next brought the teacher into an adjacent room and introduced him or her to the shock machine. The instrument panel on the machine consisted of a dial, with clockwise increments starting at "Slight Shock" and ending a few increments after "Danger: Severe Shock," indicated by XXX. The experimenter instructed the teacher to begin testing. For each test question, the teacher read one word followed by four possible paired words. If the learner answered correctly, giving the appropriate word pair, the teacher moved on to the next question. If the learner answered incorrectly, the teacher delivered a shock. For each incorrect answer, the experimenter instructed the teacher to give an increasingly stronger shock. Again, the explicitly stated goal of the experiment was to determine whether punishment, in the form of shock, improved learning, as many studies of rats and pigeons had already demonstrated. Unbeknown to the subjects, Milgram was actually testing for obedience—an individual's willingness to yield to authority.

Based on a preliminary assessment of what people actually said they *would* do under the circumstances, Milgram expected subjects to stop delivering shocks at a moderate level of pain, stopping at around 9 on a dial that went up to 30. But what people say they will do, and what they actually do, are often different. In the Milgram experiments, the mismatch was extraordinary. With either some or no prodding from the experimenter ("Please continue" or "The experiment requires that you continue"), subjects continued shocking the learner to an average maximum intensity of around 20–25, equivalent to an extreme intensity shock. With voice feedback from the learner, Milgram observed a negligible change in the level of shock delivered. Subjects willingly zapped the learner in the face of such feedback as "Let me out of here . . . [*Agonized scream*] . . . My heart's bothering me. You have no right to hold me here." Bringing the learner in closer proximity to the teacher caused a 20 percent increase in disobedience. This suggests that when the victim is in view, the teacher's empathy rises closer to the surface; when such emotions are closer to the surface, they are more likely to influence action, especially disobedience in the face of authority. Nonetheless, even when the teacher could see the learner squirming and hear him screaming, most subjects went up to level 20 on the shock meter, an increment corresponding to "Intense Shock." An extraordinary 30 percent of the subjects went up to the highest level of shock (30 on the dial, 450 volts, and a label of XXX), even though the learner no longer responded verbally and was virtually listless.

Milgram's reflections on these experiments are poignant, revealing his conception both of authority as a powerful molding agent and of moral development as a process of *learning* about principles of harm: "Subjects have learned from childhood that it is a fundamental breach of moral conduct to hurt another person against his will. Yet, almost half of the subjects abandon this tenet in following the instructions of an authority who has no special powers to enforce his commands. To disobey would bring no material loss or punishment. It is clear from the remarks and behavior of many participants that in punishing the victim they were often acting against their own values. Subjects often expressed disapproval of shocking a man in the face of his objections, and others denounced it as stupid and senseless. Yet many followed the experimental commands" (p. 41). Like drones, these normal human beings—young and old, male and female, welders, academics, and social workers—marched to the beat of authority.

Milgram's experiments capture a core element of human nature. Breaking with authority is hard. To break with rules imposed from on high is to inhibit a typical or habitual pattern of action. It is a control problem that even we—adult humans with free will, a rich theory of mind, and a long history of education in moral behavior—have difficulty overcoming. In the context of the Milgram experiments, it also shows that normal humans are willing to inflict pain on another for what seems like a meaningless end. If this is the state of affairs in an artificially contrived experiment, then the control problem must surely be of horrific proportions outside of the psychologist's lab, where inflicting pain through violence pays off in the arena of competition, and where authority figures have much greater charisma. If you need convincing, recall any of history's most noted dictators.

Milgram's experiments show something else, directly relevant to the theme of this section: obedience to authority is universal, but the degree to which authority rules varies between cultures. Following on the heels of

Milgram's research, labs across the globe replicated his exact design, including studies in Germany, Italy, Spain, Austria, Australia, and Jordan. On the side of universality, subjects in all of these countries were willing to send high levels of shock to their inept learners. But among these countries, there was considerable variation: 85 percent of German subjects were willing to follow the experimenter's authority and send shocks at the highest level, whereas in the United States and Australia, the proportions dropped to 65 and 40 percent respectively. Culture can alter the gain on the rule of authority or the obedience of a culture's members, but the capacity to rule and follow are evolved capacities of the mind, shared with our primate relatives and numerous other species.

The fact that cultures can play with the nasty side of our nature, and in some cases perpetuate norms of violence even when it no longer pays to do so, shows the power and resistance of our belief systems. In the case of honor cultures, and, in particular, the American South, a macho response to insult is fueled and perpetuated by a psychology of rage and revenge. When someone challenges a Southerner's resources, the only appropriate, nonembarrassing, nondisgraceful reaction is a violent one. In one version of the bump-and-insult experiments, the experimenter told the subject that a bystander observed the incident. Southerners, but not Northerners, reported that the bystander would think less of them, taking away macho points. If Southerners believe that their reputation is at stake, then a perfectly rational response to insult is a violent one. For Southerners, the temptation to attack lies closer to the surface, while the systems of control are suppressed. If people believe that everyone in their culture fights insult with aggression, whether or not they do or don't, then a norm of violence will stick around. These attitudes can then force a shift from the descriptive level of what is to a prescriptive level of what ought to be. Southerners not only respond with violence to insult. They think this is what people ought to do.[15] The stability of honor cultures is therefore maintained by a tight coupling between descriptive and prescriptive systems and, more concretely, by creating a self-reinforcing feedback—a cycle of violence. As the social psychologists Vandello and Cohen suggest, "Cultural patterns become internalized scripts and habits that are rarely consciously noted: if noted, rarely questioned; if questioned, rarely energetically refuted."[16]

If this characterization is accurate, then the only way out of the loop is to break the misperception that others believe in violent responses to insult. This can happen in several ways. Subcultures can develop, providing an antiviolence voice. Regardless of their size, outspoken individuals who break with conformity can overturn long-standing traditions as dozens of social-psychology experiments illustrate, and as the swift elimination of the thousand-year-old tradition of foot-binding in China revealed. Laws can have a similar effect by making explicit the views held by a majority. Although an insult would still be annoying, it would not be embarrassing or a disgrace, because there is no imagined audience. Without an audience, inaction or nonviolent action are not judged. Without judgments, there is no social stigma. Without social stigmas, the cycle of violence might break.

SLAY THE ONE YOU LOVE

On the fifteenth of March in 44 BC, Julius Caesar walked into the Senate chambers of Rome. This would be his last entrance. Before delivering his speech, Caesar's trusted friend Marcus Junius Brutus joined other members of the Senate to launch a brutal attack on Caesar. Within minutes, they had stabbed Caesar a couple dozen times. Before dying, Caesar looked into Brutus's eyes and uttered those infamous words, *"Et tu Brute?"* or "You, too, Brutus?" History suggests that the original conspirators pulled Brutus into their plot by convincing him that Caesar had become too greedy and powerful, and was therefore a threat to the prosperity of Rome. Following the murder, the conspirators ran through the city, dripping with blood and shouting the cry of freedom. Brutus defended their cause, explaining that Caesar was no longer a friend of Rome but a foe. Since Brutus and his coconspirators were, in effect, the law, there was no court to try their violent acts. Nowadays, we would convict them of murder, slapping on a life sentence or, in some places, a lethal injection. We classify Brutus's act as murder because it was premeditated. But what if Brutus had been provoked by Caesar, or repressed and abused to such a point that murder was inevitable, spontaneous, involuntary, unplanned? Would Roman law have responded in a different way? Would it count as retaliation, an act of revenge that, as the Italian proverb suggests, "is a dish that the man of taste prefers to enjoy cold"?

In societies with formal laws, there is a crisp distinction between whether an act of violence is premeditated or not. Premeditated acts can receive life sentences or the death penalty, whereas acts that result in the same consequences—death of an individual—but are the result of self-defense, negligence, or ungovernable passions, typically yield lighter sentences or none at all. This distinction provides a window into what counts as a permissible act of violence, admissible in a court of law as a form of defense. It also highlights the distinction between the intuitions that emerge from our moral faculty and the far more complicated world of moral performance that ultimately provides the data for our legal systems.

Culture-of-honor psychology has nasty consequences for some married women. Depending upon the specific culture, violations include seemingly innocent conversations with another man, flirting, refusing a prearranged marriage or requests for divorce, rape, and voluntary sexual intercourse; although rape would appear to be a clear case of involuntary sexual intercourse, those in power—read men—consider it an act caused by female temptation. To give a sense of the violence, and the hold that men have over women in such cultures, consider a rather typical and horrifying example that occurred in Jerusalem in 2001:

> About thirty men and women gathered at the house that evening. After being greeted by Mr. Asasah, they formed a circle around Nura, his thirty-two-year-old daughter, who stood frightened, her swollen belly showing under her dress. She was five months pregnant. She was single. Holding a rope in one hand and an ax in the other, her father asked her to choose. She pointed to the rope. Asasah proceeded to throw her on the floor, step on her head, and tie the rope around her throat. He then began strangling his daughter while she, in turn, did nothing to protest. The audience—so the Jerusalem paper reports—clapped their hands, yelling, "Stronger, stronger, you hero, you have proven that you are not despicable." Following the macabre ceremony, Nura's mother and sister, who had witnessed the gruesome scene, served coffee and sat with the guests.[17]

Now consider some international figures on honor killings.[18] The United Nations Commission on Human Rights reports five thousand honor killings per year, with cases from Bangladesh, Brazil, Ecuador, Egypt, India, Israel, Italy, Jordan, Pakistan, Morocco, Saudi Arabia, Sweden, Turkey, Uganda, and Yemen. In the Arab world, more than two thousand women die every year from honor killings. In the West Bank, Gaza Strip, Israel, and Jordan, nearly all murders of Palestinian women are the result of honor killings. Most occur in public places, and sometimes in front of exuberant crowds. They are designed to both shame the victim and exonerate the murderer and his family. Many of the murders are carried out by a relative of the female victim, often her father or brother, and especially the youngest brother, because the legal consequences are much lighter for juveniles. In his address to the 2000 Convention on Human Rights and Dignity, General Musharraf of Pakistan proclaimed: "It shall be the endeavor of my government to facilitate the creation of an environment in which every Pakistani can find an opportunity to lead his life with dignity and freedom. . . . The government of Pakistan vigorously condemns the practice of so-called 'honor killings.' Such actions do not find any place in our religion or law . . . [Honor killing] is murder and will be treated as such."[19] This announcement came as welcome news, given the history of staggeringly high rates of male violence toward women in Pakistan, including rape cases every two hours, and honor killings ranging in the hundreds per year. Sadly, there were 461 reported cases of honor killings in Pakistan in 2002, representing a 25 percent increase over the previous year. Although honor killings are less common in Europe than in the Middle East, Pakistan, or India, the numbers are on the rise due to an increase in emigration and attempted marriages across traditional and more constrained cultural boundaries.

Honor killings are planned. They are premeditated. More often than not, they are excused. Sometimes, the crime is defended on the grounds that the woman's violation of social and sexual norms creates rage, uncontrollable anger that necessarily leads to violence. Most often they are defended on the grounds that the woman has violated part of the culture's heritage, their social norms, their way of life. In response to these crimes, most non-European legal systems either ignore the cases or dismiss them with virtually nonexistent penalties. Complementing the discussion in the previous section, both men and women in societies that engage in honor killings actually support these homicides by encouraging and even daring members of the harmed family to take revenge. As many recognize, response to honor killings constitutes a family issue, discussion of which falls outside of the law. In Islamic cultures, there is nothing in the Koran that would allow for or encourage honor killings. However, deep within Islamic tradition is the attitude that women are property. The owner has the power to do with property as he pleases. Property can be traded, bought, sold, and destroyed.

Honor killings create two interesting, albeit sociopolitically depressing, twists on the moral psychology of killing. From the male's perspective, any indication of a threat to the family's honor is sufficient to trigger violence. The local cultural norm fails to provide a mechanism of control. In fact, the social norm suppresses any system of control by allowing the impulse to kill to come forward. For such cultures, killing is not only permissible but expected if there has been a social or sexual transgression with respect to the married couple's exclusionary relationship. From the female's perspective, the possibility of honor killings serves as a control mechanism, forcing her to confine all of her dissatisfaction with the marriage to the recesses of her memory. A woman accused of dishonoring her family has no voice and little protection from the law. Knowing this, some women turn to extreme defensive acts to avoid death. In Jordan, for example, some women check themselves into the local prison. Other women have illegal operations to reattach their hymen in order to ensure that they are perceived as virgins prior to marriage. Among many Arab cultures, if a bride's hymen is torn, or if she fails to bleed during the couple's supposed first intercourse, the husband can call off the marriage. This brings shame to the family, and the only recourse is to kill her. Even in cases where a hymen exam results in a clear diagnosis of abstinence, 75 percent of the examined Palestinian women in Jordan are murdered due to persistent disbelief. In Turkey, as recently as 2003, a legal proceeding evaluated the possibility of exonerating a rapist if he agreed to marry his victim. An advisor to the minister of justice argued against this policy change, stating that men would only marry virgins.

These are horrific statistics and ideologies, at least from the perspective of some human cultures and societies. They illustrate how social norms can perpetuate violence by tempting a man to defend his honor, while controlling a woman's sexual and romantic interests by threatening her with the prospect of extreme violence or murder. Honor cultures act as psychological chastity belts, or, following an Arab expression, a man's honor "lies between the legs of a woman."[20]

Honor killings and crimes of passion show parallel psychological signatures.[21] Both tend to be supported by a local culture, a social norm that views violence as an expected response to transgression. Both are associated with an inherent gender asymmetry. To put it starkly, a man can kill his adulterous lover and walk away as a reputed hero, whereas a woman carrying out the same act is vilified as a cold-blooded murderer and put away for life or sentenced to death. And both honor killings and crimes of passion are typically associated with either no sentence or a significantly reduced sentence. The crucial difference between these two types of violence is that crimes of passion arise from an apparently uncontrollable rage, triggering an involuntary act of aggression, often in the form of a lethal attack; sometimes the murderer expresses remorse or guilt upon reflection. Honor killings, as discussed, are cool, calculated, and rarely associated with feelings of guilt. Crimes of passion are most often triggered by the sight of one's lover caught in flagrante delicto, or in response to a lover's description of the adulterous act and her partner's sexual inadequacies. The novelist Milan Kundera wrote: ". . . because [man] has only one life to live, [he] cannot conduct experiments to test whether to follow his passions or not."[22] We are sometimes inevitably forced to follow our passions in an experiment that is running on autopilot.

While honor crimes are anchored in a culture that views women as a commodity, crimes of passion are anchored in a view of human nature that sees the emotions as uncontrollable, at least in certain situations. Here is where an understanding of the interface between our moral faculty and the systems of temptation and control comes into play. Darwin's theory of evolution shows that selection has favored competition between males for access to females. It has also favored female choice, or pickiness, with respect to finding, choosing, and remaining with a mate. Once a couple has agreed to marriage, at least in cultures where marriage is between one man and one woman, this legally binding arrangement acts as a mechanism of control on the temptation to seek other partners. It sets up an expectation with respect to the moral faculty that a promise has been made. Breaking the promise by engaging in romantic acts with another constitutes an im-permissible or forbidden act. This expectation holds whether the marriage is legally or only informally binding. Both partners should be on the look-out for threats to the relationship. Seeing one's lover in bed with another partner leaves little to the imagination. Hearing one's lover speak about making love with another partner leaves little to the imagination, but at least raises some doubt concerning the report's truth. Provocation defenses based on verbal taunting have often been rejected in courts of law, includ-ing a case where a man stabbed his wife nineteen times after she ridiculed his sexual competence and threatened divorce; on the other hand, up until 1977, the American state of Georgia considered it justifiable to kill some-one to prevent an adulterous act from occurring.

It is hard to imagine anything more threatening, more irritating, and, for some, more deserving of murder then visual evidence of an affair. This image triggers the moral faculty's judgment of a forbidden act. Al-though it might seem transparent that the principle guiding our judgment here is simply a violation of a promise, things are not that simple. If a husband finds his wife in bed with another man, is this necessarily a for-bidden act? What if the couple has an open marriage? What if the wife is trying to get pregnant with another man because her husband is sterile? What if her husband has just announced that he is impotent and thus she is free to have sex with other men so long as he can watch? There is no sci-entific evidence to speculate one way or the other about these possibilities. They reveal, however, why we need to break down events into a sequence of actions, together with their causes and consequences. This is the input to our moral faculty. This is the material it chews up prior to spitting out a judgment concerning permissible, forbidden, or obligatory action. This is the stuff of a Rawlsian creature.

The emotions should click in as well, pushing us from a moral judg-ment to an immoral act of violence. Although the Humean component of our moral faculty may well trigger this judgment, for thousands of people throughout the world, this kind of threat does not automatically lead to murder. For those where it does, however, there is a long history in many countries of recognizing, both casually and legalistically, the power of the emotions to overwhelm our moral faculty's judgment and the rational conclusion that should follow. This kind of override provides a defense or excuse for the act of violence. By excusing certain actions, however, the law has necessarily developed a specific view of human nature, one anchored in a set of beliefs about temptation and control. Why is the sight of one's lover naked and in bed with another person sufficient to tip the scale of control and make killing an excusable or permissible act? Even if men and women have different brains and different capacities for control, should courts of law allow for different levels of responsibility? Is there a universal boiling point, a threshold of temptation beyond which all control systems fail? Or is the boiling point specific to each context, and to each individual? What is provocation, such that it draws in temptation and blinds control? The law isn't terribly clear or consistent on these issues, but history shows how changing views of human nature—especially sexuality and the sexes—has entered into legal discourse. The legal scholar Victoria Nourse[23] summarizes the state of affairs in America up to 1997:

> . . . the doctrine of provocation stands at a crossroads . . . The doctrine is in extraordinary disarray . . . Although most jurisdictions have adopted what appears to be a similar "reasonable man" standard, that standard has been applied in dramatically different ways, with jurisdictions borrowing from both liberal and traditional theories. Some states require "sudden" passion, others allow emotion to build up over time; some reject claims based on "mere words," others embrace them. Today we are only safe in saying that in the law of passion, there lie two poles—one exemplified by the most liberal MPC [Model Penal Code] reforms and the other by the most traditional categorical view of the common law. In between these poles, a majority of states borrow liberally from both traditions.

Legal debates about crimes of passion center on the difference between murder and manslaughter. Dating back at least as far as the twelfth century, legal cases were decided on the basis of whether an action was intentional, accidental, or defensive. Accidents, by definition, are blameless, with the caveat that those who are negligent or carefree with their actions can be held accountable for an accident. Someone who keeps loaded guns around the house, fools around with them while drunk, and then shoots and kills a child, may have done so accidentally, but is negligent and irresponsible and unlikely to be cleared in a court of law. A verifiable accident that leads to someone's death is, therefore, excusable. Self-defense, on the other hand, is justifiable. If a thief is about to shoot if you don't hand over your money, shooting back is not only appropriate but morally justified even if it leads to killing the thief. These properties of the law mirror the principles and parameters that our moral faculty implements on its way to making an unconscious judgment. We can break self-defense down into factors that feed into our moral faculty. The agent's act of shooting results in a negative consequence (killing the thief). The agent's intention, however, is not to kill the thief but to defend himself from being killed. The act—shooting a gun—has the foreseen negative effect of killing the thief and the intended positive effect of saving the agent. Killing is a means to an end, but, in this case, a permissible one, because of a prior threat to the agent. In the context of self-defense, therefore, our moral faculty judges that the act (shooting a gun) that leads to killing is not only permissible but perhaps obligatory if there are no other options. The law supports our intuitions in this case. The doctrine of provocation provides an exception to the deontological view that killing is forbidden.

At the end of the sixteenth century, the law took an interesting turn, in part because of the idea that cases of self-defense were defended on the basis of impulsive emotions, and, specifically, a fear response. When extreme fear strikes, triggered by the threat of death, self-defense—in any form—is justified. If this argument works for self-defense, it should also work for other impulsive emotions, especially anger. As the legal scholar Jeremy Horder writes, "By the end of the sixteenth century, the distinctive character of voluntary manslaughter as a less grave form of homicide than murder had been formed in the shape of what was and is still called a concession to human infirmity, a concession to the strong retributive impulse of great anger upon provocation."[24] Thus was born the doctrine of provocation, designed to minimize the number of cases unnecessarily downgraded from murder to manslaughter, and to recognize four legitimate kinds of provocation, or what I consider *triggering acts*: insulting assaults, seeing a friend or relative attacked, seeing a countryman deprived of liberty, and seeing a spouse in an adulterous act.

For most of our history, passion crimes have been a male affair, though with some cross-cultural variation; in France, during the nine-teenth century, only about a third of all murders were by men, but almost all of the murders by women were crimes of passion.[25] When men kill in the heat of passion, they target male competitors. The underlying belief, either implicit or explicit, has been that women are incapable of making rational decisions and thus cannot be blamed for their inability to control temptation. Although this sexual asymmetry continues into the present, there are important reversals, especially the increase in husbands killing their wives and wives killing or seriously injuring their husbands. The first reversal reflects a change, evident in some cultures at least, that women are not property. As such, they are fully rational and are perhaps more capa-ble than men of resisting sexual temptation. Men have therefore shifted their targets, allocating responsibility to women. The second reversal re-flects the increasing independence, economically and psychologically, that women experience in some cultures. Whereas crimes of passion by men are triggered by the possibility or actuality of an affair, the same crimes by women are often triggered by a history of abuse; the affair may be the final straw, but there is a longer history leading up to the crime. These differences often lead to different legal judgments, with men con-victed of manslaughter due to the passion defense, while women are con-victed of murder because it appears more premeditated. Evidence of this changing landscape of violence comes from several Asian cultures, espe-cially China. The recent increase in independence among women, together with the rising disparity in wealth, has led to an increase in the number of polygamous relationships and illicit affairs, and, concomitantly, an in-crease in the number of unmarried young men. As a result, murder rates are soaring, with crimes of passion taking center stage, and domestic vio-lence present in approximately 30 percent of all families. In China, women want to make adultery a criminal offense. Overall, however, 82 percent of the polled population in 2000 were opposed to this legal move, and only 25 percent considered prostitutes, concubines, and forced marital sex as morally forbidden. Dissatisfied women, no longer willing to remain silent in their relationships but economically able to live alone, are allowing anger to percolate and trigger violence against their husbands—and against years of tradition.

From a legal perspective, crimes of passion—and the provocation defense more generally—raise interesting questions about the human mind and its capacity for control in the face of temptation. When the passion defense is effective in the courtroom, it relies on the *immediacy* of the violent action. From the perspective of our moral faculty, we might ask whether judgments of permissible killing are more likely when they follow on the heels of a triggering action such as an affair than when they are delayed.

In a 1949 U.S. case, here is how the judge instructed the jury to evaluate the observation that the defendant had killed her husband following a history of brutal treatment:

> . . . [the] circumstances which induce a desire for revenge are inconsistent with provocation since the conscious formation of a desire for revenge means that a person has had time to think, to reflect, and that would negate a sudden, temporary loss of self-control which is the essence of provocation . . . provocation being therefore as I have defined it, there are two things in considering it to which the law attaches great importance. That is why most acts of provocation are cases of sudden quarrels, sudden blows inflicted with an implement already in the hand, perhaps being used or being picked up where there has been no time for reflection . . . [26]

The idea here is that a particular situation provokes an automatic and violent response, akin to a reflex. Any delay counts against this defense in the same way as if a doctor tapped your knee with a tiny mallet, and you didn't move, she would think something was wrong, a normal reflex converted into a pathology or a case of willful inaction. The implication of this line of thinking is that the law would exclude from the doctrine of provocation any case, presumably extremely common, in which a battered housewife retaliates by killing her abusive husband. In this case, killing is forbidden, as it is based on retaliatory outrage.

In February 2003, Clara Harris was charged with murdering her husband David Harris. At the time of the trial, there were three well-established facts: Clara had hired a private investigator to check on her suspiciously acting husband; David had been cheating on Clara who, on the day of the killing, had found him and his girlfriend at a hotel; and

Clara ran over David with her car at least three times. Clara received considerable sympathy both from the jury and the public at large. The case centered on whether the crime was committed as an impulse or with forethought, planning, and intent. On the side of impulse, the jury observed that the crime was committed soon after Clara first saw her husband with his girlfriend; up until this time, she had suspicions but no personal evidence. On the side of intent, the jury noted that Clara never saw the couple having sexual intercourse; that she had waited until David was in the parking lot of the hotel and only then drove her car at him, first striking him down and then backing up to run him over a couple of times. The jury voted in favor of manslaughter, invoking a passion-crime defense. However, when it came to judging the relative severity of the crime, the jury allocated the maximum allowable sentence of twenty years in prison, knowing the minimum was two years. Although the jury recognized Clara's emotions, and the challenge of controlling them in the face of seeing one's spouse with another person, they effectively weighted all of the other evidence more heavily toward the intent side of the action. Running her husband over once would have yielded a lighter sentence. But doing it multiple times implied a plan with the intent to kill. The jury penalized Clara based on the psychology of intent.

As many legal and feminist scholars have pointed out, this twenty-first-century case spotlights the inherent gender bias that continues to prevail in many legal systems. Based on three hundred years of passion-crime cases, if Clara Harris had been the victim and David Harris the murderer, he undoubtedly would have received only two years and would now be enjoying his freedom and presumably a different wife.

Passion crimes also raise questions about what constitutes a normal, average, or modal response to a particular situation. As Aristotle put it, we must evaluate how to feel and act "towards the right person to the right extent at the right time for the right reason in the right way." To say that killing someone else is excusable when a man or woman sees their partner in bed with another is to say, in Arisotle's sense, that it was the right action at the right time. It is to say that a significant proportion of men and women are incapable of controlling their anger in this context. In Aristotelian terms, our judgments are based on a doctrine of the mean, a kind of yardstick for evaluating an action. Legal defense might rest, therefore, on the idea that certain emotions or actions necessarily trigger other kinds of emotions or actions; for example, seeing one's lover with another person triggers anger, which necessarily triggers violence. In his book *Wise Choices and Apt Feelings*, the moral philosopher Alan Gibbard has pushed this idea further, basing it on the biological notion of a norm of reaction: the range of phenotypes expressed by a single genotype in a suite of different environments, the classic example being a type of corn that grows to different heights depending upon altitude. Gibbard's intuition is that there are also emotional norms—apt feelings—that lead to particularly relevant and appropriate actions—wise choices.

In terms of the law, therefore, it might be useful to have population statistics on how people, within and between cultures, respond to adultery. There are two relevant statistics. First, in many countries, crimes of passion constitute a significant proportion of homicides; for example, in 2002, passion crimes accounted for most of the homicides in Cuba. Second, although thousands of people throughout the world are confronted each year by an adulterous challenge, extremely few convert the provocation into lethal violence. Nonetheless, the courts continue to deem adultery a sufficiently powerful provocation to warrant a shift in punishment. Like discussions of the death penalty and its capacity to deter future crimes, we should also question whether our more lenient views of passion crimes might increase the rate of domestic violence. I know of no statistics addressing this possibility.

The view that emotions dominate our reasoning capacities is essential to the passion-crime defense, and to the Kantian creatures' fight with the Humean. It is at the core of the Aristotelian and Hobbesian view of human nature, which conceived of anger as a desire to overcome the competition, a desire that must be consummated. For Hobbes, "Neither is the freedom of willing or not willing, greater in man than in other living creatures. For where there is appetite, the entire cause of appetite hath preceded; and, consequently, the act of appetite could not choose but follow, that is, hath of necessity followed."[27] This view continued well into the eighteenth and nineteenth centuries, where one finds dozens of legal cases describing the defendant as consumed by a "brief storm" of passion, being "beside himself," "transported out of body," or "out of balance." These metaphorical descriptions all imply a hierarchical view of the mind, with

our capacity for reasoning sitting above the other faculties, especially including the emotions. Sometimes the emotions are so violently triggered
that they displace our reasoning machinery, as passion wreaks havoc with
our action system, both our perception of what is permissible and what we
actually do following this judgment. This view of the human mind places
the legal distinction on the side of excuse as opposed to justification.

Beginning in the middle of the nineteenth century, legal analysis of
crimes of passion changed. The main impetus for this change was a critique of the then-dominant perspective on self-control. Like Groucho
Marx, who commented that he would wait for six years to read *Lolita* because she would then come of age, we can resist the temptation to kill off
an adulterous spouse or lover if we are reasonable. Thus, in one crime-of-
passion case, a judge noted that "though the law condescends to human
frailty, it will not indulge to human ferocity. It considers man to be a rational being, and requires that he should exercise a reasonable control over
his passions."[28] Current thinking in American law is, therefore, based on
the agent's judgment of wrongdoing as opposed to the appropriateness of
their response. Did the defendant perceive an appropriate injustice (e.g.,
given that adultery is wrong, did the defendant have evidence for adultery?), and if so, was their reaction proportional to the degree of provocation and reasonable, given that they lost control? The proportionality
aspect of the defense is, of course, difficult to evaluate. Is killing ever an
appropriate level of response? On a general level, the answer is certainly
affirmative, given the cases discussed above. Is it appropriate in the face of
witnessing an adulterous act (kissing, intercourse) or hearing about it sec-
ondhand? The law is unclear on this point, with cases yielding a variety of
answers.

Honor killings and passion crimes illustrate the power of social norms
to both set the principles and parameters of permissible killing, and to
convert them from descriptive to prescriptive principles. In some cultures,
men are allowed not only to engage their sexual appetites but to perceive
them as obligatory, as important as eating and sleeping. Equally powerful are attitudes concerning a woman's freedom—or lack thereof. In cultures where men are sexually promiscuous, often taking on several wives,
women are sexually oppressed, shackled to their partner by the threat of
violence. A potential danger is that the reduced or negligible sentencing

associated with such crimes may, in fact, fuel a higher level of violence, leading to spousal infidelities becoming hotbeds for lethal attacks. What these systems also show is that homicide, which is often seen as individual pathology, is better understood as an act fueled by powerful cultural beliefs. Depending on the cultural climate, killing is not only permissible but justified, excusable, and expected. The biology that underlies human nature and describes what happens (triggering action → permissible counteraction → anger → rage → permissible killing) has made its way into some of our cultural and legalistic prescriptions of what should happen (triggering action → permissible counteraction → anger → rage → obligatory killing). Social norms therefore have the power to convert a description of what happens into a normative view of what *ought* to happen.

NATURE'S COUNSEL

In many twelve-step programs, from Alcoholics Anonymous to Sexaholics Anonymous, individuals enter the final steps by admitting that they did something wrong. In the third episode of the 2004 season of *Desperate Housewives*, several of the main characters found themselves in a situation in which they either explicitly or implicitly acknowledged some wrongdoing: Susan apologized to her ex-husband for being rude to his new and younger wife, Bree sought counseling after announcing at a dinner party that her husband cries during sex, and Lynette's husband acknowledged the challenges of taking care of the kids after he was busted for being at a party instead of a serious business trip. Many people presumably know that they have done something wrong based on reactions by others, but don't admit to the wrongdoing or take responsibility. Some of these people are excessively narcissistic, a disorder that can bleed into the presidency, as when President Bill Clinton failed to acknowledge his affair with Monica Lewinsky and President George W. Bush failed to admit to the public that he went to war with Iraq for reasons other than the one concerning weapons of mass destruction.

Admission of wrongdoing entails acknowledging an impermissible act, a violation of some moral principle or social convention. In some of the cases above, however, it is not entirely clear what these wrongdoers understand about the nature of their social crime. They can see the negative consequences of their acts based on how other people respond, from one-on-one interactions à la *Desperate Housewives* to national and international outcries à la Presidents Clinton and Bush, respectively. What the linguistic analogy forces is a reevaluation of the nature of this knowledge and the extent to which we, in our mature state, have access to the underlying principles organizing this knowledge. When we judge others or evaluate our own actions, what do we know about the nature of our judgments? Do our explanations fully describe the principles that guide what we say and what we do in the moral domain?

If the arguments in linguistics hold for morality, then we will have to ignore what appears obvious at the surface in terms of our descriptions of behavior, turning to a layer underneath, which contains the moral faculty's codes. What we see in terms of moral behavior and justification is most likely a poor representation of the moral faculty's output. The reason is simple: In between the computation that generates intuitive judgments about morally permissible actions and our actual actions and justifications lie many different steps, interfacing with many different psychological processes, including those involved in emotion, memory, attention, perception, and belief.

When people attempt to explain their moral judgments, they are often dumbfounded, appealing to hunches or conflicting accounts. Some, like Haidt, argue that such dumbfounding arises because we are not reasoning about these moral dilemmas but rather delivering flashes of insight based on unconscious emotions—the signature of the Humean creature. For Haidt, the Humean creature has the first say, whereas the Kantian comes in as cleanup, rationalizing the judgment delivered.

I have no doubt that our emotions and capacity for principled reasoning play some role in our moral judgments. But what both of these processes minimally need before they can work out a moral verdict is an appraisal of the causes and consequences of an action. Both systems need the Rawlsian to step up to the plate first, and deliver a structural analysis. I am also suggesting something much stronger than this minimal addendum. The Rawlsian creature may be the essential system for generating the moral verdict, with both Humean and Kantian creatures following in its wake, perhaps triggered by it. I have not yet given sufficient evidence one way or the other to adjudicate on the stronger version. This evidence will arrive in two installments, first in our discussion of human development and second in our discussion of human evolution.

To set up the evidence ahead, consider once again the trolley problems. From the facts that we observe at the surface of these cases, we uncover important psychological factors that are part of the mind's deep structure, and ultimately responsible for why we judge some actions as permissible and others as forbidden.[29] We can also begin to see some interesting dynamics between the notions of obligatory, permissible, and forbidden judgments:

a. If an action is permissible, then it is potentially obligatory, but not forbidden
b. If an action is obligatory, it is permissible and not forbidden
c. If an action is forbidden, it is neither permissible nor obligatory

It is permissible for Denise to flip the switch, but I doubt anyone would say that it is obligatory. It would be obligatory for her to flip the switch if there was no one on the side track, converting this case into an opportunity for cost-free rescue. When there is an opportunity to help someone in danger, and there are no personal costs, then we perceive rescue as obligatory. If we described a variant of the Denise case with no one on the side track, and said that she let the trolley run over the five people, most—presumably all—would judge her action as forbidden and most likely punishable. In the original Denise case, however, my hunch is that those people who said that it was not permissible for Denise to flip the switch, perceive *responsibility* as the key issue. It is not Denise's responsibility to decide the fate of the trolley. Denise doesn't get to choose who dies, even if the numbers differ. Each life is of value, and the person on the side track is not in harm's way given the trolley's trajectory. Those who judge the case in these terms would presumably say that it is forbidden to sit back and watch if the side track is empty. And if responsibility drives their decision, then they should use this as a guiding parameter for other moral dilemmas. If the numbers changed radically, such that redirecting the trolley caused one person to die on the side track but enabled an entire city to

survive, would it still be impermissible for Denise to flip the switch, because it isn't her business? At some point, the responsibility card won't
work, and this is where the push to consider other parameters enters.

A nice extension of this general approach comes from work by the social psychologist Philip Tetlock. Although inspired by a different line of
thinking than the linguistic analogy presented here, his experiments fit
beautifully into the theme of universal principles and culturally config-
urable parameters. In developed nations in the Western world, everyone
with children and even those without would be offended if a salesman offered $1,000 for each child. What if he increased the offer to a million
dollars? A billion dollars? What about any price you like? Everyone is
likely to maintain their offended, even disgusted response to this question. But some may well pause at the higher sums. Those who do, often
feel extreme guilt and attempt to do what they can to redress the moral
balance. We are offended and sometimes disgusted by opportunities for
exchange, not because we think they are unfair but because these are
taboos, commodities that just don't enter the market. What is interesting
about these taboos is that all cultures have them. When pushed for an explanation, people are dumbfounded, incapable of explaining why it is forbidden to exchange certain commodities. Each culture, however, has the
freedom to decide which commodities enter into legitimate trade and
which are off bounds—taboo tradeoffs. Pursuing the linguistic analogy, I
would say that each culture has a principle of fairness in the context of exchange, with a parameter involving exchangeable goods that is set by the
local culture.

The general point here is that there are hidden parameters underlying
people's intuitions about these cases. When we ask subjects for justifications, they are hopeless. If people really don't have access to the principles
underlying their judgments, then this has important implications for how
we think about our moral faculty. For one, it makes the linguistic analogy
all the more striking. It strengthens our characterization of the Rawlsian
creature with its inaccessible but operative principles. On the other hand,
it is possible that even without any access to the underlying principles, or
with only limited access, the nature of our moral judgments shifts once
we *become* aware of these principles. Once again, the linguistic analogy is
relevant. The fact that a linguist like Chomsky has access to some of the

underlying principles guiding his knowledge of language doesn't impact
on his performance. When Chomsky communicates with others, his comprehension and diction are as good as anyone else's. If they are different,
it is not because he knows something about the language faculty that the
rest of us don't. In contrast, it may well be that once we have an understanding of some of the operative principles of our moral knowledge, and
broadcast these principles to an interested public, they will impact on performance. Keeping in mind the distinction between intended and foreseen consequences may influence our judgments of others and our own
actions. Keeping in mind that killing is sometimes permissible may influence how we think about harm. The main point is that the moral faculty
may guard its principles but once its guard has been penetrated, we may
use these principles to guide how we consciously reason about morally
permissible actions. If this captures the link between operative moral principles and our moral actions, then we will have identified an important
difference between the moral and linguistic domains of knowledge.

PART II

Native Sense

4

THE MORAL ORGAN

*Many are really virtuous who cannot explain what
virtue is . . . But the powers themselves in reality perform
their several operations with sufficient constancy and
uniformity in persons of good health whatever their
opinions be about them . . .*

—F. HUTCHESON[1]

DO WE BEGIN LIFE with some default personality, born evil or good? The sixteenth-century mathematician and philosopher Gottfried Leibniz claimed that our knowledge of morality and arithmetic is innate. He noted that it "is no great wonder if men are not always aware straight away of everything they have within them, and are not very quick to read the characters of the natural law which, according to St. Paul, God has engraved in their minds. However, since morality is more important than arithmetic, God has given to man instincts which lead, straight away and without reasoning, to part of what reason commands."[2] In the Book of Numbers, Moses claimed to have obtained his intuitions about morally permissible behavior from God, who told him to "gird his loins . . . kill his neighbors . . . rape his women . . . starve his children and steal his land." And Darwin also leaned in the direction of God's wisdom, claiming that "A man who has no assured and ever-present belief in the existence of a personal God, or of a future existence with retribution and reward, can have for his rule of life, as far as I can see, only to follow those impulses and instincts which are the strongest or which seem to him the best ones."[3] In other words, without God's leading light, there is no moral guidance.

The analogy between language and morality provides a different and, in my opinion, more informative way of looking at this age-old problem. In chapter 1, I mentioned that the first step in understanding the engineering of our moral psychology is to describe the unconscious and inaccessible principles underlying a mature person's judgments concerning right and wrong. Here I use current knowledge to explore how the principles guiding our intuitive judgments are acquired and how they are represented in the brain. I will characterize the moral organ, based on some fascinating findings from studies of infants, normal adults, and individuals with mental disorders arising from nature and nurture. I will later show how this system generates intuitions in the context of forming and maintaining competitive alliances.

First, a few thoughts about what is at stake and what counts as evidence for or against the explanation I favor. A common move among scientists of the mind is to document the universality of a trait and then claim that it is Mother Nature's handiwork. Universality is one of Mother Nature's signatures, but there are potential impostors that we must eliminate before assigning credit. Take the fact that night follows day. This belief doesn't come from Mother Nature. It comes from the fact that we are all exposed to this pattern, early in development, and thus register this information into memory. The information is learnable independently of who you are or where you live. The learnability of knowledge doesn't mean that all knowledge is learned from experience. Rather, it is a warning to keep in mind when moving from the observation of universal beliefs to the inference that such beliefs are part of the brain's hardware.

A second of Mother Nature's signatures is the early appearance of a trait in development. When a behavior emerges early in development, it is tempting to conclude that its foundation lies in nature. The reason is simple: based on timing, there has been insufficient experience for nurture to construct the details. Consider the fact that newborn babies, no more than an hour old, can imitate the facial expression of an adult sticking out his tongue or opening his lips into an O-shaped configuration. Imitation is part of nature's gift. It is not something that could have been learned within the first hour of life, especially since much of this time is spent on the mother's breast. Nurture, in the form of exposure to a silly person sticking out his tongue, is critical. Without it, the baby would not have protruded its own tongue. But we do have the ability to watch someone do something and then replay this obser-vation from memory as a guide to producing an exact replica. But some-times, an individual does something early in life that can be explained by nurture. This happens, for example, when infants acquire new facts or labels for objects, based on a single exposure. Consider a child who knows the words "duck" and "truck," and uses them appropriately to label ducks and trucks respectively. Place one toy duck and one toy truck on a table, along with a novel object. Now ask this child for the "blicket." She will bring you the novel object. By means of exclusion and the mapping of sound to mean-ing, she can learn the label for a new object and store this information in memory. Thus, the early appearance of a behavior—verbal naming, in this case—need not indicate nature's signature. Rather, all normally developing human children are born with the capacity to map sounds to meaning. Their native environment provides the lexical ingredients for building a massive vocabulary. In addition, traits that appear late in development are not necessarily the result of nurture's handiwork, as evidenced by the emer-gence of facial hair in men and breasts in women, secondary sexual charac-teristics that arrive as a result of maturation and the achievement of puberty.

If there are moral universals, then there must be capacities that all normally developing humans share. There are at least three ways that this might play out, again using parallels to language. On one end of the spec-trum is a nativist position that puts precise moral rules or norms in the newborn's head. She is born knowing that killing is wrong, helping is good, breaking promises is bad, and gratuitously harming someone is evil. On the opposite end of the spectrum is the view that our moral faculty lacks content but starts us off with a device that can acquire moral norms. With this view, there are no rules and no content, only general processes for acquiring what nurture hands us. In the middle is the view that we are born with abstract rules or principles, with nurture entering the picture to set the parameters and guide us toward the acquisition of particular moral systems. The middle view is the one I favor. It comes closest to the lin-guistic analogy. It makes the obvious point that something about the hu-man brain allows us to acquire a system of moral norms. And it makes the equally obvious point that dogs and cats that grow up with humans never acquire our moral norms, even though they are exposed to them at some level. The challenge then is to characterize what this initial state looks like, what each culture hands down, and what limitations exist on the range of possible or impossible moral systems.

The moral instinct perspective I favor does not deny cross-cultural variation. But acknowledging the observed variation does not constitute a rejection of constraints. We must look hard at the cross-cultural variation and ask whether it is boundless or limited in interesting ways. It is common, perhaps especially in the social sciences and humanities, to counter claims about human nature with examples from indigenous peoples living in areas with little or no Western contact. For example, the philosopher Jesse Prinz calls upon a number of counterexamples from anthropologists to argue against the idea that there are universal, innately specified norms against harm and incest.[4] In discussing harm, for instance, he cites the anthropologist Colin Turnbull's work on the Ik of Uganda, noting that young men would merrily pry open the mouth of an older person to extract food, and watch with anticipation and glee as a young child reached out, unaware of the danger, to touch the burning embers of a fire. Prinz also notes that incest has been much more common than we might wish to acknowledge, with evidence of brother-sister sexual intercourse occurring since ancient Egyptian times. These cases strike us as barbaric, bizarre, and disgusting. We must also remember that these scenarios are not the dominant pattern for our species. It would be like arguing that Mother Teresa and Mahatma Gandhi are emblems of human nature, of our unfaltering desire to help others. Humans do harm to others and engage in sexual behavior among kin. But there are constraints within our cultures on whom we harm and what counts as an off-limits sexual interaction. It is precisely these details that are of interest when we attempt to understand the nature of our moral faculty.

GREAT EXPECTATIONS

We expect parents to care for their children, friends to be supportive and loyal, members of a team or organization to cooperate, store owners to sell products at a fair price, cheaters to be punished, and individuals to defend their property against others. In some class-based societies, including Charles Dickens's nineteenth-century England, members of the lower class, such as Pip, were expected to marry members of the same class even if, like

Pip, they were deeply in love with an upper-class lass, such as Estella. In the eyes of many, marrying outside of your class is wrong, with the lower class polluting or infecting the upper class as if its members were diseased. When individuals violate these patterns of action or interaction, we often expect others to respond by attempting to reestablish the customary patterns, sometimes by punishing the offenders. When an individual breaks with expectation, we usually consider their action wrong. When individuals act in a way that is consistent with expectation, we assume their action to be right, even if we don't openly label it as such. Of course, the relationship between expectation and notions of right and wrong is far more complicated than this. From a normative perspective, actions that are expected may nonetheless be judged as morally wrong. Similarly, actions that are unexpected may nonetheless be judged as morally right. Like the notion of expectation itself, the concepts of right and wrong are always framed relative to some standard, whether handed down by God, Darwin, or our legal bodies. In colonial America, wealthy people were expected to have slaves, and the poor and jobless from Africa and the Caribbean were expected to be slaves. When Abraham Lincoln and other visionaries stood up against the concept of slavery, they violated expectations in the service of promoting a normatively consistent doctrine for treating members of our species. And as I mentioned in chapter 3, if a white non-slave-owner had killed a white slave-owner following repeated attempts to end slavery, we might consider this act expected and morally permissible.

The idea I wish to introduce here is that we consider thinking about the origins of our sense of right and wrong by starting with the process of generating an expectation.[5] Before human infants can run, climb, eat with a fork, discuss their impressions, and understand humor, they can form expectations about patterns of action in the world. Loosely speaking, an expectation is a belief about some future state of affairs. Expectations arise when an individual uses her knowledge of prior events to predict the reoccurrence of the same or similar events in the future. In some cases, the individual will be aware of the expectation. In other cases, the expectation will form sub rosa, but nonetheless influence behavior. This sense of expectancy is closely linked to probabilities, like the odds of rolling snake eyes or calling out blackjack when a card is placed on top of an ace. A different sense of expectancy arises in the moral domain when we think about what we expect others to do or what we should do. These are normative expectancies and they refer to obligations, promises, and commitments.

When an expectation is violated, a response ensues. When an action or its consequence matches expectation, the response is often positive. Positive emotions are rewarding and reinforcing. When an action violates expectation, a negative emotion often ensues. Negative emotions are aversive. I propose that one branch of the root of our moral judgments can be found in the nature of expectation concerning action. This exploration requires an understanding of how we build expectations, distinguish between accidental and intentional causes, respond to violations, and forge a relationship between expected actions and emotions.[6] To appreciate the potential force of this move, let's start with a simple example, detached from the complexity or our moral psychology.

Physical laws capture highly regular phenomena. Nothing could be more certain and expected than the fact that two solid objects can't occupy the same physical space at the same time, and that an object lacking support will naturally fall straight down until it contacts another physical structure. The first principle is true of all objects, while the second is true of all nonliving things and most living things with the exception of flying animals. Given that these regularities have been around since the first multicellular creatures roamed the Earth, it would make sense for evolution to have wired our brains, and the brains of other animals, with this information. If this is the case, then their behavior should show the mental signature of these physical principles. These organisms should have expectations about the physical world that together constitute a kind of naïve or folk theory. This folk theory will serve them well, most of the time, generating accurate expectations about objects and actions. But sometimes their folk theory will backfire, because something has changed in the environment. Behaviorally, individuals will make errors. Errors—if they are genuine—are informative, as they, too, provide evidence of expectation. Repeated errors provide the telltale signature of a theory that is immune to counterevidence; the theory generates one expectation and no alternative solutions. To break with tradition requires trying a new action that violates expectation. It requires breaking with conformity.

Newborn babies lack the ability to move on their own, can barely see, communicate by crying and gurgling, and lack a sense of humor. But underlying such incompetence is an exquisite system of perception linked to a skeletal database of knowledge of the world. It is a system that generates expectations about the world, both physical and psychological. It is available to them even before they can act on the world, reaching for objects, manipulating them, and talking about their experiences. How can I be so confident about this knowledge if I can't ask these immobile blobs? Let me tell you by way of a magic trick.

A magician's goal is to violate physical reality without revealing his trick. When he pulls a rabbit out of an apparently empty hat, this is magic, because something can't come from nothing, and because rabbits are too large to hide within a top hat. When the magician hides an audience member behind a cloak and makes him disappear, he has violated the principle of object permanence: out of sight is not necessarily out of existence. Piaget observed that for young infants under a year old, out of sight *IS* out of existence. If the magician shows the child his magic rabbit, she will reach for it. If he puts it back inside the hat, she will stop. No more reaching. No more expectations about the object. Rabbit out of sight = rabbit out of mind. A child at this age wouldn't play hide-and-go-seek, and wouldn't attempt to find a thief who just disappeared behind the bushes. The child's failure to reach tells you about her expectations: the concealed object is gone.

Soon after their first birthday, when they develop the ability to reach for a hidden object, you can play a different game with them. Show a child two opaque screens, A and B, and hide a toy behind A. Once she successfully and repeatedly retrieves the toy behind A, switch sides, hiding it behind screen B. Although the hiding game is the same, and although she knows that out of sight is still in her mind, she searches behind A, not B. This search error reoccurs over many tries. Screen A is like a magnet, pulling her back to the source of earlier success. This error arises in all infants, independent of socioeconomic background or culture. It is an error that reveals a signature of the developing mind.[7]

During his studies of the A-not-B problem, Piaget made an intriguing observation that has since been reported by other developmental psychologists. Sometimes, a child will look toward the B screen while reaching toward the A screen. It's as if the child's eyes reveal one system of knowledge while her action or reaching reveal a completely different system. These observations have led some researchers to conclude that Piaget's framework for understanding the child's cognitive development was flawed. Piaget believed that by measuring the development of the action system, he would gain an understanding of what children know. What he failed to measure was the child's knowledge more generally, with or without action.

There are many things that we know but fail to use in action; often, when we use it to act, we do so incompetently, at least when contrasted with the depth of our unconscious and often inaccessible knowledge. If this sounds familiar, it is: It is precisely the same argument I made in chapter 1 when I mentioned both Piaget and Kohlberg's research programs on moral development. Both used verbal justification and discussion to assess the child's stage of moral refinement. Both failed to consider the possibility that what children say doesn't necessarily coincide with what they know. And nor did they consider the possibility that the knowledge driving their judgments of morally appropriate actions is unconscious, and thus any moral justification is bound to be incomplete and possibly incoherent.

Developmental psychologists responded to this new theoretical position by using magic and the child's willingness to stare. The developmental psychologist Renée Baillargeon showed four-to-five-month-old infants—

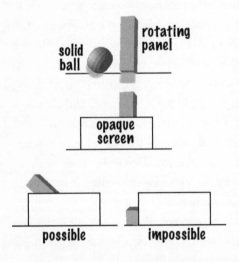

individuals who, by Piagetian standards, are months away from grasping object permanence—a solid ball sitting next to a solid panel (see figure). She then concealed part of the panel and all of the ball with a screen and showed infants two different actions: in one (left), she rotated the panel in such a way that it appeared to stop on top of the ball, and in the other (right), it appeared to rotate through the ball. Infants looked longer in the second condition. Although the ball was out of sight, the infants must have continued to think about the ball's spatial location. To detect the violation—the magic—children must remember that the ball lies in the path of the rotating panel and therefore halts its path upon rotation. When the panel appears to rotate through the ball, this represents a violation of solidity, one of the core principles of objecthood. The fact that infants have such knowledge, early in life, suggests that the lesson isn't learned but handed down through evolution as part of our standard equipment. Returning to Piaget, infants know that objects continue to exist when out of sight well before they can act upon such knowledge.[8] This conclusion leaves open the possibility that the knowledge guiding early looking is different from the knowledge guiding later reaching.

If we want to characterize what infants know and expect, we must not rely on their patterns of action or explicit behavior as the sole metric. Rather, we should look to their patterns of looking as a different or possibly complimentary source of information concerning what they know and expect. This point maps onto the discussion of our moral faculty in the previous two chapters: Even with adults, we want to distinguish between what an individual does and how he or she judges the same situation. What adults say is the morally right or wrong thing to do may be different from what they would actually do in the same situation. And for both their judgment and their actions, they may have little understanding of the underlying principles. In the same way that we attempt to distinguish between an adult's competence and performance in the moral sphere, as well as the operative versus expressed principles, a similar logic applies to our exploration of infants and children.

THE ABCs OF ACTION

Biology has given us knowledge of objects, which divides the world into things that can move on their own and things that can't. This knowledge fuels our moral faculty as it sets up the distinction between things that can hurt and help us, and things that can't.

Using looking as a measure of knowing, the developmental psychologist Alan Leslie presented infants with several scenes involving animated interactions between one red and one green block. In each scene, infants watched as the red block moved into view and then approached a stationary green block. Infants showed little interest when the red block contacted the green block and then the green block moved forward. In contrast, when the red block stopped short of the green block, and then the green block moved forward, infants looked for a long time—their eyes signaling an impossible event. In this second scene, the red block appeared to have the power of a remote control, capable of moving objects like the green block without contact. The infants' looking also suggests that they treated the green block as a nonliving, inanimate object, incapable of the kind of self-propelled motion that is characteristic of living creatures.[9]

In another series of experiments, nine-month-old infants watched as a green square started moving as soon as a red square stopped; in a second sequence, the green square started moving while the red square continued its approach. Once the infants expressed boredom with these films, the experimenter presented the same animation played backwards. Human adults see the second sequence as social, interpreting the red square as the aggressor or chaser and the green square as the reactive wimp, fleeing from its opponent. The first sequence is either not social at all or ambiguous with respect to the assignment of social roles. If infants attribute similar roles to these objects in sequence two, but not one, then during the reversal of the second sequence, the green square is chasing the wimpy red square. Infants looked longer at the reversal of the second sequence, as if social roles mattered. Significantly, this sensitivity to social roles coincides with what developmental psychologists call the nine-month revolution, or miracle in social sophistication. At this age, children also understand how triadic interactions work and that people are intentional agents. Experiments like these and others set up what I call the first principle of action:

PRINCIPLE 1: *If an object moves on its own, it is an animal or part of one.*

Before technology created robots and other similarly self-sufficient artifacts, the earth was populated by two kinds of objects: animals capable of self-propelled motion and everything else—the animates versus inanimates. Inanimates move but only when something or someone is responsible—action without contact is impossible. Most of you have likely never thought about this principle, and yet you unconsciously apply it all the time. When you walk into a supermarket, you don't worry about the possibility of a watermelon jumping off the stand and hitting you in the face. But you do worry about a person rounding the neighboring aisle and accidentally bumping into you.

Sometimes an object is already in motion by the time we spot it, and sometimes the object is stationary. Even without cues to self-propelled motion, human adults are readily able to pick out which objects are animals and which are not. But how? What makes animals different from rocks, fruits, and artifacts? The second principle of action builds on the first by adding the direction of motion:

PRINCIPLE 2: *If an object moves in a particular direction toward another object or location in space, the target represents the object's goal.*

In the movie *Microcosmos*—a documentary that takes a bug's-eye view of the world—there is a wonderful scene in which a dung beetle repeatedly attempts to move a piece of dung up a slope. Like Sisyphus, it never gets there. Nonetheless, we immediately infer its goal: get the dung up the hill. Now consider the illustration below. An animation of the circle's path on the left—a random walk—reveals no goal. On the right, we perceive the circle's animated path as goal-directed. The strength of our belief about what we see is enhanced if the object is self-propelled—an agent with goals. We also perceive an object's goal even if it never achieves the goal.[10]

The developmental psychologist Amanda Woodward ran an experiment with infants to see whether goal-directed movement is sufficient to trigger expectations about the object's goals. Infants watched an experimenter's arm emerge from behind a curtain and then reach for Toy A on a stage, ignoring Toy B. After the infant watched this event over and over again, the experimenter switched the position of the two toys. If the experimenter reached

to the same position but for Toy B, infants looked longer than if the experimenter reached for Toy A on the opposite side. This looking pattern suggested that while the infants watched the reruns of the first reaching
pattern, they generated an expectation: The arm's goal is to reach for Toy A.
Reaching for Toy B violated their expectation. When Woodward reran this
experiment using a rod or a hand that merely flopped on top of the object
as opposed to grasping it, the differences in looking time disappeared. Apparently, grasping hands have goals, but rods and floppy hands don't. Only
reaching hands are connected to intentional agents with goals.

The third principle of action, which builds on the first and second,
states:[11]

> PRINCIPLE 3: *If an object moves flexibly, changing directions*
> *in response to environmentally relevant objects or events,*
> *then it is rational.*

Only self-propelled objects can reposition themselves with respect to
ongoing or anticipated events. The object's flexibility is an indication of
rational behavior. It is rational in that it takes into account environmentally significant constraints. If a lion chases a gazelle, it is both rational
and reasonable for the gazelle to evade the lion. A gazelle standing still to
feed as a lion marches forward shows irrational inflexibility—stupidity that
natural selection should extinguish. If a gazelle sees an approaching boulder and leaps over it to evade the lion's attack, that, too, is rational. If the
gazelle leaps in the absence of a boulder, that is irrational. Just before reaching their first birthdays, and possibly well before, infants show that the
third principle guides their expectations.

The developmental psychologists Gergely and Csibra provided a key
test to show that infants are born with the third principle in place, ready to
guide their expectations. As illustrated below, infants watched a computer
animation involving two balls and a physical barrier between them. The
small ball rolled up and over the barrier (flexible change of directions),
joined the large ball, and stopped. Infants watched several reruns of this
animation. An experimenter then presented each infant with two new animations, involving the same two balls but no barrier. In one animation,
the small ball followed the same route forward, rolling a bit and then up

in an arc, landing and joining the large ball. In the second animation, the small ball rolled straight to the larger ball.

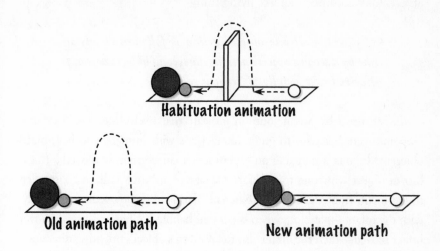

Habituation animation

Old animation path

New animation path

If the novelty of seeing the small ball take a new route is most interesting, then infants should look longer at this animation. In contrast, if infants perceive the oddity of the small ball rolling up and over *nothing*— an energetically costly and inflexible move—then even though the route is familiar, it is a bizarre choice. If infants' expectations are guided by the third principle, then the small ball takes a rational path in the first animation and an irrational path in the second. Irrationality should trigger greater gawking than rationality.

Infants look longer when the small ball takes the old, familiar route. Based on the first animation, they extract the first, second, and third principles: The small ball moves on its own, flexibly, and toward a goal. In the absence of the barrier, it makes no sense to take an arcing path. A rational object would roll straight ahead. This is expected. An arcing path is unexpected, a violation of the third principle of action. Infants gawk at irrational action, even when it comes from a faceless disk on a computer screen.

The fourth principle adds the dimension of contingency, a volleying back and forth of action-reaction. In one version of the Gergely and Csibra experiments, the animation starts out with the small and large balls pulsing, back and forth, as if they were having a digital chat across the screen. Since the two objects are separated in space, we—human adults, at least—don't see the large ball's pulsing as causing the small ball to pulse. We perceive something more interactive or social, even though the objects look nothing at all like living things.[12]

> PRINCIPLE 4: *If one object's action is followed closely in time by a second object's action, the second object's action is perceived as a socially contingent response.*

Experiments by Susan Johnson indicate that twelve-month-old infants use contingent behavior to guide interactions with humans and inanimate objects.[13] Infants watched as an experimenter surreptitiously moved a fuzzy brown object with one round part the size of a beach ball and a smaller round bit in front the size of a baseball. If this object moved contingently with the infant—babbling when the infant babbled or lighting up when the infant moved—then the infant also followed this object's orienting direction; the infant followed what appeared to be the object's gaze direction. If, however, the object moved randomly—uncoordinated with the infant's movements—then the infant showed no evidence of gaze following. For young infants, contingent behavior is sufficient to trigger a socially important behavior—joint attention. By looking where something else looks, infants share knowledge. Contingent behavior triggers a sense of the social, a feeling that there is a mind behind the object, which intends to communicate.

The fifth principle of action moves us much closer to the moral domain, linking action with emotion:

> PRINCIPLE 5: *If an object is self-propelled, goal-directed, and flexibly responsive to environmental constraints, then the object has the potential to cause harm or comfort to other like-minded objects.*

This principle emerges from an experiment by David Premack and Ann Premack,[14] aimed at understanding whether infants read emotion into or off of action, even with such unemotional objects as geometric shapes on a monitor. The Premacks showed infants different movie animations, which adults perceive as positive (*caressing, helping*) or negative (*hitting, hindering*) interactions. The *caressing* movie showed a black circle

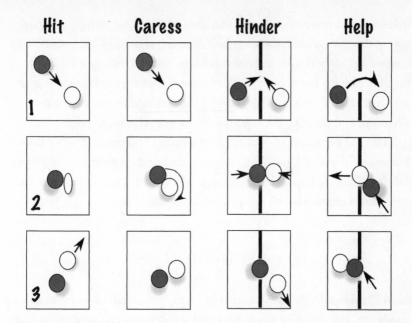

move toward a white circle, make contact, and rotate around it. The *hit-ting* movie showed a black circle move toward, make contact, and deform the white circle, transforming it into an ellipse. The *helping* movie showed a black circle move through a gap between two lines, approach, move be-low, and then push a white circle up through the gap. The *hindering* movie showed the black circle move up to a gap between two black lines and block the advance of the white circle moving toward the gap.

Infants first watched reruns of one of these movies until they were bored. Then they watched one of the other movies. If the second movie was emotionally similar to the first (e.g., caressing, then helping), infants lost in-terest; if the second movie was emotionally dissimilar (e.g., caressing, then hitting), they regained interest and watched. Like adults, infants perceive these shapes as agents, and attribute emotional coloring to their interactions. Since emotional coloring matters in terms of our preferences, these primi-tive principles of action put the child on a proper path toward developing normal social relationships. And for both the Humean and Rawlsian crea-tures, they provide an early emerging ability to link emotions and actions.

The fact that infants look longer when the emotional category of the movie changes doesn't necessarily mean that they perceive, classify, or experience these actions in terms of their emotional attributes. Perhaps infants judge these displays on the basis of purely structural principles, seeing the movies we call *helping* and *caressing* as permissible because they contain the same causes and consequences, such as the two circles joining up and staying together; forbidden actions, in contrast, involve agents that meet and then part. This is how a Rawlsian infant would perceive the situation, and more recent experiments support this idea.[15]

The five principles of action I have discussed guide children's comprehension of the world, providing some of the building blocks for our moral faculty. They are in play before the end of the first year of life. Their early appearance represents one signature of an innate system.

EVENT FULL

John Lennon's death was an event that many mourned. As an event, it was similar to, but different from, the death of Princess Diana. Both are different from events such as the annual Christmas party, the Superbowl, and the Last Supper. Actions and events are entities—things that we recognize and distinguish from other things. As entities, they have an identity.[16] Unlike that of objects, their identity is associated with a particular location in space and a change over time. Like objects, actions and events have boundaries. Objects have physical boundaries. Actions and events have spatial and temporal boundaries; what we usually think of as beginnings and endings. Sometimes these boundaries are fuzzy.

Think for a moment about the event of killing someone and the action that led to the person's death. Think, in particular, about the trolley examples from chapter 3. When Denise the bystander flips the switch, her action causes the trolley to turn onto the side track and run over one person hiking on the track. Let's say that this hiker is rushed to the hospital where, after twenty-four hours of surgery, he dies. When did the event of killing start and stop? Can we say that flipping the switch killed the hiker? Not really. We can't say that flipping a switch is equal to killing, because, in fact, the hiker didn't die on the spot. There is a delay. As witnesses, we might say that the act of flipping a switch starts when Denise's hand contacts the switch, and ends when the switch is moved into a different position; for some, the start may be earlier, as early as when Denise has the intention to move her hand toward the switch. Killing is more ambiguous. We perceive Denise's act as causing the hiker's death, even though we can acknowledge that this was not her goal; her goal was to save the five other people. In between the act of flipping the switch and the event of dying, nothing relevant happens. The doctor's surgery consists of actions that constitute an event, but neither the actions nor the event play any role in the final outcome. The doctor did not kill the hiker. The doctor's actions have a different goal: to save the hiker. If the doctor saved the hiker's life, we would still apply the same meaning to Denise's action: Flipping the switch caused the trolley to run over the hiker. We don't place much weight on the time elapsed between flipping the switch and the hiker's death. We perceive a connection between the act and the consequence. Most of us judge the act as permissible even though the consequence of the act is death.

Like sentences, events consist of multiple components and their relationships: actions, causes, consequences, and their arrangement over a period of time. When we hear a sentence, we extract meaning from the whole. We don't consciously decompose the sentence into individual words, and the words into syllables, and the syllables into phonemes. If we wanted to, we could segment the sentence into these parts. The same is true of events. Making these distinctions is important, because judgments of permissible, obligatory, and forbidden actions depend, in part, on the pieces and, especially, on understanding which actions cause which consequences in which order.

The following example from the philosopher Alvin Goldman[17] illustrates one way to think about the segmentation of events: "John squeezes his finger, thereby pulling the trigger, thereby firing the gun, thereby killing Pierre, thereby preventing Pierre from divulging the party's secrets, thereby saving the party from disaster. By killing Pierre, he also drives Pierre's lover to suicide." We can capture this complicated story, with multiple subplots and a larger number of actions, as a diagram that reveals both the pieces and the connections between them.

Squeezing a finger is like a phoneme. It has no real meaning. Squeezing a finger to pull a trigger on a gun has meaning, in that it is an intentional action with a goal. Pulling a trigger, as an action, has one meaning if the gun is loaded and another if it is empty. It has yet another meaning

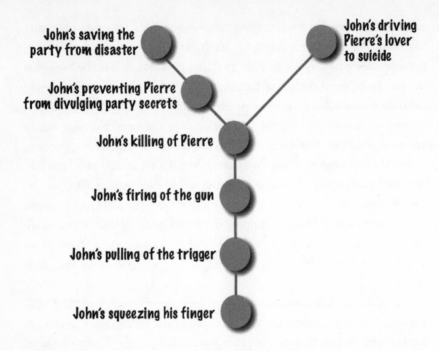

John's saving the party from disaster

John's driving Pierre's lover to suicide

John's preventing Pierre from divulging party secrets

John's killing of Pierre

John's firing of the gun

John's pulling of the trigger

John's squeezing his finger

if the gun is loaded and pointed at something. If that something is a person, the action's goal is different than if that something is a set of concentric circles pasted onto a wall. Not only are the goals different, but so are the consequences. The fork in the diagram is critical. It shows that John's killing of Pierre has two consequences. It blocks Pierre from divulging secrets and it leads to the death of Pierre's lover.

The parts of an action sequence or event are different. They might have a specific location in space, but the parts emerge over time rather than all at once. American football has predefined periods that we consciously perceive as parts of the overall game. Within these defined parts, with clearly established beginnings and ends, there are other parts—combinations of meaningless actions that seamlessly merge into meaningful actions and events: A quarterback throws to a receiver (the quarterback pulls his arm back, moves his head to find a receiver, and then moves his arm forward and releases the ball from his grip), the receiver catches the ball (the receiver runs, turns his head to see the ball, open his arms and hands, and then closes them around the ball) in the end zone for a touchdown, and then spikes the ball (raises arm with ball, then rapidly brings arm down and releases ball from grip) as the crowd cheers (individuals clapping, yelling, whistling). All of these parts unfold over time. The entire event seems salient even if we never bring the individual pieces up into our awareness. And how we divide an event into pieces depends on our familiarity with the event.[18]

Are infants built with the machinery that perceives actions with respect to a hierarchy, unconsciously recognizing the infinite potential to combine and recombine meaningless actions into meaningful actions and events? As a small step in the direction of answering this question, the developmental psychologists Dare Baldwin and Jodie Baird presented infants with a video of a woman walking into a kitchen, preparing to wash dishes, and then washing and drying the dishes.[19] As part of this sequence, they saw the woman drop and then pick up a towel. After watching this video over and over again, infants looked at one of two stills extracted from the sequence: a completed action of the woman grabbing the towel on the ground or an incomplete action of the woman reaching for the towel, but stopping midway, bending at the waist. Infants looked longer when the woman appeared frozen at the hip, suggesting that they carve up a continuous stream of actions into smaller units that represent subgoals within larger goals. Like human speech comprehension, in which we glide over the smaller phonemic units to achieve a higher level of understanding with respect to words and phrases, event perception in infants is similarly processed.

Infants are equipped with the capacity to carve up continuous events into discrete action phrases, and to interpret an object's actions in terms of five core principles. Although neither of these abilities is specific to the moral faculty, they provide infants with the essential capacity to generate expectations about objects classified as agents, to attribute intentions and goals to such agents, and to predict patterns of affiliation based on actions associated with positive or negative emotions. In the absence of these capacities, infants would develop into creatures that only focus on consequences. They would fail to distinguish intended harm from foreseen or accidentally caused harm. They would judge all actions leading to negative consequences for one or more individuals as forbidden. Such creatures would never make it into the moral arena. But even with these capacities in play, others are necessary. Classifying objects as agents is one thing. Classifying themselves and others as moral agents—individuals with responsibilities, an understanding of ownership, a sense of impartiality, and the capacity to empathize—is another.

REFLECTIONS ON SELF

We live in an obsessive culture of beauty. We are bombarded with images of handsome men, women, and children. This beauty culture tempts us to look like something we are not: with a tuck here and a rearrangement there, you, too, can fit the Hollywood image—or think you do. In the United States alone, there were 7.3 million cases of cosmetic surgery in 2002. In a particularly telling documentary on MTV, a camera crew followed the lives of four individuals, each scheduled for plastic surgery. Two women in their early thirties already had tummies tucked, thighs sculpted, and breasts enlarged; one was now going in for a nose job, the other for a hip reduction. One man, rippling with muscles, was dissatisfied with his calves, and decided to go for a calf implant. The fourth person, an obese woman who was unhappy with her weight and her inability to shed it, was going in for gastropexy, a radical intervention that involves cutting out a large part of the stomach. Each of these people would soon look different, some with a bit less of what they started with, others with a bit more. Their bodies would be transformed. Each of these people decided on plastic surgery because he or she felt that a change would make them feel better by feeding their desire to look different.

The plastic surgery not only transformed their outward appearance but their sense of self as well. The plastic surgery gave them each a renewed sense of self-control, self-confidence, and self-esteem. Self-control and a more general sense of self are essential systems linked to the moral faculty, though not exclusively. Many moral judgments involve issues of responsibility. Agents with minds, goals, and the capacity to act voluntarily are the kinds of objects that have responsibility—for themselves and others. Individuals with a sense of self can feel guilty when they have done something wrong, awed by the accomplishments of others, proud when they have helped someone in need or achieved some great goal, and envious when they see that others have something they want. What is the anatomy of individual identity—of self?

Two humans can never have the same identity. Cloning wouldn't create the perfect clones. Identical twins, even when reared apart, prove this point.[20] On some measures, the effects of the environment on identical twins is almost nil. For example, IQ scores of identical twins reared apart and together are virtually the same. Although it is possible that some of the twins reared apart were actually placed in similar environments following the separation, this is unlikely to explain the extent of overlap— there are hundreds of twins reared apart, many of which have grown up in wildly different environments. But these twins also exhibit different preferences for certain foods, activities, friends, and lovers. Bottom line: Genes constrain the range of variation. Caveat: With respect to identity, genes represent only one of several ingredients that go into the recipe for creating each individual's unique signature.

What are the other necessary ingredients? What makes me different from you is something about the continuity of my own personal experiences. I can share experiences with others in terms of seeing, tasting, and smelling the same thing, but they are my experiences. What makes me the same person over time is that although my views may change, it is my experience of the world that is changing them. I own my experiences, even if they are foisted on me by someone else, and even if I am unaware of what I own. This characterization of the self is universally preserved, even though other aspects vary across cultures.[21]

The identity problem is important to our understanding of self-control, as well as to the problem of responsibility and its link to our moral faculty. Self-knowledge is a prophylactic,[22] a protective skin that can empower us to avoid temptations or, more mundanely, avoid saying or doing the wrong thing at the wrong time. A sense of self enables us to step away from our own self-interest, recognizing that an altruistic action benefits others and that we should take responsibility for the well-being of some segment of the community of others. A sense of self and other allows us to imagine being in someone else's shoes, feeling their sorrow or joy. A sense of self allows us to build an autobiographical sketch, storing and recollecting memories in the service of guiding future behavior.

Some of the earliest evidence for a sense of self appears at about two months of age.[23] This is by no means a self-reflective sense, but one rooted in an understanding of self-control over one's own actions. Although these barely conscious creatures can't move their own bodies from one place to another, they can learn to suck and kick in order to turn on lights and make stuffed animals move, thanks to the magic of an experimenter. They recognize that their own actions can influence the *behavior* of other things in their world.

Further evidence for a sense of self in humans comes from studies modeled on Narcissus's dilemma: When an infant faces a reflective surface, what does she see? Self or Other? The mirror test was designed to assess when individuals distinguish between "my" reflection and "someone else's." This ability emerges at around eighteen to twenty-four months. For many parents, this comes as a surprise. Some close friends of mine told me that their ten-month-old daughter recognized her image in the mirror. Now, I do not typically peddle my science with my nonscience friends, but in this case, peddling was in order: "Your daughter can't recognize herself in the mirror, at least not yet." They were diffident and somewhat defensive. I said, "Look, let's do a test. Deborah can go put on some red lipstick, come back, and give Iona a kiss on her cheek. We can then put her in front of the mirror and watch." As we brought Iona in front of the mirror, she looked at us, smiling, watching us laugh. But there was no evidence at all that she recognized herself. Had we waited eight to fourteen months, she would have wiped off the lipstick from her cheek and, most likely, begun using the mirror to play games and to see parts of her body that are normally hidden from view. What is important to note here, and in all of the actual studies of infants that reveal the same finding, is that children less than eighteen to twenty-four months recognize other people and things in the mirror. Iona recognized me, my wife, her dad, and her mom in the mirror. But from her perspective, that baby in the mirror was a foreigner.

The mirror test is silent with respect to what the child of two years believes or feels about herself. When a child recognizes the lipstick on her cheek, we don't know what she thinks about these red marks, whether she wonders how they got there, who is responsible, or whether they are permanent. Some have argued, however, that because mirror recognition emerges in development at about the same time as other aspects of the self—the expression of embarrassment, early forms of empathy, pretend play—that the mirror test does reveal more than a visual-body sense or the capacity to recognize contingent responses (when "I" move, so does that similar-looking image in the mirror).[24] These pieces are essential to the Humean creature's design: A child who feels embarrassed has recognized her own inappropriate action in the midst of an audience; a child who empathizes can not only recognize another's pain or joy but have some sense of what others feel, and then act accordingly. A strong proponent of the connection between self-recognition and self-awareness is the comparative psychologist Gordon Gallup, who thinks that the ability to identify the image in the mirror requires an awareness of self. Moreover, mirror recognition not only correlates with the child's capacity to attribute beliefs and desires to others—to have the beginnings of what psychologists call a "theory of mind" (see pp. 244–53)—but also correlates with the breakdown of this ability in children with mental retardation, some autistics, and patients with frontal-lobe damage. This is an interesting idea, and, if correct, would elevate the mirror procedure to a litmus test for assessing the capacity for self-awareness.

To put the litmus test to the test we need to see how tightly related self-recognition and self-awareness are in the human brain. One test comes from individuals who can't recognize their own image in the mirror but have no difficulty recognizing that others have beliefs and desires that may or may not differ from their own. They fail the mirror test but pass theory-of-mind tests.[25] These individuals have a disorder known as prosopagnosia. Prosopagnosics not only fail to recognize their own face in a mirror but fail to recognize the faces of familiar people, while showing completely normal abilities to recognize the same people by their bodies or voices. Confronting their own deficit is horrific, precisely because of their sense of self, their awareness of what this means, what they are missing, and how absolutely different the world has become. Prosopagnosics show that self-recognition and self-awareness are not inextricably linked. They are separable. An individual who fails to recognize her mirror image may well have a richer sense of self. The mirror test is not a litmus test for self-awareness.

Whatever knowledge prosopagnosics had about familiar faces is either gone or inaccessible. Either the information has been zapped, as if someone entered a library and removed all the biology books, or the injury severed the connection between the knowledge and the system that accesses it, bringing it forward to conscious awareness—the biology books are still on the shelf but the card-access privileges have been canceled for this section of the library. Distinguishing between these two possibilities is essential for characterizing the actual status of self-knowledge, and the extent to which it guides both perception and action.

Results of several studies now show that prosopagnosics have a card-access problem.[26] When prosopagnosic patients see a familiar face, they show a much stronger arousal response, as measured by the sweatiness of their palms, than they do to an unfamiliar face. When a prosopagnosic patient meets someone who, on the previous day, used a joy-buzzer to send a mild shock during a handshake, the patient won't recognize this person but will refuse to shake hands. Like babies, who seem to know that a solid object continues to exist behind a screen, but fail to reach for this same object once it disappears behind the screen, prosopagnosics unconsciously know that certain faces are familiar and others are not, but they can't report this information—they don't know that they know. This is a different sense of unconscious than that in our discussions thus far, of the principles underlying the moral faculty. In normal humans, knowledge of our own and others' faces is available to conscious reflection. Brain damage doesn't destroy the knowledge, only the access code.

Capgras delusion highlights the distinction between recognition and emotion, two key features of the Rawlsian and Humean creatures.[27] These patients have no problem recognizing familiar faces, including their own. But once they recognize a familiar face, including their own image in the mirror, they believe they are mistaken. When a person suffering from Capgras sees a familiar face, such as the face of his mother, he recognizes the person as "Mom," but then claims that she must be an impostor. The same happens when they see their own reflection, which often leads them to remove all mirrors from their house. The recognition system seems to be intact, but the feeling of what it is like when we see someone familiar is missing.

What this disorder reveals is that knowing is often accompanied by feeling. Confidence is not only about what one knows but about what one feels. When we are confident, we are in control. When we are in control, we often feel good about our actions. For Capgras victims, confidence has been eroded by a breakdown in the connection with the circuitry of feeling.

Without appropriate feelings, the mind creates an alternative narrative to account for what is perceived. The storytelling part of the brain is in control, trying to make sense of what has been seen without any relevant emotions.

Taken together, these patient cases show that self-recognition and self-awareness are different computations. In normal humans, these computations typically work hand-in-hand. When I see people I recognize, I am often aware of seeing them and the sensations that brings. I often reflect upon what I know of these people and how they make me feel. This knowledge, and the feelings that can emerge from it, play a central role in our actions.

It is this sense of self that is ultimately anchored within a particular political climate, favoring either individual freedom and autonomy or the squelching of them. In a democracy, individuals must have access to basic liberties. In Rawls's account of justice as fairness, the self is a political conception, endowed with two moral powers: a sense of justice and a capacity to understand the good. Individuals must be given, by the governing politics of their society, a voice and access to basic values that make life worth living. Freedom for each citizen comes from the capacity to express personal interests and, to the extent that it is possible, cash in on these interests. What political bodies must worry about is extinguishing individual expression. When slavery was banned in the United States, it was not because individual slaves were able to cash in on their claims to individual freedom. Rather, it was because society as a political entity decided that slavery was an injustice to the individual, to a sense of self.[28] A just society recognizes when it has squelched the individual, deleting the political conception of self.

HEARTACHES AND GUT REACTIONS

On September 11, 2001, United Flight 93 took off for San Francisco. Unbeknown to the passengers, this plane was actually heading to the nation's capital on a suicide mission led by terrorists. Only one thing distinguished this plane from the two that had crashed into the World Trade Center a short while before: Some of the passengers, who had been alerted

to the earlier attacks, decided that they would bypass their own selfish interests and fears and make a run at the terrorists who now commanded
their plane. Their plan worked. Although the plane crashed, killing all of
its passengers, they saved the lives of an even greater number of people.
Like the trolley problems described earlier, these passengers may have *decided* that it was morally permissible to risk killing all of the passengers on
the plane as a foreseen consequence of trying to save a potentially much
larger number of equally innocent people. In the eyes of many, these passengers were heroes, ordinary mortals acting quite extraordinarily.

What drove these passengers to act in this way? It was certainly not a
thoughtless impulse. There was a plan, as revealed by phone conversations
that some of them had with their families on the ground. But their emotions must have been abuzz, a richly textured stew of the unknown, with
hints of fear and anger. Presumably, at the time of planning, there was a
sense of rational deliberation, of cool-headed thinking. But what pushed
these passengers to implement their plan was, presumably, their emotions—
a feeling of commitment and loyalty to each other, of excitement at the
prospect of taking back control of the plane, and, most likely, a surge of
testosterone-guided rage that propelled them forward. For some passengers on the plane, fear presumably predominated, inhibiting action. But
for all of these passengers, and for all normal humans, emotions inspire
action. Emotions work like well-designed engines, propelling us in different directions depending upon the task at hand. Sometimes it is best to
move, as when anger motivates attack; at other times it is best to freeze, as
when fear places its leash. Our emotions are thus biasing agents that work
together with our perceptions of planned or perceived action. They can
also work against us.

The passengers aboard flight 93 had to trust each other, assuming that
no one would defect, that each would go along with the plan. Prior to the
actual attack on the terrorists, some if not all of these passengers must
have thought about the commitment they just made, of the possible guilt
or shame they might feel if they held back at the last moment. Some may
even have felt contempt for those passengers who were frozen in fear. In
contrast to fear and anger—basic emotions that most scholars consider
universal—there is no consensus concerning the universality of contempt,
guilt, and shame, emotions that are deeply rooted in our sense of self and

other. For example, in most if not all Western cultures, individuals experience shame when they violate a norm and someone finds out about it. In contrast, in several non-Western cultures, such as Indonesia, individuals experience shame if they are in the presence of more dominant members of their group. Shame, in this non-Western sense, is not associated with any wrongdoing, and, unlike in the Western sense, is not associated with any self-destructive feelings.[29]

Recall from chapter 1 that Humean creatures solve moral dilemmas by appealing to their emotions. Emotions cause judgments of right and wrong. Sympathy triggers a judgment that helping is permissible, perhaps obligatory. Hatred triggers a judgment that harming is permissible.

Emotions can also turn a Kantian creature into a puddle, dissolving quickly. But as David Hume intuited, and as the neuroscientist Antonio Damasio has emphasized more recently, rational thought often relies on an intimate handshake with the emotions. For some, like the utilitarian moral philosopher Jeremy Bentham, our emotions play a central role in both the descriptive and prescriptive analysis of moral behavior: "Nature has placed mankind under the governance of two sovereign masters, *pain* and *pleasure*. It is for them alone to point out what we ought to do, as well as to determine what we shall do . . . They govern us in all we do, in all we say, in all we think." For others, such as Socrates, invoking emotions as a source of insight into our moral sense is a path ill-selected: "A system of morality which is based on relative emotional values is a mere illusion, a thoroughly vulgar conception which has nothing sound in it and nothing true."[30] At stake, I believe, is not whether emotions play some role in our moral judgments, but how and when they play a role. What is an emotion such that it can influence our moral evaluations? And what insights should we derive from their action in terms of characterizing both descriptive and prescriptive principles?

It may seem odd to ask for a definition of emotion. It may seem obvious that emotions are the things we experience when we step on a nail, have an orgasm, watch a horror movie, pass gas at an elegant dinner party, and gaze into the eyes of a newborn child. But what are the *things*? What kind of experiences are they? Some of these things are happenings or events in our bodies and on our body surface. We tingle following an orgasm, retract our foot when a nail penetrates, and blush when flatulence strikes at an inappropriate moment. Some of these experiences are universal and involuntary, others absent or muted in some cultures and under voluntary, conscious control. Sometimes, something inside our body changes, causing us to act in a particular way, even though at the time we are unaware of the change: heart racing, palms sweating, eyes dilating. Suddenly, we leave the premises—afraid. Sometimes we recognize an emotion in another, which may or may not trigger a response in kind—empathy—or some other kind—sympathy, jealousy, hatred, envy, or lust. Sometimes nothing at all is happening to our bodies or in our immediate sensory world, but we can stir up a thought, recollect a moment, or anticipate an event, allowing these mental mind games to trigger a feeling of sorrow, guilt, or excitement. Everything here seems reasonably familiar.

Do we have any experiences without emotions? Do we have any experiences with only emotions and no thoughts? When we decide that an action is wrong, do our emotions come before, during, or after the judgment, at each point, or not at all? Does a fetus have emotions and, if so, which ones? If the fetus lacks emotions or some of the emotions in an adult's repertoire, what parts ultimately develop, enabling anger and angst? Are we taught to recognize emotional expressions or is this built into the system from birth? How does the child's increasing understanding of facts, beliefs, and uncertainties mix with her emotions, changing her values and choices?

Many of these questions fall under what the philosopher Jesse Prinz calls the "problem of parts."[31] Crudely, it is the split between emotion and cognition, a dichotomy that is as unfortunate as the split between nature and nurture, or culture and biology. Prinz's problem of parts is either a subtraction or an addition problem, depending on where you start. Consider the animation sequence below. Start on the far left panel. Seeing this image alone presumably leaves you cold. Geometric shapes are not the kinds of things that elicit feelings, unless they have some prior history, an association constructed during a geometry class that now causes you to hate circles and love triangles. This is pure perception and thought. Now move to the right, frame by frame. As new shapes appear, the image is transformed, our perception altered, and, for some of us at least, our emotions engaged as we see Mr. Potato Head and smile. This transformation changes what we see, think, and feel. It engages the problem of parts as an

addition operation. Assuming you have the happy-smile response when you see Mr. Potato Head, what would you have to take away from this experience to lose the emotion? This is a subtraction version of the same problem. The problem of parts identifies the fact that some things invoke emotions and some don't, and asks, what part of the experience is the emotional part?

The problem of parts also connects us back to the three main characters in our moral play—the Kantian, Humean, and Rawlsian creatures. The Kantian makes decisions without appealing to the emotions. Or, if the emotions emerge, he tries to jettison them, favoring the purity of logic over the messiness of feeling. If we asked a Kantian whether it would be morally permissible to smash Mr. Potato Head, he would decide by running through the five-point method, consciously reasoning through each step, thinking about relevant principles, and then presumably answering that it is permissible—universally so—because inanimate objects don't fall within the moral circle, to use the philosopher Peter Singer's apt phrase. The Kantian may go further, probing the conditions under which it is permissible to destroy inanimate objects, considering legal or religious issues concerning property, rare pieces of art, and sacred creations. At no point do the emotions intrude.

Humeans call on their hearts to judge the situation. Smashing Mr. Potato Head is morally permissible because it doesn't induce feelings of wrongdoing. No one should feel guilty about smashing *him*, because inanimate objects are not the kinds of things with feelings. This evaluation may arise without any explicit awareness of what is happening. Mr. Potato Head never invokes any emotions, or, if he does, they are not the kinds of emotions that couple with the moral sphere. We may feel bad about smashing him, but are unlikely to feel that our act was morally bad or forbidden.

Note, however, that to introspect and call upon our emotions to adjudicate on an action, a Humean creature must perceive or imagine it, and either consciously or unconsciously recognize the kind of situation it is and the objects and events involved. There must be some kind of appraisal, even if it is as simple as recognizing that the target of our actions is inanimate. This recognition can happen rapidly and unconsciously, and so can its connection to a particular emotion.

Rawlsian creatures are appraisers, but their appraisals are unconscious and emotionally cool. The appraisal can either trigger a feeling or not. The appraisal triggers an emotion before a judgment is delivered. Alternatively, the appraisal triggers the judgment, which then triggers an emotion. When smashing Mr. Potato Head, we recognize this as an intentional act, designed to cause a physical change in the target object. The target object falls outside the sphere of morally relevant objects, recognized by appealing to the five principles of action. Although the action is intentional, it does not cause harm or negative consequences, because the target is not an agent with goals, beliefs, and desires. To see the force of this point, consider what happens when we ask whether it is morally permissible to smash Mr. Potato Head, a present just received by a little boy on his birthday. We now immediately code the act as morally wrong, a gratuitous act of harm to the child's well-being, with no positive consequences. And, of course, all of our moral creatures would judge this act as wrong. The change is in the target. Smashing Mr. Potato Head comes closer to smashing part of the little boy. A different moral calculus is needed. For the Rawlsian, the sequence of actions, causes, and consequences add up to generate a moral judgment. At stake is whether anything we would want to call emotional intervenes on the way to generating the judgment as opposed to emerging out of the judgment. Do we unconsciously judge that it is morally permissible to smash Mr. Potato Head and then have a positive feeling that pushes us to smash him? Or do we evaluate the actions and have the emotions before deciding that it is permissible to smash him? Answers to this question come from studies of child development, brain imaging of normal individuals, and patient populations with damage to areas of the brain involved in the processing of action and emotion.[32]

During the third trimester, when the fetus can hear, a speaker playing back its mother's voice has a different effect on heart rate than it does if it plays the voice of another woman. Hearing the mother's voice has a calming effect, causing a deceleration in heart rate. For some, this change constitutes an emotion, even if we can't be sure what kind of experience it is. Birth breaks this scientific impasse, at least a bit.

Newborns, within the first few hours of life, cry in response to hunger and pain. Soon thereafter, spanning a period of a few weeks, infants make noises and contort their faces in ways that we interpret as happiness, anger, fear, sadness, and even disgust. For many, these are the core emotions, present in all cultures, and evident soon after delivery from maternal Eden. Their early emergence provides an additional signature of an innate system in play.

Of perhaps greater social relevance is how infants, even within the first few weeks of life, experience emotions that appear to provide the building blocks for forging social relationships. After only a few days from delivery, newborns will cry in response to teasing, as when an experimenter repeatedly gives and takes back their pacifier. The newborns can register their distress by crying, record this information into memory, and potentially fend off future distress by turning the other cheek when the Teaser tries again.

Within the first few hours of life, newborns cry in response to hearing others cry. The developmental psychologist Nancy Eisenberg interprets the newborn's response as evidence that we are born ready to experience a "rudimentary form of empathy."[33] What is the nature of this experience? Empathy starts off when an individual recognizes particular states in another. Which states? As defined, they are emotional states, often sadness or pain, but also anger, as when we see another individual unjustly victimized and then feel anger toward whomever caused the victim's suffering. Both empathic sadness and anger cause an increase in the odds of helping, with the goal being to reduce unhappiness or relieve the victim's suffering by attacking the transgressor. In each of these cases, the first step in the appraisal is at the level of actions, their causes and consequences. When others feel a certain way, and express their feelings, they do so by means of actions, displays including changes in the face, body,

and voice. These changes on the outside represent an approximation of
what is happening on the inside, whether consciously or unconsciously.
Early in infancy, and as implied by Eisenberg and others, seeing an action
displayed in another may trigger a replay of the same action in the observer. Empathy is thus a matching up of emotions in the displayer and
observer. It differs from sympathy, in which the observer notices an act or
emotion in someone else but does so without experiencing the same emotion or repeating the same action. It also differs from personal distress, in
which seeing someone else in some state triggers a feeling of distress rather
than a matched emotional response. What differentiates empathy, sympathy, and personal distress is, then, how the observer responds to the displayer.

For this form of empathy to work, especially early in life, the newborn
must have a replay button, involuntarily triggered by seeing another act.
By replaying the action, the newborn effectively replays the emotion. This
assumes, of course, that there is a built-in—or rapidly learned—relationship
between certain actions and certain emotions. Thus, seeing someone cry
triggers a replay of the gestures used to create this sound, which, in turn,
sets up the emotion associated with the gesture and sound. In the mind of
the observer, perception and action fuse, setting up a channel for commu-
nally experienced emotions.

If we remained in the infantile mode of empathy, we would look like
yoked puppets to the rest of the world. We would cry every time we saw
someone else crying. Back and forth like yo-yos, we would dip from one
emotion to the other, driven entirely by what others feel and how we feel
for them. This helps reveal a design flaw in the perception-action perspective. Either something blocks this form of empathy or, once triggered,
something blocks its behavioral enactment. Something must intervene to
inhibit action. Something must tip the scale, devaluing the value of another's current state and needs. Something must cut off the impulse to
help.

Part of the solution to this problem comes from thinking about empathy in light of our discussion of moral dilemmas and conflict. Dilemmas, as I have discussed them, always represent a battle between different
duties or obligations, and, classically, between an obligation to self versus

other. What regulates, in part, the reflexive triggering of empathy is our sense of self and self-interest. To help, feeling empathy toward those in need, we must delay gratification. We must put off our own immediate needs for the needs of another.

As the infant develops, the earliest form of empathy mutates into a richer experience. Correlated with her developing sense of self and other, the child of about two years can begin to model the world, imagining what others experience. Some describe this as a shift from pure perception-based empathy to cognitive empathy. Empathy is no longer automatically triggered by seeing someone who is sad or in pain because the child has gained control over her own actions and thoughts. Further, the older child can imagine situations that are sad, imagine being the person experiencing the sadness, and then reason through the arguments for and against helping them. As Hoffman and Kohlberg said early on, this form of emotional musical chairs provides the basis for perspective-taking, for thinking about what it is like to feel that someone else is an important mediator of altruistic behavior. Those who score high on personality tests of empathy are more likely to show contagious yawning and more likely to help those in need.[34] Doesn't all of this show that Humean creatures are alive and well and running the moral show? Not yet.

Recall that a Rawlsian creature doesn't deny the significance of emotions in moral action. Rather, Rawlsian creatures challenge both when emotions arise in the course of making a moral decision, and whether they follow on the heels of an unconscious judgment or are triggered by them. All of the work on empathy, including studies of its neural foundation discussed up ahead, only show that empathy influences altruism. For the argument I have been developing, however, there is a difference between emotions playing a role in our moral judgments and emotions playing a role in our moral behavior, or what we actually do. Studies of empathy unambiguously show that our capacity to take another's perspective influences our behavior. It is nonetheless possible that prior to the altruistic act, our moral faculty has *decided* that the dilemma under consideration warrants, perhaps obligatorily, an altruistic act. Our emotions then kick in, either elevating the probability that we will actually help or diminishing it. It is not yet time to resolve this debate. I raise it here to keep the argument in the forefront, and the battle between Humean and Rawlsian creatures alive.

YUCK!

If empathy is the emotion most likely to cause us to approach others, disgust is the emotion most likely to cause us to flee. Unlike all other emotions, disgust is associated with exquisitely vivid triggers, perceptual devices for detection, and facial contortions. It is also the most powerful emotion against sin, especially in the domains of food and sex.

To capture the core intuition, imagine the following scenarios:

[i]   Eating your dead pet
[ii] Having intercourse with a younger sibling
[iii] Consuming someone else's vomit
[iv] Picking up a dog's feces in your bare hands

I assume these four images are disgusting for most if not all readers. When we judge something as disgusting, a negative emotion reigns us in, leashing any tendency we may have experienced to act otherwise. This constraint is highly adaptive. In the absence of a disgust response, we might well convince ourselves that it is okay to have sex with a younger sibling or eat vomit, act with deleterious consequences for our reproductive success and survival respectively. Humans with no pathology experience disgust in response to food, sexual behaviors, body deformities, contact with death and disease, and body products such as feces, vomit, and urine; humans with some pathologies, such as Huntington's disease, lack a typical disgust response, neither recognizing the emotion in others nor classifying typically disgusting items as such.[35] Although there are cross-cultural and age differences in the conditions eliciting disgust, the facial expression—typically a wrinkling of the nose, gaping of the mouth, and retraction of the upper lip—is highly recognizable and unique to our species. Together, these observations indicate that disgust emerges from a biological substrate that may be both unique to our species and unique among the emotions we experience.

Darwin defined disgust as "something revolting, primarily in relation to the sense of taste, as actually perceived or vividly imagined; and secondarily to anything which causes a similar feeling, through the sense of smell, touch and even eyesight."[36] Over a hundred years later, the psychologist Paul Rozin[37] refined Darwin's intuition, suggesting that there are different kinds of disgust, with *core* disgust focused on oral ingestion and contamination: "Revulsion at the prospect of [oral] incorporation of an offensive object. The offensive objects are contaminants; that is, if they even briefly contact an acceptable food, they tend to render that food unacceptable." What makes Rozin's view especially interesting is that many of the things that elicit disgust are not only stomach-churning but morally repugnant. Thus, once we leave core disgust, we enter into a conception of the emotion that is symbolic, attaching itself to objects, people, or behaviors that are immoral. People who consume certain things or violate particular social norms are, in some sense, disgusting.

Vegetarians who don't eat meat on moral grounds, including the inhumane conditions for farm animals, often find meat disgusting, looking down on those who eat it as immoral carnivorous barbarians. Vegetarians who don't eat meat for health reasons don't have these emotional responses and don't generate moral labels for those who enjoy their T-bone steaks or chicken breasts. Though disgust and morality are clearly intertwined, what comes first in this egg-and-chicken problem? Do moral vegetarians first experience disgust when they see meat and then develop a moral stance toward the piece of dead red flesh sitting on their plate? Or do they first work through the moral rationale against eating meat and then develop a feeling of disgust as they imagine how much suffering goes on in a slaughterhouse? Is disgust cause or consequence? Is disgust first or second?

The anthropologist Daniel Fessler provides a simple way to begin answering this question. Consider the observation that people differ in how easily and intensely they are disgusted by different objects and behaviors, but that if you are highly reactive to one kind of object, you will also be highly reactive to others. If you think that vomit is extremely disgusting, chances are you will also think that feces, the sight of a person cutting a finger, and an open facial wound are equally disgusting. If one thing readily elicits an intense feeling of disgust, other things will as well. If disgust is a cause, then moral vegetarians should be more reactive to other disgusting things than health vegetarians or nonvegetarians. In contrast, if disgust is a consequence, then moral vegetarians will be as reactive or unreactive as health vegetarians and nonvegetarians. Fessler found no relationship between sensitivity to disgust and the reasons for either eating or abstaining from eating meat. Moral vegetarians first take a stance on eating meat and then develop a profound feeling of disgust toward meat and meat-eaters. Disgust—in this specific case, at least—is second, a consequence of taking a moral stance.[38] The Rawlsian creature is driving the Humean response.

Disgust carries two other features that make it a particularly effective social emotion: It enjoys a certain level of immunity from conscious reflection, and it is contagious like yawning and laughter, infecting what others think with blinding speed. To see how this works, answer the following questions:

1. Would you drink apple juice from a brand-new hospital bedpan?
   YES ☐ NO ☐

2. Would you eat chocolate fudge that looks like dog feces?
   YES ☐ NO ☐

3. If you opened a box of chocolates and found that someone had taken a small bite out of each one, would you eat any?
   YES ☐ NO ☐

4. If your mom served you a plate of your favorite food decorated on the side with sterilized dead cockroaches, would you eat the food?
   YES ☐ NO ☐

Most people answer "no" to these questions. If they answer "yes," they do so after a noticeable pause. These answers are a bit odd, on reflection. There is nothing unsanitary about apple juice in a brand-new, sterilized bedpan. And the shape of chocolate fudge plays no role in its palatability. But our sensory systems don't know this: Bedpans are for urination, and things that look like feces typically are. Our minds have been fine-tuned to detect features in the environment that are causally and consistently connected with disease. Once detected, a signal is sent to the systems in the brain that generate disgust, and once generated, the action system is commandeered, driving an evasive response. The cascade of processes is so rapid, automatic, and powerful that our conscious, cool-headed, rational minds are incapable of overriding it. Like visual illusions, when our sensory systems detect something disgusting, we avoid it even if we consciously know that this is irrational and absurd. Disgust engages an automated sequence of actions that leads to tactical evasion.

A second component of disgust is its capacity to spread like a virus, contaminating all that comes in its path. How would you feel about wearing Hitler's sweater? Most of Rozin's undergraduate subjects at the University of Pennsylvania rated this as a highly disgusting thing to do. People who respond in this way think that Hitler was a morally repugnant character and that wearing his sweater might transmit some of his most horrific qualities.

Disgust wins the award as the single most irresponsible emotion, a feeling that has led to extreme in-group–out-group divisions followed by inhumane treatment. Disgust's trick is simple: Declare those you don't like to be vermin or parasites, and it is easy to think of them as disgusting, deserving of exclusion, dismissal, and annihilation. All horrific cases of human abuse entail this kind of transformation, from Auschwitz to Abu Ghraib.

Although core disgust has its origins in food rejection, its contextual relevance has mutated to other problems, and, in particular, sexual behavior. Up until the early 1970s, homosexuality was described as abnormal behavior in the clinician's bible, the *Diagnostic and Statistical Manual of Mental Disorders* (DSM)-III. Carried along with this classification was the belief, held by many cultures, that homosexuals were disgusting. Hearing about pedophilia and incest evoke much the same stomach-churning emotions, accompanied by moral indignation. Incest is of particular interest, given the universal taboos against it.[39]

The anthropologist Edward Westermark[40] argued that "there is an innate aversion to sexual intercourse between persons living very closely together from early youth, and that, as such persons are in most cases related, this feeling displays itself chiefly as a horror of intercourse between near kin." Westermark's intuition has been supported by numerous studies of Israeli children raised on a kibbutz, prearranged Taiwanese marriages, and even American college students.[41] Among children reared on a kibbutz, sibling marriages and sexual intercourse are extremely rare. Further, there is little sexual interest among unrelated children reared together, even though sexual relations are not explicitly prohibited. In Taiwan, when baby girls are placed in the home of their target spouse, and the children are reared together, their marriages frequently fail, often due to extramarital affairs. Among American college students, feelings of repulsion toward incestuous relationships are stronger among opposite-sex siblings that spent a large part of their childhood in the same household than siblings that spent relatively little time together. That familiarity breeds *yuck* would seem to suggest that explicit, culturally articulated taboos are unnecessary. Incest avoidance falls out of our biology, and the biology of other animals as well. It is an avoidance mechanism that was designed to prevent the deleterious consequences of mating with close relatives—inbreeding. The anthropologist James Frazer,[42] writing at about the same time as Westermark, made this point exactly: ". . . law only forbids men to do what their instincts incline them to do; what nature itself prohibits and punishes, it would be superfluous for the law to prohibit and punish." The role of the law is to target cases of cousin marriages, where the *yuck* response is relatively muted.

OUT OF MY MIND

Steven Spielberg's movie *Minority Report* is based on the premise that we could eliminate violent crime if we could infer intentions prior to action, literally seeing into the future. The futuristic mind readers in the movie are three "pre-cogs," mutants who can see violent acts unfolding before the perpetrator even has the conscious thought. When they detect such actions, their bodies physically react, images pop up on a screen, and a little red ball inscribed with the perpetrator's name alerts the precrime unit. Using a database of images from the pre-cogs, the precrime unit locates and arrests the individual for violent intent.

Like all science fiction, *Minority Report* is not entirely fictional. And like any good piece of science fiction, it raises interesting ethical questions. At one level, we are pre-cogs. Though we can't see into the future, we often infer the intentions of others before they act. Our hunches about others—what they believe and desire—are often correct. We are mind readers, forming beliefs about others' thoughts even if we have never spoken with the person or watched them interacting. We see someone with a crew cut and army fatigues, and we immediately set up certain beliefs about this person's interests, goals, and ambitions; and these beliefs are different from the ones we set up for people whom, upon seeing their dress code and hairstyle, we call hippie or yuppie. We have rich, unconscious theories of what other people think. Though these are not formal theories, inscribed in a personal diary or the periodicals of science, they are informal, often unconscious beliefs that enable us to navigate in a social world, cashing in on desires while resisting others because they may cause harm. From the privacy of our own minds, we simulate the world, breaking into the privacy of other minds.

We will always be faced with the ethical issue that the pre-cog story presents. A person who thinks about murdering someone can tell his friends that he is going to do it, plan each step, visualize the weapon, the dead body, and the arrival of the police, and then stop—or be stopped—before the final act. As the legal scholar Clarence Darrow once pointed out, "There is no such crime as a crime of thought; there are only crimes of action." We have free will, the ability to reflect upon what we and others desire, and then a choice to act on this thought.[43] At least this is the story that some of us believe. Those who don't are not stubborn, dimwitted, or uneducated. They are individuals, including young children and humans with certain forms of brain damage, who lack access to what others think, failing to realize that their hidden beliefs and desires mediate their overt actions.

To understand how our developing capacity to infer what people believe and desire interfaces with our developing capacity to deliver moral verdicts of right and wrong, let's turn to the stage, the puppet stage of Punch and Judy. In a typically playful but profound essay, the philosopher Daniel Dennett asks the reader to imagine a classic show between these two historical puppets. Having once again abused Punch, Judy walks away and then accidentally trips and falls into a box. Punch, seeing the opportunity for ending his abusive relationship, plans to push Judy off the stage, ending not only her career but her life. As he prepares his murderous act, he turns around to get some rope. While his back is turned, Judy opens the box and sneaks off. Punch returns, ties up the box, shoves it off stage, and rejoices over his perfect homicide.

Depending on your ability to read minds or simulate what is going on in Punch and Judy's heads, there are two interpretations. Punch believes that Judy is trapped inside the box. When he pushes the box off stage, he believes that he has killed her. Punch's belief is, however, false. If children only pay attention to consequences—as Piaget and Kohlberg believed was characteristic of children up to about nine years old—then Punch didn't do anything wrong. Judy is fine. If children pay attention to beliefs and intentions, then Punch did do something wrong. His plan failed, but this is not what he intended, nor is it what he believed. If we had children of different ages on a jury, how would they vote? To answer this question, let me provide you with a time-lapsed version of the child's mental development.[44]

Before their first birthdays, infants make judgments about the outcomes of action that are based on an initial distinction between living and nonliving things, and use features such as contingent and flexible behavior to make inferences about goals and agency. At around fourteen months, children begin to show a choreographed pattern of eye movements with other humans. When the child sees someone looking, he looks, too, cooperating in an act of joint attention. This is not only a shared moment of perceiving, but also of knowing and feeling—a fusion of thoughts. This ability allows children to appreciate when someone knows what they know while simultaneously providing a platform for understanding referential gestures, such as pointing and the utterance of words. For example, when children at this age are uncertain about a situation, such as a novel object or a stranger, they look back and forth between their caretaker and the object. By looking at the caretaker's eyes and expression, the child gains deep information about the caretaker's thoughts. This information provides a mental bridge between what the caretaker might do and what the toddler plans to do. Joint attention is not, however, necessary for developing an understanding of what other individuals know, nor is it necessary for understanding what words mean. Blind children acquire both of these capacities.[45] Losing sight or hearing is almost unthinkable for most of us. However, these natural experiments show why we must not assume that a particular modality is essential for understanding what others know, believe, or feel.

Several months after joint attention develops in normal children, another behavior emerges: pretend play.[46] Pretense provides a unique window into the child's ability to think about alternative realities. When a child starts to pretend, she has taken her first steps down the path of mental simulation, of imagining alternative worlds, of doing a bit of science fiction. This ability provides essential input to the moral faculty, even though it is not specific to the moral domain. To engage with any moral dilemma, it is necessary to imagine one world in which an action is taken and consequences follow, and a second world, where no action is taken and a different set of consequences follow. From a playspace that allows for a contrast, young children can draw conclusions about which option is preferable, which is permissible, obligatory, or forbidden.

When children engage in pretend play, they will readily pick up a toy phone and chat with a family member, use a spoon to feed a doll, develop imaginary friends without the delusional consequences of Jimmy Stewart and his rabbit friend Harvey, and, most critically, recognize when someone else is pretending and join in the game. The combination of joint attention and pretense are stepping stones to the next critical step: making proper inferences about what someone desires, knows, thinks, or intends. From age two to five, these inferences grow significantly, and children understand that beliefs can cause emotions (she believes spiders are scary and that is why she is afraid), that one can see things for what they *really* are or what they *appear* to be (milk is really white but appears red when viewed from behind a piece of red cellophane), that actions can be intended or accidental (intentionally throwing a pie in someone's face as opposed to accidentally hitting someone in the face with a pie after tripping),[47] and that deception works by making someone believe something that is false (there are no more cookies in the jar). Following his fourth birthday, the child's mind-reading abilities move up another notch, hailed by the capacity to understand such lines as "Gordon *knows* what Deborah *believes* about their friends Marc and Lilan" and "Nancy *believes* that John *believes* that good documentaries must be educational, even if the television networks don't *believe* this." One hallmark of this age is that children can distinguish between what they believe and know to be true of the world and a false belief of another. Show a child a bag of M&M candies and ask her what's inside. She will answer, "M&Ms." Open the contents and show her that you have replaced the M&Ms with pencils. Now ask this child what her mom, who is waiting outside, will say. Five-year-olds say "M&Ms," revealing their understanding of false beliefs. Three-year-olds, in contrast, say "Pencils," failing to grasp the distinction between their own current beliefs, their prior beliefs before the experimenter revealed the pencils, and the beliefs of someone who was not privy to the switch. Beliefs about others' beliefs develop some time after the fourth birthday.

The developmental path I just sketched is not as simple as it appears, and nor is it clearly coupled with our moral judgments. I first introduce a few complications, show why they matter, and then connect our capacity to read minds with the system that delivers verdicts of right or wrong.[48]

Along the road to acquiring the rich conceptual systems that children use to infer what others believe, desire, and intend, other capacities develop as well, some critically important to linking social concepts with moral action. Perhaps the most important, due to its significance in morally relevant behavior, is how children develop the capacity to inhibit certain actions and release others. Consider once again the child's capacity to infer false beliefs. Running alongside the M&M test as a classic probe of children's mental states, developmental psychologists have developed what is fondly and famously referred to as the "Sally-Ann" task, named after the two puppets that appeared in the original experiments.[49] Though there are numerous variants, each task is designed to pick apart why children either fail or succeed, and when they do so. The core narrative runs as follows. A child watches as Sally and Ann play with a ball. Sally then puts the ball in a basket and leaves the room. While Sally is away, Ann takes the ball out of the basket and places it in a box. Sally then returns to the room. The experimenter now asks the child, "Where will Sally look for the ball?" The classic result, mirroring the results of the M&M test, is that three-year-olds point and say the box, while four- to five-year-olds point and say the basket. The older children understand that Sally has a false belief. She must believe that her ball is in the basket, because this is where she left it and she didn't see Ann move it to the box. Since she didn't *see* Ann move the ball, she can't *know* or *believe* that it is in the box. In the absence of a critical perceptual event—seeing—Sally lacks a critical mental state: knowing or believing. From their third to their fifth birthdays, children undergo a conceptual revolution, akin to those seen in science when one theoretical perspective or paradigm replaces another.[50] Faced with the fact that Sally searches in the basket and not the box, three-year-olds are handed a piece of counterevidence. They predicted that Sally would look in the box. Their prediction was wrong. Over a period of one year, their conceptual system changes, as they grasp a critical fact about human minds: Sometimes we believe things that others don't. Sometimes, others believe things that we know to be false.

At the beginning of this chapter, I mentioned the idea that what a child appears to know as revealed by her eyes may well be different from what the child knows as revealed by her reaching. In the Sally-Ann test, the experimenter asks the child to point or say where Sally will look for the ball. These are explicit actions that require access to explicit knowledge. But before the child points or expresses her opinion in words, what are her eyes doing? To address this question, the cognitive scientists Wendy Clements and Josef Perner ran a Sally-Ann test with three- and four-year-olds. In addition to asking them to point and say where Sally would look, they also filmed their eyes. Three-year-olds looked to the basket and then pointed or said the box; four-year-olds, predictably, looked and pointed and said the basket. Three-year-olds appear to have knowledge of what others know or believe, but they can't access it consciously.[51] What blocks access to this knowledge? Understanding this question is important, as it provides a key connection to the moral domain of norms and considerations for the other. Like Ulysses and his visit to the sirens, as soon as the child understands her own weaknesses and strengths, she can use this knowledge to overcome temptation, or at least recognize when she is vulnerable to it.

When we point, we point to where something is. But to demonstrate an understanding of Sally's false belief in the puppet show, the child must point to the basket where the ball is *not*. To succeed, she must inhibit the natural, habitual tendency to point to where something is (the box), pointing instead to where something is not. If an experimenter runs three-year-olds on the classic Sally-Ann test, but instead of pointing asks them to place a sticker on the relevant container, they succeed, putting the sticker on the box. Unlike pointing, *stickering* is not a habitual response. We can place stickers wherever we like.

What these wonderfully simple and clever experiments show is that young children may understand false beliefs but this particular task prevents them from displaying their abilities. They may have a certain level of competence with respect to what they unconsciously know, even if some aspects of their performance or behavior hide this knowledge.

The fourth year of life hails a celebration, a victory in mind reading. Pieces of the inhibitory problem dissolve, and the child not only understands that beliefs can be true or false, but can act on these beliefs in a wide range of contexts.

We now have the pieces we need to hook back into morality. When we assign blame for wrongdoing and praise for helping, we are either explicitly or implicitly recognizing that the agent has acted intentionally and with the proper motivations. If my hand hits your face and you cry, your judgment about this action and its consequences depends on your reading of the cause of my action. If I intended to hit you and make you cry, then my act is morally reprehensible. If I intended to hit you in order to move you out of the way of a hurtling rock, then my act is morally praiseworthy—I saved you from greater injury. If I intended to reach for a glass but slipped and hit you in the face, I am clumsy but my action shouldn't evoke moral derision. What I did wasn't right or wrong but the result of an accident. If I intended to reach for a glass on a shelf near your face, and was also aware of the slippery floor next to you, my hitting your face was accidental but reckless or negligent, given the circumstances. I should have taken greater note of the odds of injuring you, asked you to move, and then reached for the glass. My action should evoke a moral appraisal. You should hold me accountable for my action, and conclude that what I did was wrong.

When do children have access to this rich psychological machinery of intentions? According to Piaget and Kohlberg, children don't have any of the machinery until they are about nine or ten years old. Younger children simply look at consequences and evaluate from that point on. If the consequence is bad (agent gratuitously hurts someone), the agent was bad. If the consequence is good (agent helps someone in need), the agent was good. This age-related diagnosis is wrong.

Studies starting in the 1980s showed that four- to five-year-olds generate different moral judgments when an individual carries out the same acts, with the same consequences, but with different motivations, with some evidence for a bias to consider actions worse than omissions. And they make these distinctions even when the acts themselves are completely neutral, lacking in any emotional or moral significance. For example, Bill turns on a hose to help his mother water the plants, while Bob turns on a hose to dissolve his younger brother's sand castle. Or Jane turns up the temperature on the oven to help her mother bake the cookies, while Jill turns up the temperature to cause her sister's cookies to burn. Or Joey puts his sister's only party dress out in the rain to get wet, while Jimmy leaves his sister's favorite party dress outside while it is raining. When children judge these events, they say that Bob and Jill are bad and should be punished, and deliver these concerns even though the storyteller never recounted the actual outcome of the events. They also conclude that Joey is worse than Jimmy, revealing that omissions are better than acts even when the consequences are the same. Of direct relevance to the theory of our moral instincts, they make these judgments on the basis of particular details of the target events (causes, actions, consequences) and do so consistently and independently of the specific content of such events. They achieve this level of competence even though their justifications are largely incoherent or, at best, insufficient to explain their differing opinions. The fact that they make their evaluations across a broad range of contexts and with actions that in and of themselves have no moral value suggests that they are operating on the basis of general principles, as opposed to rules for specific actions. Of further interest, the timing of this competence coincides almost perfectly with the emergence of a well-developed theory of mind. As they approach their fifth birthday, children appreciate not only that others have beliefs, desires, and intentions, but that these mental states play a role in judging whether someone is good or bad.

IT'S ALIVE!

What's the difference between:

1] An earth tremor that causes a rock to roll down a hill and kill a man
2] A dog dislodging a rock on a hill that rolls down and kills a man
3] A chimpanzee throwing a rock at a man and killing him
4] A woman throwing a rock at a man and killing him

Without thinking too hard, 1 and 2 carry no moral weight, 4 does, and 3 might. Why? Cases 1 and 2 are physical accidents with no psychological cause. The dog dislodged the rock and caused it to roll, but given the description, the dog was not aiming at the man, and was certainly not aiming to kill the man. The act of throwing is intentional, typically with a goal in mind. Though chimpanzees are quite inept at throwing, visitors to the zoo may have experienced their fecal attacks on the glass window separating ape from human. Chimpanzees and humans are the kinds of objects that can have intentions. Both the chimpanzee and woman intended to throw the rock but may not have intended to either hit or kill the man.[52] Throwing is the kind of action that is done intentionally and presumably with the goal of making contact with someone or something. But from the description, we don't know if either the chimpanzee or woman intended to make contact or kill the man. All four cases end in the same negative consequence, but what differentiates these cases on moral grounds is a distinction between intentional and accidental actions, the motives underlying the intentional actions, the relationship between foreseen and intended consequences, and the characteristics of the agent and target.

What gives case 4 relatively unambiguous moral status is that we think of women and other adults as having responsibilities toward others. It is debatable whether human infants and children should be seen as morally responsible agents or whether, as some would argue, they are moral patients—organisms that deserve our moral concern even if they have no sense of rights, wrongs, and responsibilities to others. As the philosopher Peter Singer pointed out years ago in a discussion of animal welfare, and as the cognitive scientist Paul Bloom more recently noted in a discussion of human-infant development, our species has increasingly widened the scope of its moral concerns, from only humans to a wide range of other species. Of interest here is how and when we acquire the capacity to discriminate among the multitude of objects in the world, developing biases in terms of our responsibilities to others, our prejudices to favor the in-group and exclude the out-group, and our capacity to judge some actions as morally relevant because they concern objects with rights.

Both motion and body cues are available to the child's eyes, and she uses these perceptual features to develop categories. Early work in both child development and some corners of philosophy assumed that this was the entire story about infant categorization: Find the relevant features, develop some metric of similarity, and then cluster those things that are most similar. For example, the child sees robins, cardinals, blue jays, and mockingbirds, and, based on this list, concludes that *all* birds have small beaks, relatively short tails, feathers, thin spindly legs, and a few toe-like appendages. This similarity set then sets up the notion of a prototypical or average bird. One problem with a reliance on similarity, however, is that depending on how one construes the relevant feature set, any two objects might be classified as similar: A blue jay is similar to a hammer in that both are subjected to the forces of gravity, weigh less than a ton, are incapable of doing differential calculus, and have no sense of humor. Further, and as pointed out by the developmental psychologist Frank Keil, these early views of the child's path from perceptual to conceptual categorization were guilty of "original sim"! They presumed that because infants use features to guide certain aspects of categorization, this is all they use. But there is more to categorization than the assembly of features at the surface of an object and the notion of a prototypical member.

At an early age, children appreciate that germs and cooties can be passed back and forth, even though they can't be seen. Appreciation of these unseen entities forms the basis for thinking that others are disgusting and, therefore, for using an unseen trait as a criterion for categorizing individuals into likable or unlikable, and extending to morally virtuous and abhorrent. Psychologists often refer to some of these unobservables as "essences." To appreciate their force in the child's thinking about categories, consider a wonderfully insightful experiment by Keil, designed to assess how different kinds of transformations might alter the child's taxonomy.

Keil told young children about a skunk that had been hit by a car and rushed to an emergency hospital. Upon arrival, the doctors determined that the skunk's fur had been severely damaged. Since they didn't have a stock of skunk fur, they decided to remove all the damaged fur and replace it with the fur of a raccoon. The operation worked perfectly, and the patient was released back into the forest. When Keil asked the children about the kind of animal released into the forest, most children said a "skunk." What trumps the categorization process are the insides, not the superficial outsides that resemble a different species.

What these studies suggest is that infants are equipped with unconscious knowledge of the biological world that enables them to draw inferences from unobservable properties of objects. Depending on the child's native environment, their early intuitions about biology might map onto a relatively small suite of organisms if they are born in a city, or to a complex web of flora and fauna if they are born in an Amazonian rain forest. This initial system has only a skeletal theoretical structure, with relatively impoverished notions of life and death, disease, growth, and kinship; it is a system that is based more on facts than on theories that can predict more generally which things die, get sick, and reproduce. For example, in Diane Keaton's movie *Heaven*, she asks a young child about sex in the afterlife. The child pauses, looks skyward, and then asks, "If there is sex in Heaven, do they give birth to dead babies?" Similarly, when I was in Uganda, and my Caucasian American wife was pregnant, a young Ugandan boy asked me whether, given my lengthy stay in his country, our baby would be born black; I am confident that he wasn't challenging my wife's fidelity. These questions reveal the signature of a theoretically impoverished system.

Although there is controversy concerning when, precisely, our theoretically more sophisticated system emerges, and what precisely fuels its development, normally developing children have it by the age of about ten years. Children with Williams syndrome, a developmental disorder, only acquire the early system of folk biology.[53] They have a loose assemblage of facts that allow them to classify animals and make some rudimentary predictions or generalizations about their behavior. However, there is little theory underlying their judgments. When asked about dying and death, they are most likely to say that a dead animal doesn't move, similar in kind to one that is sleeping. They do not appeal to the richer theoretical notions that normal children expose, including the cessation of breathing, the lack of future biological needs, and social relations. Developing this richer system is an essential part of the support team for our moral faculty, given that our judgments about innumerable moral dilemmas—euthanasia, abortion, infanticide, to name a few—rely on such factors as seeing the significance of death and the importance of certain survival needs.

While the child's folk biology matures, she is also acquiring a system that makes countless discriminations along more social dimensions—a crucial part of forming a sense of the in-group and out-group. Within the first few days of life, newborns distinguish the face, voice, and smell of their mother as distinct from other women. The child then meets other individuals, including close and distant kin or non-kin, male and female, friends and foe, young and old, high and low rank, same and different race. For some of these categories, psychologists and anthropologists have carried out the same transformation studies as those described for animals, arriving at similar conclusions. For example, though race is a human creation, as distinct from a system of classification that is rooted in biology, it is part of our evolved psychology, a system that picks up on cues that identify members as being inside or outside one's group. As the anthropologist Lawrence Hirschfeld[54] has demonstrated, if you show a young child a picture of two potential fathers, one black and one white, and then present a picture of a black child dressed in the same clothes as the white man, children will point to the black man as the father. Children realize that skin color is the relevant predictor. Similarly, they recognize that a white child raised by black parents will remain white. Again, these studies don't show that the child's concept of race is as rich and, in many cases, morally weighted as that of an adult. What they show is that early in life, children are sensitive to some of the factors mediating race, and that some of these unobservables cannot be transformed.

One of the downsides to essentialist thinking, and certain aspects of categorization more generally, is that we readily develop stereotypes and prejudices.[55] Stereotypes represent one form of categorization, a process that pigeonholes people into social groups, and serves as a foundation for morally abhorrent behavior. Prejudices represent the attitudes we form toward these groups, be they consciously or unconsciously active. How we form stereotypes and prejudices bears directly on how our moral faculty interfaces with other bits of the mind, shaping both our judgments and our behavior. When countries or particular social classes have attempted to dominate other groups, they have classically done so by moving the other group further and further away from what is prototypically human. For the Nazis, Jews were vermin, parasites, nonhuman others. Similar attributes have been assigned by the rich to the poor since antiquity. In *Émile*, one of Jean-Jacques Rousseau's longer essays on human nature and development, he rhetorically asks, "Why are kings without pity for their subjects? It is because they count on never being human beings." Though neither class nor race is a biological category, our mind is equipped with the hardware and software to pick out cues that identify the other, or *l'autre*, as the continental philosopher Emmanuel Levinas discussed. Although we can appreciate alternative ways of carving the world, we can't erase the constraints that our mind imposes on our perceptions, and this includes dividing up the world into dominant and subordinate, black and white.

In part I, I discussed categorization in the context of crimes of passion, and, in particular, about what the average or prototypical person would do in similar situations. The key legal issue is whether the provocation defense can be invoked when a lover is caught in flagrante delicto with another. Legally, a jury, together with a judge, must decide whether or not other people, in a similar situation, would have acted similarly. This raises some notion of a prototypical person with some suite of prototypical emotional responses. Similar issues have emerged in legal cases of obscenity, which derive from a view both of the average person and of the emotions she or he might have in response to things obscene. In the United States, this perspective derives from a 1973 ruling, in which Justice Warren Burger argued that obscenity be defined based on whether a piece of work fuels disgust in "the average person, applying contemporary community standards."[56]

The problem with notions of prototypicality is that, like any other generalization, they may blind us to the variation, causing us—unconsciously—to deliver highly biased judgments. In fact, as the social psychologist Mahzarin Banaji has pointed out in a series of studies, we may often make judgments about moral character with no awareness of the depth of our prejudices.[57] Millions of people claim that they are not racist or sexist or ageist. But when confronted with rapidly presented photos of faces and asked to judge whether the person is good or bad, people show strong biases against all of the out-groups. People hiring for a position also exhibit prejudices against these out-groups, even if they have stellar records that far outclass those of their in-group competitors. Many of these unconscious or implicit attitudes arise from prior statistical patterns of association; others arise from word of mouth, completely lacking in empirical support. Nothing could be more rudimentary: latch on to frequent co-occurrences of traits with people and build a caricature or stereotype of the class. Our minds are seduced by the opportunity to form categories, turning the messiness of the world into an orderly array of classes, setting impartiality to the wind. Once formed, these categories are then bound to our emotions, with negative feelings attached to members of the out-group and positive feelings toward those in the in-group. And because these prejudices lie tucked away in the unconscious recesses of the mind, they are difficult to break apart. How can you tamper with biases that you are generally unaware of?

Although we form unconscious attitudes toward those in the out-group, and such prejudices lead to caricatures, this process is neither specific to the moral faculty nor operative in the same way. More debatable is whether these biases have consequences for our moral judgments or competence. With respect to the underlying process, although our implicit attitudes operate outside the grasp of awareness, the principles are not at all inaccessible. They can be grasped by anyone looking carefully at their own patterns of judgments. And once we are aware of such biases, it is even possible to override them. In contrast, if my characterization of the Rawlsian creature is right, then we don't have access to the principles that underlie our moral judgments.

GRATIFYING PATIENCE

In Ambrose Bierce's *Devil's Dictionary*, patience is defined as "a minor form of despair disguised as virtue." Advertisers prey on this weakness, filling our heads with a just-do-it attitude. Our competitive, workaholic mentality often pushes us to cut corners, and cheat if need be. We enroll in multiple credit-card programs in order to purchase what we want, when we want. Realizing the seductive power of a credit card, some of us reject the plastic and use cash. We score fast-food hits whenever we can, craving immediate fulfillment. Some of us, on principle, refuse the large fries and double cheeseburger in a box in favor of slowly cooked curry, at home, with candles and a glass of wine. Moral systems ultimately rely on forward-looking individuals who can bypass, for self and other, the temptation to feed immediate self-interest. The eighteenth-century essayist William Hazlitt got it just right when he said, "The imagination, by means of which alone I can anticipate future objects, or be interested in them, must carry me out of myself into the feelings of others by one and the same process by which I am thrown forward as it were into my future being, and interested in it."[58] The capacity to wait, exert patience, and fend off temptation is a core part of the support team associated with our moral faculty.

For almost forty years now, the social psychologist Walter Mischel has conducted experiments with children and adults in order to characterize the anatomy of our patience—specifically, the capacity to delay gratification.[59] His goal has been to understand how and why a child's capacity for delayed gratification changes over time, and whether the capacity to wait is an innate personality trait, an individual signature of temperament that predicts intellectual competence later in life as well as other control problems, including gambling, eating, sexual promiscuity, and alcohol consumption. Paralleling Aesop's fable, Mischel's work has characterized which of us are grasshoppers who discount the future and which are ants who save up for the possibly challenging times ahead.

In the classic delay-of-gratification study, an experimenter brings a child into a room, sits her down in front of a table, and explains that she can have either a small treat immediately or a larger treat later; the treats vary depending on the age of the child: marshmallows or cookies for the little ones and money for the adolescents. The experimenter then tells the subject: "You can have this reward now or, if you wait until I return, you can have several more of the same." If, during the unspecified waiting period, the subject wants the smaller reward, she may ring a bell to bring the experimenter back into the room. Since the experimenter never specifies when he will return, subjects don't know how long they will have to wait for the larger reward. The consistent finding, across cultures and socioeconomic classes, is that children under the age of four years have little to no patience. They take the immediate and smaller reward instead of waiting for the larger reward. Gradually, children acquire the capacity to block their impulses, keeping temptation at bay as they wait for the larger reward. But this assessment of the child's growth is more complicated and more interesting. The critical measure is not whether children do or do not take the smaller reward, but how long they wait to do so.

The young child's inability to delay gratification between the age of two to four years covers up considerable differences between children at this age. Some of these differences represent the signature of an innate personality trait, a character that comes through before culture has had a significant opportunity to inscribe its own, add-on signature. Some children, at the age of two, wait more seconds or minutes than others.

The child's genome generally creates a style of engaging with the world that either *internalizes* or *externalizes* actions. Children presenting the internalist signature take greater personal responsibility for what happens. If someone gives them ice cream, they think, "I was good. I deserve the ice cream." If a friend stops playing with them, they think, "I must not be playing nicely." The signature of an externalist is exactly opposite. When someone offers ice cream, it is because the person offering is nice. When a friend stops playing, it is because the friend is tired. When an experimenter tests these two personality types on the classic delayed-gratification task, the internalists wait longer for the larger and more desirable reward. These same internalists are also less likely to violate their mothers' prohibitions ("Don't play with that object"), and less likely to cheat in a guessing game with an experimenter.[60] Self-control predicts the tendency to transgress the unstated rule. The capacity to delay gratification meets moral behavior in an intimate handshake.

The number of seconds a two-year-old waits is like a crystal ball that predicts her future moral behavior; her ethical style, if you will. Watch how long she delays gratification, and you can extrapolate what she will be like as an adolescent and even a thirtysomething. Sixth-grade boys who showed impatience on the delayed-gratification task were more likely to cheat in a game than boys who were patient. Children who pick the immediate reward are more likely to end up as institutionalized delinquents than children who pick the delayed reward. And in a longitudinal study of subjects tested as toddlers, those who delayed longest were more likely to cope with negative situations in adulthood, to achieve better job security, to obtain higher Scholastic Achievement Test scores, and to maintain stable, nonviolent romantic relationships; both men and women who, as children, immediately grabbed the smaller reward were more likely to respond with severe anger and aggression toward their partners than those who delayed for the larger reward.[61] These studies suggest that impatience or impulsivity on the delayed-gratification task is an excellent predictor of who will transgress the mores of the culture.

The capacity to delay gratification plays out in another corner of the moral arena: reciprocity. If I give someone a gift at a personal cost, and live in a society governed by the Golden Rule, then I must wait for the gift to be returned. Altruism often requires inhibiting selfish desires. A reciprocal relationship requires waiting for returns on the gift given.

To test the relationship between age, prudence, and altruism, the developmental psychologist Chris Moore ran a series of experiments with three- to five-year-old children.[62] The key insight underlying these experiments was the idea that both prudence and altruism involve thinking about someone's mental states in the future. In the case of prudence, we imagine ourselves transformed in time and what our future state will be like if we can just wait it out. In the case of altruism, we imagine what someone else's mental state will be like if we do something nice for them. Each child played four games with a research assistant. For each game, the child selected one of two options:

GAME 1—*Sharing without a cost*: take one sticker for self or one for self and one for the assistant.

GAME 2—*Sharing with a cost*: take two stickers for self or one for self and one for the assistant.

GAME 3—*Delayed gratification*: take one sticker for self now or two for self later.

GAME 4—*Sharing with delayed gratification*: take one for self now or one sticker each for later.

Four findings emerged. First, younger children selected the immediate reward more often than did older children, on both delay conditions. Second, children of all ages were equally likely to share, with or without the costs of delay. Third, among the youngest children, individuals who picked the larger but delayed reward for self were more likely to pick the shared delayed reward. These results show that a child's capacity to wait for something good sets boundaries or constraints on her capacity to be nice to others. Fourth, and linking back to our earlier discussion, children who did well on theory-of-mind tasks, including versions of a false-belief problem, were more likely to share even if they had to wait for their own sticker. The correlation between sharing and understanding someone's future beliefs and desires is interesting, but does not allow us to pick out the causal arrow. Changes in a child's motivation to share could drive her awareness of what others believe. Alternatively, the child's appreciation that others have beliefs that may or may not differ from her own may drive her willingness to share.

In follow-up work, Moore showed that the child's capacity to care for others and share with them is highly variable. What drives this capacity is not children's ability to infer what others believe and desire but, rather, intricacies of their social environment, including the number of siblings. These studies show a change in performance, linking the ability to understand other minds with their ability to help. What they don't show is whether children at this age or earlier would deliver moral verdicts while watching another child play one of these games. For example, although a three-year-old might favor the immediate reward, with or without the option of sharing, she might nonetheless judge another child as bad if she favors the immediate reward. Highly revealing would be a study in which children both played the game and watched others playing. Are their rulings for others consistent with their own behavior?

The capacity for delayed gratification reveals something profound about a person's personality. Patience is not only a virtue but a marker of success in life. In Aristotle's ethics, the virtues are referred to as *hexeis*, and a *hexis* represents a disposition that is relatively fixed or resistant to change. Walter Mischel's work on delayed gratification provides one of the most robust examples of Aristotle's concerns with the psychology of ethics, and what later became known as virtue ethics.[63] But Mischel's work on delayed gratification also presents a puzzle. Given the fact that those who wait longer on the delayed-gratification task are more likely to have good jobs, stable romantic relationships, and low anxieties, why hasn't natural selection gradually eliminated the impulsive types? Why don't females prefer to marry the patient males, since they are less likely to go postal? I think there is a relatively simple answer: Patience is not always a virtue. It is a feature of human nature that can be tuned to different settings, often varying within and between cultures. Impulsivity can pay off in many walks of life: athletes, soldiers, debaters, and, often, individuals who respond in heroic ways—grabbing a child from a burning building, jumping in after a drowning crew member, and so on. It can also pay off in more mundane but evolutionarily relevant situations, as when resources are slim and an immediate opportunity to eat, mate, or gamble arises.[64] Given these advantages, think of impulsivity as a knob on an old-fashioned radio, one that can be turned to increase the volume from highly self-controlled to wildly impulsive. The massive self-controllers are frozen, stuck on the couch of life. The reflexive types jump off the couch at every opportunity, never planning, never thinking of the consequences of their actions. But since both personality types have their strengths, there is no consistent advantage to acting one way or the other.

Our species will always consist of both cool-headed, self-controlled members and hot-headed, impulsive types. We are born with a factory setting for impulsivity, which places limits on our capacity for control and our ability to do what's right. Some have it set high, some low. This variation doesn't influence our moral judgments, but it does influence our moral behavior. We will always be tempted by sin. As Oscar Wilde[65] noted, "What is termed Sin is an essential element of progress. Without it the world would stagnate, or grow old, or become colourless. By its curiosity, sin increases the experience of the race. Through its intensified assertion of individualism it saves us from monotony of type. In its rejection of the current notions about morality, it is one with the higher ethics."

CLOCKWORK ORANGE

Anthony Burgess explained the title of his novel *A Clockwork Orange* as follows: "I mean it to stand for the application of a mechanistic morality to a living organism oozing with juice and sweetness." Thanks to a series of discoveries, we can now identify pieces of this moral clock, what makes it tick, and what happens when a spring or cog goes down. Here is where we finally attack head-on the question of whether, like language, the moral faculty is backed by an organ that has been specially designed to handle problems of right and wrong. And as in the case of the language organ, the anatomical detective work lies not in finding a circumscribed region of the brain that is as discrete as the heart or the liver. Rather, the work lies in finding a circuit, specialized for recognizing certain problems as morally relevant and others as irrelevant, and then generating intuitions about possible moral outcomes. If there is a moral organ, then signs of design should be obvious, as obvious as the design of the heart and its function in pumping blood to the other parts of our body. And if this organ exists, then in the same way that a dysfunctional heart can lead to death, so, too, can a dysfunctional moral organ lead to actions that are forbidden by the local culture, as well as inaction in situations where it is morally obligatory to act.

In what follows, I have two goals. I want to pinpoint areas of the brain that appear to be directly involved in moral judgments, be they based in Kantian reasoning, Humean sentiments, Rawlsian grammars of action, or some combination of the above. I also want to assess whether some of this circuitry is not only involved in moral judgments but restricted to this domain of knowledge. For example, if there is a grammar of action, then there must be circuitry dedicated to the analysis of events into morally relevant causes, actions, and consequences. This system may not, however, be dedicated to the moral domain alone. The same circuitry that enables many of us to judge the American military atrocities at Abu Ghraib as morally forbidden may also engage when we enjoy the beautiful arabesques in a ballet or the punches in a boxing match. Somehow, of course, the brain assigns moral relevance to our military actions, but not to ballet or boxing, even though boxing entails harm.

Let me start with an imaging study that, although by no means the first, allows me to discuss the relationship between brain activation and some of the moral dilemmas I discussed in previous sections. The philosopher–cognitive scientist Joshua Greene scanned subjects' brain activity while they read a series of dilemmas, especially different permutations of the trolley problem, together with Sophie's choice, smothering a crying baby during wartime in order to avoid detection by the enemy, and removing a healthy person's organs in order to save five patients.[66] Greene started out from the position that there is more to our moral judgments than deliberate reasoning. Like Haidt and others leaning in the direction of moral intuition, Greene set up his experiments to directly explore the relative contributions of emotion and reasoning to our moral judgments. In essence, Greene was hoping to see the Kantian and Humean creatures in action, perhaps with differential contributions, depending upon the scenario.[67] As I have argued, he would undoubtedly also activate the Rawlsian creature, as no moral judgment is possible without some appraisal of the causes and consequences of action.

Neuroimaging is at its best and most useful when there are competing psychological theories, with one side proposing a unitary mechanism and the other proposing multiple mechanisms. If you are a Kantian creature, you think that the key process underlying moral judgment is deliberate reasoning based on clearly articulated principles. Consequently, the areas involved in such reasoning, be they specific to morality or not, should not only turn on but have the most active voice. If you are a Humean creature, you think that only our emotions play a role in moral judgments, and, thus, the circuitry underlying the production and perception of emotions should turn on. There are, of course, many other possibilities, including one that Greene favors: Both Kantian and Humean creatures have a voice, and sometimes they are in harmony and sometimes they are at war. When they are fighting, some process must resolve this disagreement, providing a psychological détente.

Subjects' behavior revealed distinctive psychological signatures for each category of dilemma. When reading a moral-personal scenario, such as Frank pushing a large person in front of the trolley, they deliberated for a relatively long time if they judged this case as permissible. If they judged the case as forbidden, then they sometimes took a long time but more often answered quickly. Recall that most people think it is inappropriate or impermissible for Frank to toss the large person in front of the trolley to save five people further on down the track. What these reaction-time results suggest is that if you go against the current and judge this case as appropriate, it takes you a while to muster the confidence to say that it is okay for Frank to toss the large person—an up-close-and-personal experience. Evidence in favor of the Humean voice comes from looking at all cases in which subjects delivered "permissible" judgments. Subjects spent almost seven seconds working out their answer to the moral-personal scenarios, but a significantly shorter four to five seconds on the moral-impersonal and nonmoral cases. This suggests that without the cry of the Humean perspective, subjects respond to moral and non-moral cases with equal swiftness. As Greene suggests, what typically creates the lengthy deliberation on the moral-personal cases is the tension between the Humean and Kantian voices.

What's happening inside each subject's brain as it evaluates a scenario and then responds? During the moral-personal scenarios, brain scans revealed significant activation in areas crucially involved in emotional processing, a circuit that roughly runs from the frontal lobes to the limbic system (medial frontal gyrus, posterior cingulate gyrus, and the angular gyrus). Furthermore, when subjects judged moral-personal cases in which the utilitarian consequences (maximize the good; saving five is better than saving one) were in direct conflict with the deontological rules that are presumed to be emotionally loaded (don't harm others! don't push the heavy man!), this conflict or tension directly engaged the anterior cingulate. In dozens of studies, this area lights up when individuals experience conflicting options or choices, such as the classic Stroop task, in which subjects must read a color word typed out in different print colors. For example, reading the word "white" printed in black is harder than reading it in white print, because there is interference between physical color and linguistic meaning. In the Frank-footbridge dilemma, activation in the anterior cingulate is not only an indication that the subject is experiencing conflict, but the extent to which this area activates is associated with the time it takes to deliver a response. Subjects who pondered over these conflicting cases showed greater activation in the anterior cingulate. Lastly, Greene found that when subjects went against the tide, stating that a moral-personal case was permissible, they showed much greater activation of the dorsolateral prefrontal cortex, an area involved in planning and reasoning.

What do we learn from this work, and the several other imaging studies designed to characterize our clockwork orange?[68] Unambiguously, when people confront certain kinds of moral dilemmas, they activate a vast network of brain regions, including areas involved in emotion, decision-making, conflict, social relations, and memory. Of course, these areas are also recruited for nonmoral dilemmas. The central issue then is to decide whether any of these areas, or others, are specifically and uniquely recruited for moral dilemmas but not nonmoral dilemmas. At present, none of these studies pinpoint a uniquely dedicated moral organ, circuitry that is selectively triggered by conflicting moral duties but no other. As in any empirical investigation, the lack of evidence for a system that selectively processes moral content is not evidence against such selectivity. Rather, the current work does not enable us to decide between a brain network that strictly delivers moral judgments and areas that are recruited for morality and other socially relevant problems. This is our central issue.

What these brain-imaging studies do show is that when we experience conflict from competing duties or obligations, one source of conflict comes from the dueling voices of the Kantian and Humean creatures. For Greene's subjects, if Frank on the footbridge is a no-brainer because utilitarian consequences win, then the moral judgment is fueled by deliberate reasoning and the emotions are suppressed, or, at least, unconcerned with the case. For a full-fledged utilitarian, Frank-on-the-footbridge isn't a moral dilemma at all. There is no conflict (anterior cingulate isn't engaged), no competing duties (no voice from the limbic system), simply one and only one choice: push the heavy man and save five people. Solving

Frank's dilemma is like judging whether the inequality $1 < 5$ is true: a trivial problem if you know the ordinal relationships between numbers on a number line.

These results provide a delicious twist on the philosophical debates over the nature of moral dilemmas: If it ain't got emotion, it ain't got moral swing. Emotional conflict provides the telltale signature of a moral dilemma. What we interpret as competing duties represents the output of the rushing blood flow in the emotional areas of the brain. Warning: This isn't a prescriptive claim. Reading blood flow is no better than reading tarot cards when it comes to working out what one ought to do. What blood flow provides, in the form of a snapshot of the brain, is an image of what happens when we detect a moral conflict. Brain imaging spits out a description of whether an individual perceives conflict in a dilemma, the sources of conflict, and the nature of the resolution. Unambiguously, all of the imaging studies to date show that the areas involved in emotional processing are engaged when we deliver a moral judgment, especially cases that are personally charged.

These studies, and the interpretations of them, are not without problems. Let's return to our three creatures. For Greene, moral-personal dilemmas put Kantian and Humean creatures into direct conflict. Depending on the subject, our systems of reasoning sometimes win out, achieving their goals in a cool and collected fashion. With the consequences in sight, and the means ignored, all of the actions entailed in the classic trolley problems are appropriate, permissible, and right. Attending to the means, however, raises the level of conflict, at least for some cases. It is not permissible to achieve the greater good—saving five—by pushing the large person over the footbridge, because this act entails harm, and, especially, the use of harm as a means to the greater good. On Greene's account, the emotions set up a roadblock. But recall that a Rawlsian would never deny the role of the emotions in some aspect of our moral behavior. What a Rawlsian would challenge is *when* such emotions engage. Resolving this disagreement will have to wait until our imaging technologies enable not only the detection of relevant activation areas but the timing of when they light up and then shut down.

When we evaluate whether an action is fair or whether inaction leading to harm is justified, we often simulate, in our mind's eye, what it would be like to be someone else. Our first-person experiences of the world are translated into their third-person experiences—what we feel and think is functionally equivalent to what they feel or think. We decide whether something is fair by imagining not only what we would exchange, but what someone might exchange with us. We decide whether it is permissible to harm someone else by imagining what it would be like to be harmed or to watch someone else engaging in a harmful act.

Over the last ten years, a group of Italian neurophysiologists, led by Giacomo Rizzolatti, have provided a description of how the hardware of the brain implements the simulation software of the mind.[69] I believe that this circuitry, known as the mirror neuron system, plays a critical role in our moral judgments, and may represent a key design feature of the Rawlsian creature. Interestingly, this system is unique neither to humans nor to morality. It may, however, provide a necessary first step in triggering the moral emotions, and thus offers a conduit between the Rawlsian and Humean creatures.

The initial findings, first discovered by recording from neurons in the premotor cortex of macaque monkeys, have largely been replicated in neuroimaging studies of humans, and with hints of a deficit coming from studies of autistics. The basic result is both straightforward and mind-boggling: Neurons in the premotor cortex show the same level of activity when the individual reaches for an object as when he watches someone else do the same, or when the individual hears a sound associated with an action or performs the same action himself. For example, if a rhesus monkey cracks open a peanut, the neurons that are active during this act are equally active when the animal hears the sound of a peanut cracking. Different neurons seem to turn on when different gestures are either performed or observed. For example, some neurons are active for grasping with thumb and index finger and others for lifting with the entire hand. Parallel findings emerge for humans, although the precise locus of the mirror neuron system is different, as is its capacity to respond to a greater range of actions. For example, whereas the rhesus monkey's system only responds when the target goal or object is present, the human system responds when subjects imagine an action or imitate someone else performing an action on an invisible object. Further, recent studies suggest that part of this system turns on when we directly experience a disgusting event or observe someone else experiencing disgust, with parallel findings for the experience of pain and empathy toward others in pain. The mirror neuron system is therefore an important engine for simulating emotions and thoughts—for getting under someone else's skin, feeling what it is like to be another human.

Nothing that I have said about the mirror neuron system suggests that it is dedicated to our moral faculty. However, the computations it runs are essential to our moral faculty—again, a part of the support team. A Rawlsian creature would be crippled in its absence, and so, too, would a Humean, given its role in emotional understanding. Judging whether our own actions are morally permissible, obligatory, or forbidden, must, if we accept Kant's universalist principle, be mirrored by an equivalent judgment for others. If it is permissible for me to break a promise, it must be permissible for others to break a promise. If I think that it is forbidden to harm another in order to save many from a greater harm, then it is forbidden for others to carry through on this action. If I think that reciprocal exchange is obligatory under the current conditions, then others must think that it is obligatory as well. To see whether we can run these computations without a mirror neuron system—a key test of its causal force—we will need to find patients with damage to this area, or take advantage of the new technology of transcranial magnetic stimulation—a device that sends magnetic current to specific areas of the brain, either stimulating activity or taking it out of action.

BRAIN-DAMAGED UTILITARIANS

When a child makes the same error over and over again, we attribute the error to youth, lack of knowledge, a problem with following rules, a failure to appreciate what is morally right. It is often our responsibility as adults to help children over such developmental hiccups, pointing out what went wrong and, if possible, why. If we are good teachers, and if the child is ready to grow, then she will discover the glory of conceptual change. Her errors will fade into the distant past as she masters her world. She will discover that where she has a will, she has a way.

When an adult makes the same error over and over again, we seek a different explanation. Although we can no longer explain the problem by invoking youth as the primary cause, it is possible that both children and adults are vulnerable to repeating an error because both lack the relevant psychological mechanisms. Consider three examples:

- Whenever Mr. X leaves a room, he walks through the doorway, turns around, walks back into the room, and then repeats this cycle thirteen times—no more, no less.
- Whenever Mr. Y speaks to someone, he uses every profanity in the book, a barrage of off-color remarks.
- Whenever Mr. Z makes a decision, he never takes into account future consequences but, rather, myopically restricts his considerations to immediate payoffs.

All three gentlemen appear to have the same general problem, at least at the surface: They lack proper inhibitory control. Mr. X can't inhibit the exiting-entering cycle, Mr. Y can't inhibit socially inappropriate commentary, and Mr. Z can't inhibit the temptation to act on an immediate reward. All three cases—real, as opposed to hypothetical—represent the behavioral signature of frontal-lobe damage. Some readers may even recognize Mr. Y as the nom de plume for the famous Phineas Gage, a well-respected railroad worker who, in 1848, was struck down by a tamping iron—almost four feet long and weighing thirteen pounds—that rocketed through the frontal lobe of his brain. Gage soon stood up, shaken but not too badly stirred, and then proceeded to have reasonable conversations with the members of his crew as well as an attending doctor. Soon thereafter, however, all semblance of moral sensibilities deteriorated, including an eruption of profanities and a disrespect for other members of his community. Lacking self-control, Gage left his town, finding odd jobs here and there, ultimately dying in 1860 of epileptic seizures and overall poor health. Is Phineas Gage proof of the Humean creature gone south?[70]

Previously, I mentioned that children are severely challenged by problems of delayed gratification, as they would rather take a smaller reward immediately than a larger reward later. Mischel suggested that children gradually learn to overcome the initial bias for the smaller immediate reward when they are able to integrate cool, rational decision-making with hot, passionate emotions and motivations. Patients with damage to the frontal lobes are like children prior to the fusion of the hot and cold systems. Not only do they fail to integrate their emotions into their rational deliberations, they appear to operate without ever consulting their emotions. Phineas Gage is but one of the more illustrative cases.[71] The others, covering damage both early in development and later in life, point to the same conclusion: to adhere to social norms requires emotional control, and emotional control requires an intact connection between the frontal lobes (especially the ventromedial and orbitofrontal regions) and the limbic system, especially the amygdala, a key player in emotional expression. The question we need to address here is whether such damage selectively knocks out some or all of our moral faculty, and, if so, which of our three creatures—Kantian, Humean, or Rawlsian—suffers the worst damage.

The neuroscientist Antonio Damasio and his colleagues developed a simple task to highlight deficits in decision-making and, importantly, to distinguish its signature from other cases of brain damage.[72] An experimenter presents each subject with a pot of money and four decks of cards, all facing down. By turning over cards from the decks, subjects can either lose or gain money. Two decks yield a net profit over the long haul, while the other two yield a net loss; to increase conflict, the two winning decks offer smaller rewards and punishments, thereby making the temptation to sample from the other decks high, because of the larger rewards. While subjects select cards, the experimenter records their emotional temperature by tapping into the sweatiness of their skin. After about fifty cards, normal adults begin to show a different profile of response from patients with ventromedial prefrontal damage. Normal subjects pick from the two winning decks and virtually ignore the bad decks. Patients consistently pick the bad decks. Patients are seduced by the high payoffs associated with the bad decks, selecting cards as if they were immune to punishment and the long-term net loss associated with their strategy. Patients are also different in terms of their skin's response to each card selection. Whereas normal subjects show a highly variegated change in sweatiness over the course of the game, with peaks corresponding to selections from the losing decks, patients look as if their sweat-o-meter stopped working: Their skin conductance profiles are relatively flat, with no differences among the four decks. Further, normal individuals show large increases in their skin response before they pick from the bad decks, before they are even aware that these are bad decks. When everything is working properly, our emotions function like hunch generators, a flittering of unconscious expectations that guide long-term decisions. For patients with ventromedial prefrontal damage, there are no hunches, and thus their decisions are short-sighted. These patients are like young children, captured by the lure of an immediate reward. Without the frontal lobes to reign in the amygdala's short-sightedness, temptation strikes. The future is irrelevant.

Based on these results, and others, Damasio suggests that our decision to take one action as opposed to another relies on the choreography between processes in the prefrontal cortex and a suite of internal states of the body, including changes in heart rate, breathing, temperature, muscle tone, and, especially, our feelings. Recurrent encounters with particular objects and events will create changes in the somato-sensory areas of the brain that function like road maps for bodily action. Imagine someone who has been stung several times by a wasp. The sting creates a change in the body. It hurts. The brain creates a memory of the association between seeing or hearing a wasp and the physiological change caused by the bite. Subsequently, simply seeing or hearing a wasp is sufficient to generate a hypothetical, as-if-you-were-bitten state of mind. Detecting the wasp is sufficient to avoid it, because the body has already created an emotional response. Changes at the body can therefore guide changes in action, either covertly or overtly, as the person consciously works out an evasive move.

What makes Damasio's work of relevance here is that patients with damage to the frontal lobes do not present a general deficit in problem-solving or general intelligence. Rather, their problem is more circumspect, as evidenced by their striking failure to take into account future consequences. To operate in the world, whether the savannas of our past or the cities of our present, requires the capacity to think about the consequences of our actions, how they influence self and other. Making such decisions often involves waiting, including many cases where we must forfeit immediate opportunities to cash in on more profitable ones that will emerge through patience. As the political scientist Jon Elster and the behavioral economist George Ainslie have pointed out, countless problems of temptation and control are linked to our discounting curves, our tendency to see immediate rewards as more valuable, more tempting.[73] Damasio's work has provided a wedge into the underlying neural machinery of this complicated problem. If the frontal lobes malfunction, an inappropriate decision is likely to follow due to a general insensitivity to consequences. Since such individuals lack the kind of feedback most of us enjoy when we act in ways that are beneficial or injurious to self or other, they make personal and societal mistakes, and fail to correct them. Neither positive nor negative actions have any consequence, because they fail to make it into the tally sheets of the mind's accounting system. This failure is not due to economic blindness. These patients know the difference between reward and punishment. But unlike people without such damage, these patients are seduced by the sight, sound, or smell of an immediate reward, and act impulsively, picking this option without any consideration for the future. The emotional conductor—guided by a now-dominant amygdala—has complete control. Irrational, impulsive action is the only option. This deficit helps explain why Phineas Gage the model citizen turned into Phineas Gage the moral deviant. In the absence of our emotional conductor, there is little hope that moral behavior will spring forth.

When Damasio and his colleagues tested these adult patients on a set of moral dilemmas developed by Kohlberg, they appeared normal. That is, they distinguished between morally permissible and forbidden actions and justified them at an advanced, adult level. As Damasio and others have argued,[74] these patients appear to have intact moral knowledge. What is damaged is the circuitry that allows emotion to collide with such knowledge and guide action. Although Damasio doesn't put it in these words, it appears that these patients have normal moral competence, but abnormal moral performance. The systems that normally guide what we do with our moral knowledge are damaged in these patients.

All of the patient data I have described thus far concerns individuals who incurred damage in adulthood. Although these subjects appear to behave irrationally, due in part to the lack of emotional feedback, these results raise several fascinating questions: If damage occurs early in childhood, is the resultant behavioral deficit the same, different, or absent due to the plasticity and reorganizing abilities of the immature brain? Is the apparent moral deficit an indication of a performance problem alone, or is there an accompanying competence problem as well? Do these patients not only act inappropriately but also make inappropriate judgments because their moral knowledge is damaged as well?

Damasio and his group have now assembled a new set of subjects who incurred damage to the frontal lobes as infants or young toddlers.[75] In many cases where damage to a higher cortical area arises in early development, the brain undergoes a miraculous process of reorganization. The damage can be monumental, as in cases where an entire hemisphere must be removed due to epileptic seizures. In these cases, if the left, language-dominant hemisphere is removed, there is an initial paralysis to the right side of the body, and loss of speech, but soon thereafter, and with the help of rehabilitation, subjects regain control over the body as well as their speech. Damasio's developmental patients show no evidence of recovery, raising profound questions about why some areas recover whereas others do not. Two of the earliest reported cases incurred damage before their second birthday. As adults, both had repeated convictions for petty crimes. The repetitive nature of the crimes implicates either an inability to learn from mistakes or a disregard for societal norms. And when tested on Kohlberg's moral dilemmas, their scores were completely abnormal, falling within the range of immature children.

In discussing Kohlberg's approach to uncovering our moral psychology, I noted that his particular line of inquiry assumes that the hallmark of moral maturity is the ability to justify, by means of Kantian principles, why some actions are permissible and others forbidden. The evidence presented thus far indicates, however, that this is only one aspect of our moral psychology, leaving unexplained the fact that for a wide range of moral dilemmas we deliver rapid judgments in the absence of coherent justifications. If emotions are involved in this aspect of our moral psychology, responsible for causally driving our moral verdicts, then patients with this kind of frontal-lobe damage should show deficits in their judgments. To explore this possibility, my students and I teamed up with Damasio and several of his colleagues. Using our battery of moral dilemmas, we tested his well-studied frontal patients, including some individuals who incurred damage in infancy, and others, who incurred damage in adulthood. We were especially interested in deciding between the Humean and Rawlsian creatures. If we are Humean creatures, making our intuitive moral judgments on the basis of flashes of emotion alone, then frontal-lobe patients should exhibit a pattern of response different from normal subjects. A Humean creature *needs* his emotions to make moral decisions. Rawlsian creatures do not, attending as they do to the causes and consequences of particular actions.

When the frontal-lobe patients read the trolley problems, they looked like normal subjects for Denise, but looked abnormal for Frank, Ned, and Oscar. Almost all patients said that it was permissible for Denise—the bystander—to flip the switch, killing one person on the side track in order to save five on the main track. In contrast to normal subjects, however, more patients said that it was permissible for Frank to push the heavy man in front of the trolley, killing him but saving the five. For Ned and Oscar, two cases that normal subjects tend to distinguish based on whether the harmful consequences are intended or foreseen, patients don't perceive a difference. When they read the philosopher Peter Unger's altruism cases—driving by the injured child on the side of the road versus throwing the UNICEF card away—they were like normal subjects: You can't drive by the injured child on the road but you can toss the UNICEF card. They were also like normal subjects for the CEO cases described in chapter 1: They attribute intent and blame to a CEO who supports a large moneymaking policy that hurts the environment, but they do not make this attribution when the policy ends up helping the environment. Last, these patients were more likely than normal subjects to show sympathy and generosity in cases involving the homeless. For example, they were more likely to say that a person walking home on a blistery cold night, one block from home, should give up his winter coat to a homeless man or collection box, an act that will save the homeless person from freezing to death overnight.

These observations argue for a more nuanced view of frontal-lobe deficits. In some cases, these patients are normal, attending to the relevant causes and consequences of the agent's actions. In other cases, they appear to focus more on the consequences, irrespective of means. If the consequences are good, the action is permissible. Although these results imply an important role for the Humean creature, suggesting that in the absence of significant emotional input we tilt in a utilitarian direction, there is more to consider. If the input from the amygdala to the frontal lobes is necessary for making moral judgments, across the board, then these patients should have shown deficits across the board. They did not. Rather, it looks as though damage to this area leads to abnormal responses for dilemmas involving particularly strong deontological distinctions. On the other hand, perhaps it is unfair to call these responses *abnormal*. Freed from the nasty ambiguities that most of us confront when we consider more than the consequences of someone's action, these patients see moral dilemmas with the clarity of a tried-and-true utilitarian! These patients lack the emotional checks and balances on their actions, but also lack some of the relevant competencies when it comes to simply judging the moral permissibility of an act.

GUILT-FREE KILLING

When Anthony Burgess sent the manuscript of *A Clockwork Orange* to the United States, his New York editor told him that he would publish the book, but only if Burgess dropped the last chapter—chapter 21. Burgess needed the money, and so went along with the suggested change. The rest of the world published the full twenty-one chapters. When Stanley Kubrick produced the film adaptation, it was hailed as cinematic genius. Kubrick used the shorter, American edition of the novel.

When I first read about the shortening of *Orange*, I immediately assumed that the last chapter would be ferociously violent, a continuation of the protagonist's destructive streak, a rampage against the moral norms. I was completely wrong. As Burgess put it in the preface to the updated American *Orange*: "Briefly, my young thuggish protagonist grows up. He grows bored of violence and recognizes that human energy is better expended on creation than destruction. Senseless violence is a prerogative of youth, which has much energy but little talent for the constructive." This change is not sappy or pathetic but, rather, a proper ending to a great novel. As Burgess acidly pointed out: "When a fictional work fails to show change, when it merely indicates that human character is set, stony, unregenerable, then you are out of the field of the novel and into that of the fable or allegory. The American or Kubrickian *Orange* is a fable; the British or world one is a novel." When it comes to violence, are humans more like characters in a novel or a fable? This question dovetails into the main themes of this chapter: the design of the moral organ, the nature of its development, and the consequences of its breakdown. Here I explore how a particular neuropathological deficit provides a window into some of the necessary circuitry for controlling violence, and thus, indirectly, for regulating moral norms.

No one teaches children to be angry or aggressive. They simply are, sometimes, in some situations. Anger and aggression are capacities, present in all animals and handed down to us from our ancestors.[76] They are part of our biological endowment, a piece of our innate repertoire. They are adaptive, playing a key role in competition both between and within groups. All cultures are aggressive, to some extent. Within most cultures, men are more aggressive than women. Such sex differences emerge early. Little boys are often rougher during play than little girls, even though girls in one culture can be more aggressive than boys in another culture. In cultures with records on driving, men are responsible for a vastly greater proportion of deaths due to aggressive driving. Cultures can curtail aggression or enhance it, but the sex differences remain, pointing to an underlying biological difference.

Everyone, at some point in his or her life, will engage in the thought of either severely hurting or even killing someone else. When these thoughts emerge, they are frightening, often because we can so vividly imagine the immoral act. Men tend to have more fantasy-like thoughts of homicidal behavior than women. Fortunately, most of us never give in to such temptations. We suppress our anger, thereby controlling our violence. Psychopaths don't.

In a *Calvin and Hobbes* comic strip, Calvin yells down to his parents: "Why do I have to go to bed now? I never get to do what I want! If I grow up to be some sort of psychopath because of this you'll be sorry!" "Nobody ever became a psychopath because he had to go to bed at a reasonable hour," his father replies. "Yeah," retorts Calvin, "but you won't let me chew tobacco either! You never know what might push me over the brink!" Calvin is probably not a psychopath. Although he may have outbursts such as these, he does not have the signature profile of a psychopath. For clinicians who work with psychopaths, it is critically important to make an appropriate diagnosis. Their diagnosis plays directly into the files of lawyers who must decide between acts of violence committed by sane as opposed to insane perpetrators. On the surface, psychopaths appear quite sane. This is their seductive power.[77]

Searching for the word "aggression" in the DSM-IV turns up several mental-disorder profiles. Accompanying the psychopath are a family of cases known as antisocial personality disorders. Psychopaths and people with antisocial personality disorder often show signs of inappropriate aggression and criminal behavior. However, whereas criminals show the signature of this disorder, many psychopaths do not. Most criminals are not psychopaths. The psychopath's trademark is a lucid mind, with often clear-headed, cool, rational justification for their behavior. What jumps out of their justifications, however, is an unparalleled egocentrism, supported by a lack of empathy that most of us find foreign and frightening. Guilt is a foreign concept to the psychopath. Without it, the emotional leash has been severed. The presence of a potential victim is as tempting to the psychopath as is a drink to the alcoholic, a slot machine to a gambler, or a piece of chocolate to a young child.

In North America alone, there are an estimated 2 million psychopaths, and most of them are male. John Gacy and Ted Bundy are but two of the many who have come to our attention through the media. In some ways, they are not extraordinary at all. Both men appeared to be solid citizens with charming personalities. Gacy, for example, was a contractor, voted "Man of the Year" by the Junior Chamber of Commerce, and frequently went to children's parties as a clown. He killed thirty-two men and buried them under his house. Bundy, a man with an equally impressive résumé, killed several dozen women; he claimed that pornography pushed him over the brink, and that, like a cancer, something had taken over his brain.

Not all serial killers are psychopaths. Edward Gein not only killed and mutilated his victims, he sometimes ate them and made household objects out of their body parts. He was diagnosed with chronic schizophrenia and sentenced to a hospital for the criminally insane. Of those serial killers who are psychopaths, most are difficult to defend in court, because they appear sane. Unlike someone who is severely retarded and lacks an understanding of either the causes or consequences of their actions, psychopaths are perfectly lucid about their actions. In the psychologist Robert Hare's *Psychopathy Checklist*, he identifies a suite of emotional and social symptoms, including: glib, superficial, egocentric, grandiose, impulsive, irresponsible, deceitful, manipulative, and guiltless.[78] Underlying all of these symptoms is a more fundamental deficit, a capacity that in most normal humans allows us to have a calibrated sense of other. As many clinicians suggest, psychopaths are uncalibrated because they lack the capacity for empathy. The reason most of us don't crush an ant, swat at a butterfly, kick a cat, or slap a baby is because we have some sense of what it might feel like to be another living creature. We can imagine what it is like to walk in their shoes. Empathy is a fundamental link in our ethical behavior. It is a missing link in the psychopath's mind. Evidence that it was always missing comes from longitudinal studies that trace back to their early childhood. Clinicians report that psychopaths were abnormally aggressive when they were young, and often physically lashed out against their own or a neighbor's pets. These developmental data suggest that at least some individuals are born with such disorders. As Calvin points out, however, some environmental situations may provide more inflammatory triggers than others, pushing individuals to one side or the other of the psychopathic divide.

Several studies also suggest that psychopaths fail to distinguish between moral and social transgressions. Moral transgressions arise when an individual's actions impact directly on the rights and welfare of others. Stealing money from a blind man's collection hat violates his right to keep what he has earned, while slapping a screaming child constitutes a violation of the child's welfare. Conventional transgressions, in contrast, occur when an individual's actions violate typical or normative responses to societally imposed rules, such as not wearing a tie to work or speaking in class without raising one's hand. Psychopaths fail to take into account the victim's welfare and tend to treat moral and conventional transgressions as the same. As a result, they often state that it is permissible to violate a victim's rights and welfare as long as an authority figure gives the nod. They say, for example, that it is permissible for one child to push another out of the way in order to get onto a swing if the teacher says this is okay. Recall that even young children understand this distinction.

The cognitive neuroscientist James Blair offers a more specific diagnosis, suggesting that the psychopath's deficit can be distilled into one problem: an inability to recognize submissive cues. Consequently, they lack sufficient control over aggression.[79] Normal adults are equipped with a mechanism that inhibits violence. This mechanism works by recognizing cues associated with distress, such as the facial expression of sadness or the sounds associated with fear and submission. Once such cues are recognized, they set in motion the parts of the brain that attribute beliefs and desires to others, and then use the inferences from this system to coordinate action with the guiding hand of emotion, especially empathy and sympathy. Among psychopaths, this inhibitory system is damaged. As a result, the lack of emotional input causes a blurring between moral and conventional transgressions, making it seem as though an authority figure can adjudicate on the permissibility of a violation. Without the feeling of aversiveness that comes from detecting distress, there may be little reason to put the brakes on an act of harm. What do studies of the brain tell us about these emotional capacities and their breakdown in psychopathy?

In a recent brain-imaging study, the cognitive neuroscientist Tania Singer brought heterosexual couples, with confirmed romantic feelings for one another, into a testing room. The goal: to isolate the circuitry underlying empathy. The experimenter attached separate shock-delivering electrodes to the man and the woman, and then the woman stepped inside the scanner. While the experimenter collected brain images, the woman could only see her husband's hand and a set of lights indicating whether he was receiving a mild shock or a significant jolt. When the woman received a shock, the scanner revealed three critical areas of brain activation: an area corresponding to physical pain in the target hand (somatosensory cortices), an area involved in emotion regulation (anterior insula), and an area involved in conflict resolution (anterior cingulate cortex). The insula and anterior cingulate are also part of the mirror neuron system mentioned earlier. Of greater interest, when the experimenter zapped the partner's hand, the somatosensory cortices were quiet while the anterior insula and anterior cingulate cortex were active. Activation in these two areas was strongest for women reporting a higher degree of empathy—a mirroring of emotions.

If a normal and healthy human brain shows a signature pattern of activation in the context of experiencing empathy, what about the brain of a psychopath?[80] Given the presumed deficit in emotional processing, one would expect to find damage to the circuitry underlying the emotions and the pathways that connect emotion to decision-making and action. Although experts in the field are still divided as to the actual psychological and anatomical deficits associated with psychopathy, some common themes are emerging. In contrast to normal subjects, psychopaths show reduced activation of areas involved in attention and emotional processing. Thus, their performance on a variety of tasks is often poor, because they are readily distracted or receive insufficient emotional input. Adrian Raine, a psychologist who has long studied the brains of murderers, reports that there are differences in the size of the hippocampus between *successful* and *unsuccessful* psychopaths. Success, in the current context, means "succeeded in actually murdering their victim"; unsuccessful psychopaths were caught before killing their victim. Based on studies in nonhuman animals and humans, there is ample evidence showing that the hippocampus plays a central role in regulating aggression. In earlier studies, individuals with a variety of disorders linked to abnormal aggressive behavior have implicated asymmetries in hippocampal size, specifically larger on the right than on the left. Raine's analyses show that unsuccessful psychopaths have a larger right hippocampus relative to successful psychopaths and a control population of nonpsychopaths. Given the connection between the hippocampus and the prefrontal cortex, these anatomical asymmetries point to a necessary coordination between inhibitory mechanisms and decision-making. Unsuccessful psychopaths are more likely to misjudge a situation, and thus are more likely to be caught. The only caveat here is that the asymmetries in the hippocampus are not specific to psychopathy, so there is not a necessary causal link between anatomy and psychological disorder.

If we return to our three moral creatures—Kantian, Humean, and Rawlsian—we can ask whether the deficit observed among psychopaths is due to deliberate reasoning, emotion, a grammar of action, or some combination. As discussed, the most accepted view seems to be that because of an emotional deficit, psychopaths blur the distinction between social conventions and moral rules, and, perhaps as a result, are more likely to engage in morally abhorrent behavior. But there are two alternative interpretations, one hinted at above. Although psychopaths clearly have an emotional deficit, their failure to distinguish moral and social conventions may result from a failure to bind emotions with a theory about which actions are right or wrong. Social conventions are relatively flat emotionally, whereas moral conventions—and especially their transgressions—are emotionally charged. Though we need to understand why this emotional asymmetry exists, and how it develops, observations unambiguously show that psychopaths lack a typical response to aversive cues, failing to unite this kind of emotional information with an understanding of why certain acts are morally wrong, as distinct from merely bad. For example, when a child falls, cuts his knee, and cries, this is a cry for help due to distress. The event is bad, but certainly not wrong or punishable.

The fact that people are able to associate different kinds of social transgressions with different kinds of emotion suggests an important link between the intuitive principles underlying moral judgment and our emotional responses. Back to the Rawlsian and Humean creatures. A central difference between social conventions and moral rules is the seriousness of an infraction. When someone violates a moral rule, it feels more serious; transgressions in the conventional domain tend to be associated with a relatively cool or neutral emotional response—eating with elbows on the dinner table is poor etiquette in some cultures, but certainly not an event that triggers passionate outrage.[81] This suggests that moral rules consist of two ingredients: a prescriptive theory or body of knowledge about what one ought to do, and an anchoring set of emotions. Recent theoretical and empirical work by the philosopher Shaun Nichols pushes this position as far as it has been taken to date, giving new life to Hume's sentimental rules and highlighting the significance of patient populations in deciding among competing theories of how the mind works.

Nichols points out that an emotional deficit alone can't explain the psychopath's deficit, nor is it sufficient to understand the conventional-moral distinction. We experience a towering number of events in our daily lives, which are aversive and indicative of distress but fail to elicit a moral evaluation. As mentioned above, when a child falls and scrapes his knee, while this is associated with distress, the child hasn't done anything wrong. When we see an accident victim, we typically experience distress and concern, but we don't accuse the victim of wrongdoing, unless he was driving while drunk. The mind may code the child's fall and accident victim's injury as *bad*, but certainly not *wrong*. When experimenters present these kinds of scenarios to young children, they never state that the child or victim should be punished. This shows that what is aversive, distressful, and bad is not necessarily morally wrong. What's missing, therefore, are the psychological ingredients that go into our perception that something is wrong and punishable. Conventional transgressions may be wrong, but rarely do we—young and old alike—think of them as punishable.

Nichols's view of these two ingredients of our moral psychology—what I consider a marriage between Rawlsian and Humean creatures—leads to two predictions. One, people will respond differently to an emotionally charged moral claim about harm as opposed to an emotionally neutral one. Second, in situations where there is a transgression of some norm that does not involve harm, the infusion of emotion will cause a shift from a conventional to a moral violation. There is ample evidence in support of the first prediction, and Nichols ran a simple experiment to test the second.

Consider the following event: You are at an elegant dinner party when Bob, one of the guests, snorts loudly and spits into his water glass. Was it okay for Bob to spit in his water? If it was not okay for Bob to spit into his water, then how bad was it? Why was it bad for Bob to spit in his water? Would it be okay for Bob to spit into his water if the host of the party had okayed it first?

Subjects answered these questions as if they were reading a moral dilemma about physical harm. Their answers were consistent with a moral as opposed to a conventional transgression. Spitting into a glass of water at a dinner party is a serious transgression that is forbidden even if an authority figure such as the host attempts to override the unstated policy. Moreover, when Nichols looked at subjects' overall sensitivity to disgust using a tried-and-true scale, those who found events such as spitting really disgusting were more likely to see infractions such as Bob's as very serious moral transgressions. In contrast, those who were less likely to experience the nose-wriggling of disgust perceived spitting as a mere convention, an act that is permissible if an authority figure says so.

Nichols's experiments show three things. First, it can't be that the psychopath's deficit is the sole result of a failed capacity to inhibit violent instincts toward those in distress. Disgusting actions don't trigger distress cues. Yet transgressions that involve disgust are treated as moral infractions as opposed to conventional ones. Our capacity to inhibit violent tendencies no doubt plays a role in our moral judgments, but this doesn't help us understand the difference between conventional and moral events. Second, because people judge stories about disgusting events as more serious and independent of authority, there is evidence that emotions can shift events from conventional to moral. This is important. Returning to the three toy models I introduced in chapter 1, these results suggest that at least some of our moral judgments—perhaps only those handling norms against harm and disgust—may emerge *from* our emotions. The Humean creature has a say in our intuitive judgments about moral transgressions. Third, emotions can't do all of the heavy lifting when it comes to deciding between conventional and moral events. There are many harmful and disgusting events that are not prohibited. The trolley problems reveal that we often judge harm as permissible if it is not intended and if it leads to a greater good. Similarly, acts such as unintentional vomiting or diarrhea are disgusting but certainly not forbidden. A Rawlsian creature—equipped with a body of moral knowledge operating over the causes and consequences of action—can explain these cases.

In sum, Nichols's simple experiments make a good case for the joint contributions of action appraisal and emotion in guiding our intuitions about conventional and moral cases. In these particular domains, both Rawlsian and Humean creatures have a say. At present, however, we have a limited understanding of the biasing effects of emotion on our moral judgments, because all of the studies to date have focused exclusively on disgust. It is possible that disgust holds a unique position in guiding our moral intuitions. Whether or not disgust has this unique role or not, Nichols's study raises the interesting possibility that norms acquire their robustness when they are tied to strong emotions. Upholding such norms makes people feel good, while violations make them feel bad, ridden with guilt, shame, or embarrassment.

Psychopaths, as an extreme case of pathology, reveal that humans are equipped with systems that control aggression, and sometimes these systems break down. Given the relatively poor success that clinicians have had with rehabilitating psychopaths—when released from prison, they show four times the level of repeated offense as that of prisoners with other antisocial personality disorders—it looks like the control problem must be attacked earlier in development, when the habit of aggression has only just begun. This assumes, of course, that psychopathy *is* treatable, and that behavioral or pharmacological treatments can undo what is learned. If a more significant component of psychopathy is due to genetic factors, then treatment may be more difficult than presumed by many clinicians. Ironically, recognizing the possibility that the psychopath's genetically endowed moral competence may be intact could provide the most promising solution for recovery, allowing the clinician to work with a moral foundation that is universally shared.

5

PERMISSIBLE INSTINCTS

*It disturbs me no more to find men base, unjust, or
selfish than to see apes mischievous, wolves savage, or the
vulture ravenous for its prey.*

—MOLIÈRE[1]

CONCEPTION IS A MAGICAL AND WONDERFUL experience for most parents. Nine months later, this experience is transformed into another—the birth of a child. Unbeknown to most parents, however, is a nasty little fact: While the fetus develops, it is maneuvering to take all the resources that a mother can give, and more. Yes, the human fetus is greedy. Yes, the fetus is not playing fair. And yes, mothers pay. The Bible provides one explanation, one imbued with moralistic coloring: "Unto the woman he said, I will greatly multiply thy sorrow and thy conception; in sorrow thou shalt bring forth children . . ." Evolutionary theory provides a different explanation, one that is descriptively powerful but prescriptively colorless.

The idea that parents and offspring are engaged in a battle is not new. Every parent knows this: scratch a teenager, find conflict. But what most parents don't know is why this battle exists. Their ignorance is partially due to the shelves of parenting books that simply and clearly describe the terrible twos, and then lay out a prescriptive recipe for combating what is, developmentally, a fait accompli. Trivers brought fresh insight into this problem over thirty years ago by showing how an understanding of the genetics of the parent-offspring relationship inevitably leads to conflict, where one side is tempted to take more than its fair share while the other is tempted to give less. Biological parents are genetically related to their children by exactly one half. But each child is related to itself by exactly one. Consequently, whereas the child wants to get as much as it can from its parents, parents need to distribute their wealth so as not to foreclose opportunities for future offspring. This simple genetic difference leads to conflict, one present in all species with sexual reproduction like ours. The outcome of such conflict is, hopefully, a carefully choreographed pattern of give-and-take. Offspring get what they need to develop into healthy, productive individuals. Parents give what they can without compromising their chances of having other, equally healthy and productive children. At some level, then, selfish offspring and selfish parents must cooperate for mutual benefit. After all, they both want, at some level, the same thing: genetic immortality.

I begin this chapter with the dynamics between parents and offspring for three reasons: It represents the first social relationship we experience in life, it is an evolutionarily ancient relationship that pits concern for self against concern for other (perhaps the most basic dimension of moral decision), and we now have an exquisite understanding of its adaptive function as well as the mechanisms that are at its core, from genes to neurons to beliefs. It is a relationship that in today's world is imbued with moral weight, including the permissibility of abortion, infanticide, bottle-feeding, wet nursing, and genetic engineering. From greedy babies and controlling parents, I turn to cooperative relationships among genetically unrelated and often unfamiliar individuals, where the temptation to defect is high, putting a premium on the development and implementation of regulatory control. Specifically, these new relationships put pressure on a moral faculty that can see past a rationally optimal, self-interested action. Humans acquire a variety of social skills that facilitate cheating and cheater-detection. We also acquire a thirst for novelty and creativity, which creates an opportunity for increased exploitation, which, in turn, creates increasing pressure for control. This conflict fuels our sense of right and wrong, and raises questions about what we ought to do. The aim of this section then is to see how the moral machinery I dissected in the last chapter does some work for an individual living and acting in a social world. Here we will see why our intuitions about permissible actions sometimes fail to align with our actual actions. Here we will see the cleavage between competence and performance, and the tug-of-war between rationality and morality. And here we will see how evolutionary intuitions drive new ways of looking at moral conflict in our own species, foreshadowing some of the arguments in part III.

FETAL ATTRACTORS

The evolutionary biologist David Haig provided a surprising twist to Trivers's account of parent-offspring conflict. Conflict begins before the child is able to look up with a seductive smile and manipulate its mother into a longer nursing bout. Although the human fetus can't see, talk, or move on its own, it knows how to work covertly, turning the placenta into a cafeteria that delivers more than the standard fare. Haig's insight emerged from a critical analysis of two surprisingly disparate animal groups and research problems: genetically engineered mice and women with pregnancy complications.

As we all learned in high school biology, the genes that make us who we are act the same way whether we received our copy from Mom or Dad. On this view, genetically engineering a fetus with Dad's genetic material should be similar to a fetus with Mom's genetic material. When this thought experiment was realized in the laboratory with mice, as distinct from humans, the results flew in the face of received wisdom. The all-Dad fetuses were much larger and more active than the all-Mom fetuses, often with bigger bodies but smaller heads; all-Mom fetuses appeared to have smaller bodies but larger heads. These results immediately suggested a parental asymmetry—and, to Haig, the signature of biological warfare between paternal and maternal copies of a gene. Biologists soon uncovered a new class of genes, called "imprinted genes." Unlike the genes we learned about in high school, these have the unusual property of having a parental label. Sometimes, Mom's copy is turned on in us, and sometimes it is Dad's copy.

Understanding how imprinting works not only helps explain the genetically engineered mice but also sets up a new way of understanding conflict between father and mother, mother and fetus, and inside the fetus's own body. It also sets the stage for thinking about the evolution of our moral faculty from the bottom up, from genetic to behavioral conflict. An all-Dad fetus is big because it has been engineered with paternally active genes. From the perspective of a male mouse sire who will most likely mate with a female and never see her again—no flowers, chocolates, or promises of a big house—it pays to produce the biggest, healthiest baby, because this increases the chance of surviving and competing with others. However, bigger is more expensive in terms of carrying the fetus around, supplying it with the necessary nutrients, and, ultimately, giving birth. As Dave Barry put it, "Childbirth, as a strictly physical phenomenon, is comparable to driving a United Parcel truck through an inner tube."[2] Given the costs, mothers want control, turning off genes from the father that are likely to make the fetus too big. In species such as mice, where females give birth to many offspring over their life span, mothers disburse resources as a function of the number of potential offspring that are left to reproduce, and the current environmental conditions. This is the cold logic of evolution: If this is her first litter and times are bad, it may not be worth investing too much, waiting instead for the next round and the potential for a better season of resources. If this is her last litter, it is worth pumping in all she has, because there are no more direct opportunities to leave a genetic legacy. With maternal and paternal perspectives entered into our account, we have the fire that ignites the conflict. Fathers always want bigger babies, mothers smaller ones, up to a point.

Conflict between mother and fetus emerges, in part, as a result of conflict between mother and father. When the paternal copy is active, the fetus has been engineered to get more resources from its mother. And, sometimes, mothers have genetic mechanisms to fight back, redirecting or changing the allocations the fetus has ordered. Lastly, the conflict within the fetus arises because each individual is some conflicted blend of maternally and paternally imprinted genes. A divided self is part of human nature's signature.

The fetus uses a militia of hormonal tricks to block spontaneous abortion, commandeer the flow of blood to the placenta, and thereby divert greater volumes its way instead of to tissues that the mother depends on for her own health. One of the fetus's best tricks, however, is handed down by the father. Men carry a gene that, if turned on in the fetus, enables the secretion of a hormone that blocks the effects of the mother's insulin. The result is an increase in the amount of sugar in the blood during the third trimester. This energy boost is fantastic from the fetus's perspective and dangerous from the mother's, since she may contract gestational diabetes. When mothers lack such complications, they have worked out a cooperative truce with their offspring, one that minimizes damage to their own bodies while maximizing the resources they have to offer their child. Complications present the telltale sign of a crafty fetus, one who has won the tug-of-war. Winning the tug-of-war can, however, have its downside.

The transition from fetus to newborn infant changes the game, as the anthropologist Sarah Blaffer Hrdy describes:

> By the time the baby is expelled by the uterine muscles, it must be prepared for its exile from gestational Eden. From hormonally empowered, firmly entrenched, fully enfranchised occupant of its mother's body, the baby's status declines to that of a poor, naked, two-legged mendicant, not even yet bipedal, a neonate who must appeal in order to be picked up, kept warm, and be suckled.[3]

How does the neonate appeal? By looking cute, vulnerable, and needy. The human neonate is equipped with design features that exploit the sensory biases of its caretakers. Neonates have developed facial and vocal signals that exploit their mothers and fathers. Why is Mickey Mouse cute? Because his head is much bigger than his body and his eyes are relatively large for his face. These *neotenous*, or juvenile, characteristics are like visual candy, deliciously attractive to our eyes.

Fatness is an honest signal of health, an indication that a parent has done the right thing by investing and bringing the baby to full term. A neonate can't fake the layers of fat. Their fatness is an indication that they have gotten what they need, at least nutritionally. It provides one level of protection against our capacity to abandon or kill infants that don't look healthy.

What about those innocent smiles, pouting lips, adorable nuzzling movements, melodious coos, and horrific cries? Are these always honest indicators of need and, if not, how do parents crack the code, discriminating between the lies and the truths? All infants cry, often upon taking their first breath of air outside the womb. Cries of hunger and pain sound different, and parents rapidly learn to distinguish between them. As children develop, they gain control over their emotional expressions, crying when there is an appropriate audience but holding back their tears when a sympathetic caretaker is absent. The design features of cries are ideally suited to catch our attention, stir our emotions, and motivate us to swing into action, attempting to stop or eliminate the causes of our offspring's hunger or pain. Loud, harsh, noisy sounds are aversive. Soft, harmonic sounds are pleasant. When humans hear an infant crying, they want to turn it off.

The evolutionary biologist Amotz Zahavi argued that signals are honest if and only if they are costly to produce, if the costs are proportional to the signaler's current condition (e.g., the same signal is costlier to produce for an individual in poor rather than good condition), and if signaling ability is heritable, passed on genetically from parents to offspring.[4] When signals meet these conditions, they are referred to as "handicaps." They are handicaps in the sense that those individuals who have managed to produce or sustain such signals in the face of selection against costly actions must be superbly fit. They have escaped the destructive hand of natural selection.

Darwin called crying "a puzzler." We can make some sense of the puzzle, however, by appealing to the logic of handicaps. Crying, especially with tears, qualifies as a handicap. It is difficult to produce on command, costly in terms of energy and the blurring of vision, and is the only emotional expression to leave an enduring physical trace after the initial incident. From the infant's perspective, these are crucial features, as they undoubtedly increase the odds that a caretaker will respond in a positive manner, even after the cause of the infant's tears has dissipated. They are a form of commitment. One can well imagine, then, a scenario in which crying with tears—a uniquely human expression—evolved from crying without tears, with selection operating to increase the probability of an unfakeable expression of distress.

This brief discussion of crying babies illustrates one simple point: Selection has equipped infants with signals designed to manipulate what their parents hear and see, often causing them to lose control and give in to the infant's wishes. Given the relatively close genetic interests of parents and offspring, this kind of manipulation is perhaps not that surprising, or unfortunate. From the parent's perspective, their sensory biases have an adaptive function: They guarantee investment in their genes. They guarantee what we perceive as the morally appropriate response to having a child. However, as soon as an infant gains some control over its own actions, including the earliest forms of reaching and moving, it must confront the interests of other individuals, who may or may not share any genes in common. From an unadulterated self-concern must grow partially adulterated other-concern. From the cold logic of myopically optimizing individual fitness there must evolve a system with more warmth and compassion for others. There is a hidden calculus here as well, but it is, of necessity, one in which direct concern for self is at least temporarily shelved for some other optimal good.

The choreography between parent and offspring raises another, more volatile topic: When, if ever, is it permissible for a parent to harm or kill its offspring? Although questions of abortion and infanticide have, for most people in most cultures, been settled one way or the other, there are other issues that link our judgments about these cases to more general principles of harm. In particular, to what extent is the logic underlying our judgment about abortion and infanticide consistent with other forms of harm? When we judge the permissibility of killing a fetus or infant— as opposed to engaging or not in the act of killing—what factors enter into our judgment? Volumes have been written on this topic, and the issues are complex and heated.[5] Here I focus on a narrow corner in order to bring the problem back to the causes and consequences of harmful actions, and the extent to which our moral faculty delivers a verdict prior to our emotions. As a footnote, let me state at the outset that the cold calculus I am about to apply to this problem does not in any way diminish or compete with the heated emotional issues that engage all human beings confronting problems of abortion and infanticide. Attempting to account for the principles that may underlie our judgments of abortion does not undermine the personal significance that each individual attaches to this problem, including, for some, the horror of hurting an innocent other.

For starters, I want to bypass all discussion about when the fetus *counts* as a person, an individual with some rights or all rights. Trying to draw a line is an exercise in futility. Following the lead of several moral philosophers, especially Thomson and Kamm,[6] I assume that at some point in the process from conception forward, there is something—fertilized ovum, fetus—that counts as an individual with rights of some sort. Given this starting point, I then ask about the calculations, conscious or unconscious, that we may or may not engage in when deciding whether this individual should have unquestioned access to its mother's body, perhaps an inalienable right, from the start.

Every person has a right to life. So the fetus does, too. A woman has the right to decide what happens to her body, inside and out. If a person's right to life trumps a woman's right to arbitrate over her body, the fetus wins in the debate: The case on abortion is closed. But now imagine the following scenario from Thomson. One morning, you wake up in your bed, medically yoked to a famous violinist who has a fatal kidney problem. The *Society for Music Lovers* decided that you were the perfect person, due to blood compatibility. If the violinist is unplugged, he will die. If you remain plugged in for nine months, he will recover. Are you morally obligated to stay plugged in? Moral philosophers discussing this case agree that while you may choose to stay plugged in, you are not obligated to do so. A virtuous act is not necessarily an obligatory act. When we posed this question on our Internet site, laypeople across the board, religious and nonreligious alike, agreed with philosophical intuition: It is perfectly reasonable to unplug the violinist.

How should we think about the psychology that enters into our judgment about Thomson's violinist case, apocryphal as it is? The violinist is clearly a person with a right to life. In theory, his right to life should trump your right to do what you wish with your body. If this was all there was to the case, then our moral faculty would deliver an obligatory verdict, forcing you to stay plugged in. But this is not how we perceive the dilemma, so what makes our verdict here different from the case of abortion? Unlike voluntary pregnancy, your connection to the violinist—and his dependency on you—is the result of an involuntary process. You didn't agree to being plugged in. You didn't make any commitment to the violinist. You may feel sorry for the violinist, and it would be virtuous of you to help out, but there is no obligation.

To draw a more direct parallel to abortion, let us change one part of the scenario: you consent to being plugged in, but then decide to remove the plug at some point. Here, our Internet-based sample of subjects indicated that it was generally forbidden to unplug. They seemed to perceive this case as no different from a surrogate mother who decides to abort after carrying a fetus for a few months. The permissibility of harm is thus linked to the issue of commitment, although this is not the only relevant parameter.

Now let's ask the harder questions: Is abortion permissible, possibly obligatory, when the fetus threatens the mother's health and potential survival? For example, consider the case where a mother develops pregnancy complications that, should they persist, will unquestionably end her life. This can happen, as discussed above, because of competition between imprinted genes and, especially, paternally active genes that cause the fetus to seek more resources than the mother can provide. Let's say that during pregnancy, the father leaves the mother. She is financially devastated, with no resources to feed her child, only enough to feed herself. What if she wanted a male baby, knows the fetus is female, and knows that she will be devastated with a daughter, enter into depression, unable to take care of her child? How does one decide which of these harms is more permissible, if any?

Thinking about the psychological ingredients associated with the Rawlsian creature helps clarify what is occurring in some of these cases. Aborting the fetus constitutes an *act* of killing. Letting the pregnancy continue constitutes an *omission* with the same general consequence: someone dies. The act kills the fetus, the omission kills the mother. We are back to a dilemma that puts harm to the fetus against harm to the mother. Who caused the current situation? The mother and her partner started the process of conception, with the goal of giving birth to and raising a healthy child, so this was not the cause of conflict. The cause of conflict lies with the fetus, or the father, if one wants to attribute cause to the imprinted genes. This looks like the fetus is to blame. But we can just as easily turn this problem around and say that if the mother had been in better condition, she would have been able to provide for a greedier-than-average fetus. What is the mother's goal? Her immediate goal is to save herself. As a foreseen consequence, she must harm another—she must kill the fetus. Unlike the

violinist case, however, the fetus is a threat to the mother, and an immediate, ongoing threat at that. By aborting the fetus, the mother intends its
death, even if it is a means to something else. In this case, killing the fetus
is the means to the mother's survival. Digging beneath the strong emotions that many of us feel when we think about abortion, we find our
moral faculty, a Rawlsian system designed to process cases of harm in
terms of causes and consequences.

The psychological factors and complications raised here only skim the
surface. As we enter an era of ever-increasing technology, designed to prolong life and eliminate traditionally complicated medical issues, we are
faced with new dilemmas, including problems that our psychology did
not evolve to solve. If a pregnant mother enters into a life-threatening situation triggered by the fetus, should the law make it obligatory for her to
have a cesarean because she is guaranteed relief from child care at no personal cost, including adoption? As we think about these new complications and the moral challenges they raise, we should keep in mind that our
moral faculty may judge these cases with more general principles of harm,
anchored in the logic of a grammar of action.

OF LORDS AND FLIES

In *A Theory of Justice*, Rawls presciently intuited a tension between evolutionary selfishness and the success of a rational moral system that attempts to go beyond this foundation, working out what would be. As the
following quote indicates, Rawls was cognizant of the sociobiology revolution even before Ed Wilson wrote *Sociobiology* and Richard Dawkins
wrote *The Selfish Gene*:

> The crucial question here, however, is whether the principles of
> justice are closer to the tendency of evolution than the principle
> of utility. Offhand it would seem that if selection is always of
> individuals and of their genetic lines, and if the capacity for the
> various forms of moral behavior has some genetic basis, then al-
> truism in the strict sense would generally be limited to kin and the
> smaller face-to-face groups. In these cases the willingness to make

considerable self-sacrifice would favor one's descendants and tend to be selected. Turning to the other extreme, a society which had a strong propensity to supererogatory conduct in its relations with other societies would jeopardize the existence of its own distinctive culture and its members would risk domination. Therefore one might conjecture that the capacity to act from the more universal forms of rational benevolence is likely to be eliminated, whereas the capacity to follow the principles of justice and natural duty in relations between groups and individuals other than kin would be favored.[7]

Seeing the tension between these two motivational forces, Rawls posed the veil of ignorance problem described in chapter 2. To reiterate, we want to understand how to construct a morally just and fair society. Imagine starting from the ground up, with no principles in place, and no formal laws to dictate the distribution of resources and rights. Assume that everyone is maximally self-interested and charged with developing principles that are just. Since these decisions must be made behind a veil of ignorance, no individual can know beforehand what kind of position he or she will obtain. Consequently, there is a built-in constraint that should limit, if not eliminate, selfishness. Every one should cooperate to lay out the best possible outcome for all.

When civilizations disappear, it is often at the hands of selfishness, of a selfish core that strives to undermine what in theory may be a well-intentioned construct of justice. What we must understand is how a selfish core can sometimes, maybe often, develop into deep regard and respect for others. In the same way the discovery of imprinted genes forced us to consider a divided self that reflects the competition between maternal and paternal interests, the realization of an evolved self in society forces us to recognize a divided self that reflects the competition between selfish- and group-oriented instincts.

Let's step into the problem of combating selfishness in the service of maintaining a stable cooperative society by considering William Golding's *Lord of the Flies*. This riveting fiction, standard reading in most intro courses to English literature, should be standard reading in biology, economics, psychology, and philosophy. Golding starts us off with a gaggle of children stranded on an island and presented with a delicious twist on the initial Rawlsian state. In contrast to the rational, sit-down-at-the-table-and-think-through-all-the-options group of adults, Golding asks us to imagine what it would be like for wild, naked, and hungry children. Many of the children are scared. Most are hungry. All are unclear about what to do next. Soon enough, however, several of the older children emerge as leaders, hiding their fears and speaking with confidence about what is needed. Ralph has charisma and emerges as the king, backed by Piggy, who has brains. They are like Christian and Cyrano de Bergerac, head of state and his staff, puppet and puppeteer. Although Ralph's rules hold for a while, resulting in group harmony and cooperation, the feisty Jack immediately capitalizes on a weakness. He speaks out, claiming that Ralph is afraid of the forest, afraid to go hunting for food and scared of the dangerous pigs. Jack is fearless, his charisma magnetic, and his offer of food irresistible. As the children begin to divide, some following Ralph and some Jack, each makes a speech that captures the challenges of stabilizing cooperation.

Jack knows that the children's hunger is permeable to the temptation of pig meat. He also preys on their sense of fairness and reciprocation. Since he gave them food, he leaves them with little option but to join his tribe. He also preys most effectively on the young children who are afraid and need protection. Ralph's countermove attempts to strike a different part of the human psyche—specifically, the emotions that mediate loyalty. As the economist Robert Frank noted early on, in a challenge to the standard self-interest models in economics, emotions can often override selfish instincts, forcing a cooperative hand in the face of shame or guilt at breaking a commitment.

Ralph expects everyone to stay with him because he is chief and because they elected him. He also attempts to counter Jack's offer of food with an offer to keep the fire going. But here Ralph fails to think through the logic of cooperation. Keeping the fire going requires more than one individual. It requires cooperation. But there will always be the temptation to defect, to take advantage of those who keep the fire going. Jack is offering something much more tempting, and with apparently no costs. Join Jack's group and gain the protection of his hunters and the food that they provide. Moreover, by joining Jack's group, the children can still benefit from the smoke signals launched by Ralph's group. And better to lose with Jack than Ralph, because at least Jack's tribe has good hunters. Joining Jack's group is a no-brainer. But there is that nagging feeling of loyalty and commitment, of empathizing with Ralph because he took the initial role of leader, was kind, fair, and considerate with respect to the group's interests. As the final jab, Jack points out the bone in Ralph's hand, emphasizing the fact that he, too, needs and wants meat, and should be grateful to Jack for hunting down the pig. He implies, therefore, that even Ralph should step down and join the other side. The rest of the story captures the rising power of Jack's tribe, the downfall of Ralph's, and the tragic death of Piggy. Cooperation crashes as the temptation to defect rises, and trust grinds down. Supporting Ralph may have appeared to be the right thing to do, but in the face of external demands, our moral faculty's judgment collides with our moral behavior, fueled by alternative motivations and temptations.

When Trivers first developed his intuitions about reciprocity—simply, I'll do X for you if you'll do X or something X-like for me at some point down the road—he merged three intellectual approaches: cost-benefit economics; selfish-gene evolutionary biology; and the psychology of fairness, including its emotional armament. Trivers argued that if the following three conditions are satisfied, reciprocity will evolve and remain stable:

1—small costs to giving and large benefits to receiving
2—a delay between the initial and reciprocated act of giving
3—multiple opportunities for interacting, with giving contingent upon receiving

Although Trivers's theory of reciprocity looked as though it would provide a solution to altruism among non-kin, almost thirty-five years of research has failed to provide more than a few suggestive examples from the animal kingdom—a conclusion I will flesh out in chapter 7. I suggest that this conclusion is unsurprising once one begins to unpack the psychological mechanisms required for reciprocity. These include, most important, the capacity to quantify the costs and benefits of an exchange, compute the contingencies, inhibit the temptation to defect, and punish those who fail to play fair. Although we are almost completely in the dark with respect to when these different psychological ingredients evolved and became available to members of our species, we do have some understanding of how such ingredients develop within our species. The critical question then is: Once these pieces evolved, did they enable a speciation event, from *Homo economicus* to *Homo reciprocans*? Said differently, although our greedy fetus looks like the ultimate outcome-maximizer (*economicus*), has this same fetus also been handed an innate sense of fairness that eventually motivates an interest in the processes underlying an outcome, be they good for the individual or some highly selective group (*reciprocans*)?[8] How we answer this question is significant, because it forms the foundation for many theoretical and practical issues in the fields of economics and law, disciplines that pride themselves on lending clarity to the prescriptive side of morality.

One answer is that both species coexist in some stable state, neither liking the other but simply tolerating each other's presence. Independently of this dynamic, both species rely on a computational logic that develops in all humans, independent of religion, sex, race, and education. To set the stage for how we acquired this logic, both in development and in evolution, I return here to some of the evidence amassed in part I describing the mature state of moral competence and attempt to account for its acquisition. The goal is to understand how we acquire the capacity to engage in stable rounds of cooperation (moving beyond the parent-offspring dance), fend off temptation, detect cheaters and punish them. In brief, I will explain how the arena of cooperation provides the psychological foundation for understanding our moral sense, its anatomy and function. Breaking a cooperative relationship is minimally a violation of social norms and maximally an immoral act that represents a breach of a legally binding agreement. The remaining part of this chapter therefore explains other aspects of our moral psychology, including capacities that may be specific to it as well as shared with other faculties of the mind.

COUNTING FAIR PLAY

Consider Trivers's first condition concerning quantification of costs and benefits. In the absence of quantificational mechanisms, it would not be possible for individuals to play most of the economists' games. What constitutes fairness if you can't compute the costs and benefits? If the relationship between systems of quantification and systems of justice appears tenuous, then listen to Voltaire, who poignantly fingers one aspect of this relationship: "Man is born without principles, but with the faculty of receiving them. His natural disposition will incline him either to cruelty or kindness; his understanding will in time inform him that the square of twelve is a hundred and forty-four, and that he ought not to do to others what he would not that others should do to him; but he will not, of himself, acquire these truths in early childhood. He will not understand the first, and he will not feel the second." In the domain of mathematics, Voltaire is dead wrong. As a species, we are born with two quantificational systems, innate machinery that enables infants to compute small numbers precisely and large numbers approximately.[9] Both systems are present before infants actually deploy them in the service of helping or harming others. And both provide the building blocks for acquiring higher mathematics and ethics.

To showcase how these different number systems work, developmental psychologists use the same set of methodological tricks that I introduced in chapter 4. In particular, they take advantage of an infant's looking and boredom, to uncover the nature of his expectations and representations. For example, show a five-month-old baby an empty stage, place one doll on the stage, conceal it with a screen, add a second doll behind the screen, then remove the screen to reveal either one, two, or three dolls. If they properly computed the addition operation, and have in mind the two dolls sitting behind the screen, then an outcome of two is boring and expected, relative to an outcome of one or three, which should be jaw-dropping and unexpected. More concretely, infants should look longer at an outcome of one or three dolls than two dolls. And they do. With this kind of method, infants showed clear and precise discrimination up to three objects. In contrast, when comparable methods were used with large numbers, discrimination was no longer precise, relying instead on the ratios and approximate estimations. Thus, infants discriminated four from eight, and eight from sixteen objects, but not four from six, or eight from twelve.

What allows us to move beyond these two systems, and when does this conceptual revolution occur in development? Most researchers agree that around the age of three years, children acquire a large *precise* number system. This system fully integrates the integer list and eventually enables the child to perform operations such as addition and division. The key developmental change appears to be the acquisition of words for numbers, and an understanding of their meaning. We can be confident that this idea is right, because children under the age of three have a good deal of linguistic competence, including an exquisite understanding of many words, together with the capacity to string them together into meaningful sentences. What they lack is an understanding of number words. For example, a two-and-a-half-year-old child can often run off the integer list with ballistic speed. She has, however, no understanding of what each number-word means. She may understand that the numbers one, two, three, and four are part of a list, but will incorrectly think that one refers to just one thing, whereas two, three, and four refer to any quantity other than one. Children growing up in cultures with distinct words for numbers associated with the integer list acquire a full understanding of their meanings at around three and a half to four years. Moreover, some cultures have languages with the expressive power and intricacies of English, French, and Chinese, but no words for numbers above two or five, at which point they simply indicate quantity by reference to the word equivalent of "many." In these cultures, number discrimination relies entirely on the small precise and large approximate systems. The development of a large precise number system does not depend on language in general. It depends on words for numbers specifically.[10]

The developmental history of the number system, though often viewed in isolation of other systems of the mind, is intimately tied to the moral sphere, and to cooperation in particular. Due to their limited number capacity, young children can't compute equalities with large numbers. With the exception of relatively gross inequities, they can't judge whether a transaction is fair. This conclusion relies, however, on a particular conception of the child's sense of fairness. If fair translates to *equal exchange*, then a precise quantificational system is necessary. In contrast, if the child considers as fair an exchange of *some* amount or quantity, then an approximate system will do fine. If the child's conception of fairness changes over development, from some to equal exchange, then the door is open for exploitation. Older children can exploit younger children by giving less than they should under an equal-exchange policy, while providing a fair exchange under the some-exchange policy.

To explore how young children think about the problem of fair sharing, the psychologist Robert Huntsman set up four tasks.[11] In each task, he asked a four-year-old child to decide on the distribution of a resource among self and other. In some tasks, the number of resources was greater than the number of potential recipients, and in some it was less. For example, the child had to distribute three ice cream cones among four children, or thirty candies among five children. In other tasks, the child had to decide whether some recipients were more deserving than others because they worked harder. When there are no constraints on the availability of resources, and no differences between recipients in deservedness, four- to eleven-year-old children distribute resources equally. When there are constraints on the distribution of resources, Huntsman found that the youngest children were selfish, taking more for themselves even when this left others with nothing at all.

In a study of three- to five-year-olds by Haidt and colleagues, two children played with blocks and then sat down with an experimenter who provided them with stickers as a reward for their play. For each pairing, one child always received fewer stickers in total than the other; to keep within this age range's numerical abilities, Haidt kept the number of stickers allocated in a round to four or less. At the end of the sticker distribution, the experimenter first asked each child to state how many stickers they had, and then asked whether things were okay or fair, looking for either verbal or nonverbal indicators of an inequity judgment. Results showed that for the individual receiving fewer stickers, even the youngest children immediately stated that the distribution was unfair; because Haidt never analyzed the magnitude of sticker inequity, and because the total number of stickers was never greater than four, it is not possible to assess whether children's responses derive from an equal- or some-exchange policy, as in the Huntsman experiments. For children receiving more stickers, a different pattern of response emerged: they seemed perfectly content with the situation. Of considerable interest, especially in terms of the competence underlying children's intuitions about fairness as opposed to their performance or what they would do if they had been in charge of distribution, is the observation that children rarely gave coherent explanations or justifications. For example, when Haidt asked one of the boys who received more stickers why this had happened, he replied, "We have to get four stickers, because my mom and dad like stickers." This suggests that children are equipped with mechanisms that enable them to rapidly and unconsciously evaluate the outcome of a distribution in terms of fairness. Children do not, however, have access to these mechanisms, and thus come up with after-the-fact rationalizations for the current outcome.

Young children often think that a fair deal consists of distributing *some* resources to everyone as opposed to *equal* resources to everyone. If this is their concept of fairness, then they are not acting selfishly. As children develop, they place greater emphasis on a recipient's need and merit. For example, children give more resources to those who work harder, and thus merit the goods on the basis of some primitive notion of justice as fairness. In these cases, however, it is unclear how children acquire an understanding of merit or what their particular conception of it actually is. A child might think, "I like those who work as hard as I do. I invest in those I like." There is no explicit sense of "merit" in this kind of thought bubble. Rather, the child taps an empathic response, one that matches feeling for feeling, action for action. Regardless of how this system works or develops, it plays a significant role in cooperation as individuals will be more likely to play with those who play like them. Such sentiments might emerge from direct experience with others (a round of reciprocation) or from indirect experience, such as watching others reciprocate. As the evolutionary biologist Richard Alexander insightfully put it in an early discussion of the biological roots of morality, "In indirect reciprocity the return is expected from someone other than the recipient of the beneficence. This return may come from essentially any individual or collection of individuals in the group. Indirect reciprocity involves reputation and status, and results in everyone in a social group continually being assessed and reassessed by interactants, past and potential, on the basis of their interactions with others."[12]

Huntsman, along with most other researchers working in this field, interpreted the decrease in selfishness over time as a consequence of socialization; in particular, the role of the child's culture, education, and peer group. Equality is like a Platonic ideal, something that children want for themselves and for others. But do they want it because of culture or despite of culture? Could a sense of fairness emerge in all children, cross-culturally, in the same way that facial hair in boys and breasts in girls emerge? Is fairness a human universal, a core part of our moral sense, a principle that operates under the radar of consciousness, but open to parametric switching by the local culture?

Most of the work on fairness focuses on children in industrial nations, typically among socioeconomically average families. With the exception of studies like Haidt's, most tend to focus on the child's actions and justifications, as opposed to her perception or comprehension of others' actions. But to understand what is developing and how, we must disentangle what the child perceives as fair from how the child chooses to act and justify her actions. We must also distinguish between the child's conception of fairness and the factors that come into play when the child acts fairly or not. For example, playing fair requires not only a conception of fairness but also the capacity to inhibit selfish desires. As I mentioned in the last chapter, however, the systems of the brain involved in inhibitory control develop slowly, on a maturational time course that continues through puberty and somewhat beyond. The inhibitory systems are not specifically a part of our moral faculty, but they interface with it and act as a constraint on its expression. The same is true of the systems that mediate our emotions, and especially those involved in regulating altruistic behavior, including both positive (empathy, sympathy) and negative sentiments (guilt, shame).

How does the child integrate its developing sense of fairness into games of cooperation, and especially reciprocation?[13] In one of the few studies designed to actually explore reciprocation with real commodities, the psychologist Linda Keil ran an experiment with seven- to twelve-year-old children. One child was told that he was watching a real-time video of a coworker sorting letters by zip code. Each sorted letter returned five cents. In some cases, the video revealed a child sorting at high speed and, at other times, a child sorting slowly. At the end of one round of sorting, an experimenter assembled all of the sorted letters into piles and asked one child to distribute the amount of money accrued. Young children distributed less fairly than older children, were more vengeful following an unfair allocation, and were less likely to reciprocate on the basis of the previous round.

The economist James Harbaugh and colleagues tested school-aged children on some of the classic bargaining games discussed in chapter 2.[14] Recall that in one-shot dictator games, adults give either half of the total starting amount or nothing at all; in one-shot ultimatum games, the modal offer is about 50 percent, with rejections of offers less than about 20 percent. When same-aged children play these games against each other, younger children make smaller offers than older children in both dictator and ultimatum games and, in the latter game, accept smaller offers as well. Even the youngest children, however, make smaller offers in the dictator game than in the ultimatum game, showing that they already think strategically about their offers and the rules of the game. Further, when children within an age-group play, shorter children make much larger offers than taller children. The developmental changes in the magnitude of the child's offer could be due to changes in the brain that underlie both how percentages are calculated—how the brain does math—and how this system interacts with the developing moral faculty. Height may provide a proxy for social dominance. This perception could be mediated by watching tall people dominate short people under certain circumstances. Overall, these results suggest that there are developmental changes in the temptation to cheat in games of cooperation, as well as changes in the perception of fairness.

Classically, the ultimatum and dictator games are played anonymously in every sense: neither player knows the other, nor will they meet in the future. It is possible, however, to give each player some information about the other that might influence their perception of trust and, thus, their strategic choice of options in the game. Trust is an important ingredient of stable cooperative relationships, and it is important to understand how this trait develops.

Harbaugh and Krause designed a reciprocation game for children, one focused on the role of trust and based on a game originally designed for adults. In the adult version, player 1 has the opportunity to share $40 down the middle or pass control to player 2. If player 1 passes control, then player 2 can either take $30 for himself and pass $15 to player 1, or share $50 down the middle. Although this is a one-shot game, with no opportunity for retaliation by either player, if player 1 passes control, he effectively trusts player 2 to do the right thing and split the $50. Player 2, however, is faced with the temptation to cheat, taking the larger pot at no personal cost, at least economically. Adults in the player 2 position typically choose the cooperative response and split the $50 pot. And the frequency of taking the cooperative split increases if adults sniff a spray of the hormone oxytocin, previously known as the cuddle hormone, due to its role in mother-infant bonding.

The pattern for children ages eight through seventeen playing a version of the trust game is generally the same as for adults, with two exceptions: The first player is less likely to pass his or her turn to the second player, and there is no relationship between the amount of money passed on by player 1 and the amount of money returned by player 2. The young child's motto: "I'll scratch your back a bit if you scratch mine a lot."

I draw two conclusions from these studies. First, children's sense of fairness is in play as early as four years old, probably earlier. Their sense of fairness is intuitive, based on an internal logic that they are only dimly aware of but that computes the payoffs of an exchange and then generates a permissibility judgment. Second, young children are more selfish than older children, even though young children have some sense of equitable sharing when there are no constraints on the distribution of resources. Moreover, the initial conception of fairness is more likely to follow the model of *some* distribution of resources, as distinct from *equal* distribution. What this means in terms of social interactions is that as soon as children are aware of their own and others' beliefs, they can both honor a commitment for fair distribution and give in to temptation by trying to cheat, or deceive another by lying. These results open the door to many interesting questions concerning the interface between our biologically endowed moral faculty and the local spin that each culture may impose on the details. For example, and as noted in chapter 2, though all cultures have some notion of fairness, as revealed by cross-cultural work on bargaining games, cultures differ in terms of where they set the different parameters. Nothing is known about the development of these cultural signatures. How much experience, and what kind of experience, is necessary before children act like the adults in their culture? Once a child has acquired the bargaining signature of his native culture, setting the relevant parameters, is acquiring the signature of a second culture like learning a second language, something that not only takes considerable time and effort but involves a process that is completely different from acquiring the first bargaining signature?

Much remains to be discovered in this corner of our moral faculty and its development. The conclusion I draw, however, is that the underlying competence mediating our sense of fairness is no different in children than it is in adults. What differs among children and adults lies outside the moral faculty. What grows in the child, and interacts with her moral judgments, are systems of self-control, emotion, numerical computation, and memory that allow for more accurate bookkeeping records.

BABY LIES

What's the difference between lying about your tax return for personal gain and receiving personal gains from your tax department due to a calculation error? Both entail personal gain. Both entail a change in the calculated numbers. But the first involves an intentional lie, while the second involves a failure to report an error. The first is a lie of commission, the second a lie of omission. Both are wrong, even though the general asymmetry between actions and omissions[15] makes the commission seem worse. Both are part of human nature, appearing early in development.

In every culture, lying, cheating, and deceiving are generally *verboten*. But as in most, perhaps all, forbidden actions, there are exceptions. Little white lies are almost certainly universally permissible. In these cases, we distort the truth without intending harm.[16] In fact, the intended outcome of a white lie is often to avoid harming someone else. Seen this way, white lies aren't lies at all, as they fit a more general rule of social communication: help, don't harm, the pool of intended listeners. Everyone has had the experience of lying to an annoying relative that the timing of a planned visit is bad because of a prior commitment; all parents have experience making up "stories" to block their child's curiosity or interest in engaging in some activity that is tedious or inappropriate; and most everyone has told a potential suitor that they are too busy, because conveying the real reason for rejecting an invitation—boring conversationalist, bad breath, heinous laugh, oily skin—is too hurtful. In order to recognize a lie or flag a cheater, the moral faculty must evaluate the causal and intentional aspects of a sequence of actions and consequences. Sometimes the outcome will be a judgment that lying was forbidden, and sometimes it will be judged as permissible, perhaps even obligatory.

The developmental psychologist James Russell suggests that deceptive "behaviour normally requires two distinct cognitive skills, namely, appreciating that false beliefs can be implanted into the mind of others, and suppressing what one knows to be true whilst expressing what is false . . . The first focuses upon the fact of deceit, the second upon the execution of a strategy."[17] Children eventually acquire both, but how?

To cheat and get away with it, the cheater must recognize when she is being watched, even if she doesn't make the connection between the perceptual act of seeing and the mental act of knowing.[18] Joint attention, as discussed in chapter 4, is an essential developmental milestone, occurring around fourteen months. It is not a specialized capacity for moral evaluation, but as a part of the support team, it plays into each individual's capacity to deliver morally meaningful decisions, including requirements to help others in need. Several months after joint attention develops in normal children, pretend play emerges, allowing them to think about alternative realities. Joint attention and pretense are stepping-stones to making inferences about what someone desires, knows, believes, or intends.

Pretense and shared attention are both pieces of the machinery employed during deception, though they, of course, are not specific to it, nor to the moral faculty. Added on to these are capacities that develop a bit later, including the ability to distinguish between appearance and reality, and to create a full-blown theory of mind. But deceptive maneuvers start early—before some of these add-ons—representing the telltale signature of a developing mind that sees differences between self and other.[19] These early forms of moral transgression turn into more serious ones, raising important questions about when children become aware of the offense, and its consequences for self and other. This, in turn, links the study of the child's developing moral psychology to the legal analysis and definition of a witness, and what it means to tell the truth. We will come back to this once we have fleshed out the psychology.

To disentangle the various factors that enter into an act of deception, and assess when they develop and color both judgment and behavior, consider an experiment with two- to four-year-old children, conducted by the developmental psychologist Michael Chandler and his colleagues. An experimenter informed each child that a puppet named Tony would help hide a bag of gold coins and jewels in one of four containers. Tony, however, was a messy puppet and always left a trail of steps. The goal was to hide the bag so that a second experimenter—presently out of the room—would not find it when he returned. Of interest was whether children of different ages would spontaneously try to mislead or deceive the experimenter.

All age groups deceived, including zipped lips about the bag's location, wiping Tony's trail, lying to the experimenter about the location of the bag, and, my favorite, destroying Tony's steps and adding in new ones to a false location. No one handed these children the Machiavellian playbook on deception. They spontaneously generated these acts in the service of misleading the experimenter. The children's behavior indicates a complicated suite of competences. For instance, they must know at some level that Tony's steps are a giveaway with respect to the hiding location. This could be a simple association, learned from their own experience of watching Tony hide the bag and then knowing where to find it. They must also know that when there are no tracks, there are no cues as to the location of the bag, as long as no one says anything. They also realize that they can use their own knowledge to mislead the experimenter, telling him to look elsewhere. And their memory must be good enough to recall where the bag was hidden, in order to point to an empty one. This suggests that they can simultaneously entertain two models of the world: a true one, associated with the bag's actual presence, and a false one, signified by the bag's ersatz location. Finally, most subjects failed to provide sufficient justifications for their deceptive acts, reinforcing the distinction between an early intuitive competence and the faculty that may consciously reason through the same conceptual terrain.

Are young children also sensitive to the circumstances in which lying is permissible? When do children distinguish between lying with malignant intent and lying to avoid harming another? By the age of four, children already understand that certain facial expressions should be concealed in certain social contexts. For example, in a study where children were promised a desirable toy for their good behavior but handed an unattractive one instead, they showed facial signs of disappointment if left on their own with this toy, but suppressed these facial expressions if the experimenter stayed in the room. These studies show an early sensitivity to communicative rules, guided in part by a desire to avoid hurting someone else's feelings.

To explore the emergence of white lies, the developmental psychologists Talwar and Lee tested three- to seven-year-old children on a lipstick task, one mimicking the logic of the mirror experiments for self-recognition discussed in the last chapter. For one group of children, an experimenter holding a Polaroid camera entered the test room with a red lipstick mark on her nose and asked whether she looked okay for a picture. Each child then took a picture of the experimenter, who left the room before the picture had developed. Next, a second experimenter entered and asked whether the previous person looked okay for a picture, both looking at and discussing the developed picture. As a control, a second group of children ran through the same procedure, but with no lipstick markings on the first experimenter.

Children of all ages were more likely to tell lies in group 1 than group 2. When the first experimenter asked about her appearance, children in group 1 lied about the lipstick, but then later told the second experimenter that this person did not look okay for a picture. The fact that children tell the truth to the second experimenter, but lie to the first, shows that they are not merely yes-kids in the face of authority.

The emergence of the capacity to tell white lies coincides nicely with the emergence of lies to cover moral transgressions, such as peeking under a cloth to see a toy when an experimenter explicitly forbids this action. Together, these studies show that our moral faculty is sensitive to contingencies, if-then rules, that allow for exceptions to moral rules about what is or isn't forbidden. These competences emerge early, presumably in every child, and without the help of teachers, parents, and other sages.

The fact that young children can sometimes engage in deception doesn't mean that their competence is fully fledged, reaching the level of sophistication that is characteristic of our mature competence. Dozens of studies show, and any parent's personal experience vindicates, the oft-sited ineptness of young children in pulling off a clearly crafted lie. In offshoots of the Tony studies discussed above, young children will often gaze at the location of the hidden goods, giving away their ruse; when asked whether they peeked under a forbidden cloth, or grabbed a cookie from the cookie jar, they will avoid eye contact and refuse to answer, failing to recognize that their cat-got-your-tongue expression is a sure giveaway. Often, when they attempt to deceive, they fail to understand the consequences of their actions—what laying down false trails does in terms of others' beliefs. These early forms of deception differ from their mature expression because of two missing ingredients. Neither ingredient is specific to deception or the moral sphere generally, and neither matures because of what parents teach their children or what they learn in school.

The first ingredient is the theory of mind module discussed in chapter 4. This core aspect of the human mind is on a maturational timetable, developing quite independently of culture and education, achieving various developmental milestones from the first year of life up until about ten years. Young children often botch their attempts at deception because they don't realize how their own actions alter the beliefs of others. They don't understand that the key to Machiavellian deception is recognizing when others have false beliefs. This capacity emerges some time after the fourth birthday, and, at this juncture, deception takes on a new complexion. The second ingredient, also discussed in chapter 4, is the mind's executive system. This system, linked to the frontal lobes of the brain, is directly responsible for controlling action, including the regulation of emotion.

The significance of linking deception to the fully fledged theory-of-mind system is that it provides the key connection back to our Rawlsian creature. When a Rawlsian creature evaluates an action vis-à-vis its permissibility, it is unconsciously and automatically assessing the causal and intentional aspects of the action and its consequences. What counts as a moral transgression depends upon the underlying cause. In chapter 1, I discussed a famous moral dilemma, originally constructed by the existentialist philosopher Jean-Paul Sartre. An individual faced two options concerning a promise to return a borrowed rifle following hunting season: keep to the promise and return the rifle knowing that its owner has been clinically diagnosed as a psychopath, or break the promise, keep the rifle, and save the potential lives of many who would otherwise be harmed by the psychopath. Most people's intuition is that lying in this situation is permissible, as it avoids the greater harm. What is central to this intuition is a judgment that the intent of the lie is a positive outcome—saving many—and not the negative one of breaking a promise. People—including young children—judging this situation would presumably also agree that if the owner was not a psychopath, it would be wrong to break the promise.

One arena where issues of competence meet issues of performance head-on is in a trial court. All legal systems are based on the capacity of some ruling body to distinguish between lies and truths. Whatever the mechanism, some presumably neutral judge must be able to question the accused and accompanying set of witnesses in order to get to the truth of the matter. In some situations, children may be the only witnesses in a case. In many countries, the number of children serving as witnesses has skyrocketed, with one U.S. estimate reporting approximately one hundred thousand per year since about 1990, and with ages as low as three years. How is the court to decide whether their reports are useful, honest representations of what actually happened?

In the United States and Canada, the court takes two steps before admitting a child to the witness stand. First, the judge asks a set of questions to evaluate whether the child understands the difference between lying and telling the truth, as well as the consequences of each. The goal is to ascertain whether the child has a commitment to the morally appropriate action of telling the truth. If the judge is satisfied by the child's answers, and is convinced that his or her memory of the case is sufficiently good, then on to step two: The child must promise to tell the truth. If both steps check out, then the child appears on the witness stand. In the absence of a reliable child witness, sexual offenders may never go behind bars.

The court's two-step procedure is based on some assumptions about the relationship between competence and performance. Is it the case, for example, that children who understand the difference between truth- and lie-telling are more likely to tell the truth? When children promise to tell the truth, do they understand what is entailed in making this promise? Does the promise work, promoting truth-telling? To test the validity of these assumptions, Talwar and her colleagues ran a series of experiments with three- to seven-year-old children. Results showed that all age groups recognized lies as moral transgressions, and often stated that the liar should come clean and tell the truth. When the hypothetical scenarios involved their own actions, children of all ages often lied to cover up personal transgressions. There are two ways to interpret this finding. On the one hand, children may have a general and early developing competence to distinguish lies from truths, but when it comes to their own actions (performance), and especially the capacity to report what they would do, they sometimes lie. The path from competence—recognizing a transgression—to performance—doing something about it—may not line up as parallel or integrated paths. Other faculties may intervene to guard against revealing the truth. Alternatively, although not necessarily exclusively of the first, children may understand the general rule that lying is forbidden, but simply not follow through on the Kantian categorical imperative: If it's not okay for others to lie, then it's not okay for me to lie. On either account, the child's inconsistency reveals a flaw in the court's procedure: The fact that a child recognizes the difference between lying and telling the truth, and sees truth as anchoring the higher moral ground, doesn't mean that they will reach for this turf when it comes to their own actions. The court's procedure is flawed. Step one provides no guarantee that they will tell the truth.

The media attention to the shocking cases of pedophilia among the clergy, which peaked at the turn of the new millennium, have raised the visibility of cases of sexual abuse. These cases also raise the stakes on the court's evaluative machinery: To what extent do young children lie to cover up transgressions by individuals such as priests and parents that they are close to, and do they recognize not only that they are lying but the consequences of their distortions for self and others? As Talwar and colleagues point out, this kind of lying often takes the form of omitting information when questioned; omission of information can take the form of pleading ignorance or providing false information to cover up the truth. The omission of information need not carry any moral burden at all. When someone tells me something in confidence, we share secretive information. Sometimes, it would be wrong to tell someone else about this information, even if questioned by an authority. For example, if a friend tells me about an idea he has for a patent, something that will make him rich, it would be wrong for me to tell others, given that I was sworn to secrecy. On the other hand, if an adult sexually abuses a child and then tells the child to keep this a secret, it seems morally permissible, perhaps obligatory, for the child to report the abuse to an authority. Once again, a rigid deontological stance is problematic, because it is sometimes permissible to lie, breaking a promise to keep a secret. The intention of the liar and promise-breaker is essential.

Consider the patent and sexual-abuse cases together. Breaking the promise is, in some sense, designed to harm the person to whom the promise was made. In the case of the patent, the promise-breaker is giving away top-secret information that will ruin the inventor's chances of striking it rich. It is hard to read this in any but the most malicious sense. In contrast, although the promise-breaking child also intends to harm the person to whom the promise was made, we read this in a positive sense—as due justice to the sexual abuser. The exception to the no-promise-breaking rule goes back to the validity of the initial promise. The sexual abuser never had the right to request a promise of secrecy, as his actions harmed the child. One can't ask another to maintain a promise if the target event constitutes a moral transgression. Even this deontological principle will have exceptions, presumably formulated in terms of modifiable parameters.

Talwar and colleagues replicated the sequence of events that typically transpire when children witness a moral transgression and are then called into court as witnesses: seeing the event, being questioned about the event by a social worker, being interviewed by a member of the court to ascertain understanding of moral distinctions, and being asked by the court to tell the truth. In the two social conditions, children between the ages of three and eleven years watched as a parent broke a puppet as soon as an experimenter left the test room. The parent then showed signs of distress about the puppet and told the child not to tell anyone about it. Once the child agreed to keeping this secret, the experimenter returned and asked a series of questions about the puppet, with the parent either present or absent. In the nonsocial condition, the same-age children entered a room, saw their mom or dad sitting next to a broken puppet, and were then told what happened, being asked to keep this a secret. The two social conditions were designed to mimic courtroom procedures, where young children are often accompanied by their parents. Is there an audience effect, such that children are more or less likely to cover up a transgression when a parent is present? The nonsocial condition was designed to remove the possibility that the experimenter might hold the child responsible for damage to the puppet. It is not uncommon for children to think that authority figures will hold them, as opposed to an adult, responsible for a transgression even when they are innocent. In a follow-up interview, a second experimenter again checked the child's understanding of lying and truth-telling, and then asked for a promise to tell the truth when discussing the event.

In the nonsocial condition, children of all ages were more likely to lie about the puppet to both the first and second experimenter. When children are clear about the responsible agent in a moral transgression, and know that they won't be blamed for the outcome, they are more likely to lie in both the social worker– and court-equivalent interviews. In this sense, lying is a selfish affair, reserved for situations where it is personally beneficial, suppressed when it may do personal harm. When children lie in the context of parental transgression, their motives may not be altruistic!

Surprisingly, having a parent present or absent did not affect what children said in the interview. Children of all ages were as likely to lie with a parent hawkishly staring at them as if this same parent was off in another room. Of further relevance to the court's procedures is the fact that in all three conditions, children were more likely to tell the truth during the second interview than the first. This sounds an important alarm: Social workers are likely to hear lies from abused children, and court workers are likely to hear the truth. The main reason for this difference seems to be that the court's competence examination, designed to show what young children understand about lying and moral transgressions more generally, is like a truth-eliciting serum. When children are forced to explicitly distinguish between lying and telling the truth, truth prevails.

The fact that children with confidence about the source of the transgression were more likely to lie should not take away from the fact that most children, independently of condition, told the truth. Most children ratted on their parents. Thus, even in the face of an authority figure breathing down on them, children told the truth. What is unclear from this work is how much it can be extended to other, more emotionally salient moral dilemmas. In these experiments, parents never tell their children about the consequences of breaking a promise to keep a secret. The negative consequences are never articulated. In the case of sexual abuse, abusers may often threaten the victim with severe consequences should they divulge the secret: "If you tell, I will go to jail and we will never be together again." Further, the child is personally involved in the secret in the case of sexual abuse, but not in the case of the puppet experiments. These caveats aside, results from studies of the child's developing moral competence indicate that our courts are putting many incompetent children on the witness stand. The mismatch between moral competence and performance is significant. Legal scholars, take note.

PINOCCHIO'S NOSE

In Jane Austen's *Mansfield Park*, Mary Crawford pronounces: "Selfishness must always be forgiven . . . because there is no hope for a cure."[20] I began chapter 4 with a similar pronouncement, but one derived from developmental and evolutionary biology. The Haigian view of the fetus is the opposite of a passive sponge waiting to absorb what Mom has to offer. Our little fetus is a warrior, a selfish, resource-sucking machine. Cooperation and reciprocal exchanges are but one of the many contexts in which we confront the temptation to cheat. But there is also a part of us that is fair, genuinely altruistic, and set up to punish those who try to destroy the inner circle of cooperators. This piece of our psychology may well be uniquely human. It evolved as a counterattack on cheaters and is a core part of our moral faculty.

The economic games of cooperation that I discussed in this chapter as well as chapter 2 are social, even when they are played only once and anonymously. As games, they show that humans are good at detecting free riders, cheaters who try to take advantage of the goodwill of others. Humans are good at all sorts of things. Simply because we are good at ferreting out cheaters doesn't mean that our minds evolved a specialized ability to do so. It could be that our ability to detect cheaters is part of a more general ability to figure out when someone or something has violated a rule. To determine if a rule has been violated, we work our way through the inferences, check the assumptions, and assess what is true or false. Solving problems in formal logic provides one entry point into our general reasoning abilities.

Logic is a beautiful form of mathematics, because it abstracts away from the noise that context and language bring to the table, yielding abstract variables and their relationships. For example, consider two statements called P and Q, respectively. If I state that P is a true statement, and P implies Q, it follows that Q is a true statement. Similarly, if I tell you that either P or Q is a true statement, and then state that P is true, you should logically conclude that Q is false. Fill in whatever you like for the Ps and Qs. Logic rules. Here's an example:

Joe is sleeping is true. *Joe is sleeping* implies that *Joe is unaware that his dog is eating food from the refrigerator. Joe is unaware that his dog is eating food from the refrigerator* is true.

Solving problems of logical inference is surprisingly not our forte. Most college students experience mental pain and anguish when taking courses in logic. Consider the following problem and diagnosis:

PROBLEM: You are a detective investigating a recently reported murder. The policeman on the scene hands you the following information. Three suspects have been detained: Fred, Bill, and Joe. If Fred is not guilty, then Bill and Joe are guilty. Considering just Bill and Joe, and the evidence at hand, only one of the two is not guilty. Either Fred is not guilty or Bill is guilty. Who, therefore, is guilty?

DIAGNOSIS: Few people get the answer right, straight off. It often takes several passes, as well as pen and paper, to work out the IFs, EITHERs, and ORs. The answer is that Fred and Bill are guilty. The second statement says that either Bill is not guilty or Joe is not guilty. Consequently, we can return to the first statement and conclude that Fred must be guilty because Bill and Joe can't both be guilty. If Fred is guilty, then Bill must also be guilty. For many, this sounds like word salad.

There is a vast literature on the deficits in our reasoning capacities, much of it focused on the ways in which our thoughts get tied up in knots when we are confronted with statistics. This suggests that unlike walking, seeing, or hearing—things we do easily, without instruction or years of education—reasoning is hard and requires experience and often explicit tutelage. This view of our general reasoning abilities is accurate. It misses out, however, on the possibility that our minds evolved more efficient reasoning abilities for specific kinds of problems. It is this kind of possibility that provides our first serious step into the world of reasoned judgments without explicit reasoning, a world that is at the core of our moral faculty.

The evolutionary psychologists Leda Cosmides and John Tooby suggest that there is one context in which our reasoning abilities are like fine-tuned machines, superficially giving the appearance that we are all wiz kids at logical inference. It is precisely the context that our hominid ancestors confronted, and most likely worried about: social contracts and systems of exchange. Back to Trivers and the concept of reciprocal altruism. If you turn the inference "if P, then Q" into a rule involving a social contract, then run-of-the-mill humans, as opposed to formally trained logicians, find such problems trivially easy. Cosmides and Tooby's insight was that our minds evolved the capacity to solve socially relevant problems, such as detecting cheaters who violate rules. Since reciprocation depends critically on fair exchange, and since stable reciprocation depends upon ferreting out those who renege on their promises or commitments and subsequently punish them, it is likely that evolution equipped us with a specialized ability to work through the cost-benefit analysis of a social contract. To test their intuition, they borrowed a task from the psychologist Peter Wason, known as the "Wason Selection Task." The task was originally designed to explore our capacity to solve logical inferences, and, specifically, conditional relationships of the form "if P then Q." To illustrate the problem, consider a classic case and one transformed to tap the logic of social contracts.

> CLASSIC: You have a deck of cards, which, unlike regular playing
> cards, have a letter on one side and a number on the other. An
> experimenter removes four cards and places them in front of
> you as follows:

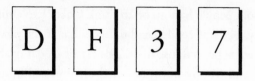

The following rule, which may be true or false, applies to these cards: If there is a D on one side of the card, then there is a 3 on the other side of the card. To decide whether the rule is true or false, which card or cards do you turn over?

> SOCIAL CONTRACT: You have been hired as a bouncer in a bar and you must enforce the following rule: If a person is drinking beer, he or she must be over twenty-one years old. The cards below represent four people at the bar. One side of the card says what the person is drinking, and the other side of the card says how old the person is. Which cards do you have to turn over to ensure that the rule has been enforced?

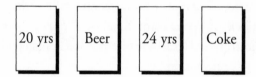

Most people find the first problem harder than the second. The answer to the first is: cards D and 7, because for any number other than 3 on the other side of the D card, there is a rule violation; if the 7 card has a D, the rule is violated. There is no need to turn over the F card, because the rule does not stipulate conditions for cards marked with an F. Similarly, although the rule states that all D cards must have a 3 on the other side, it doesn't say that all cards marked with a 3 must be D and only D; when people make errors on this kind of problem, they neglect to select the 7 card.

The answer to the social-contract question is: cards "20 yrs" and "Beer." We violate the rule when individuals under twenty-one years drink beer and when individuals drinking beer are less than twenty-one years. The rule does not stipulate what twenty-four-year-olds can drink, nor does it stipulate how old people have to be to drink Coke. Explaining this second rule seems silly, perhaps condescending. But that is the point. The social-contract problem appears transparent. The classic case is opaque. Why?

Cosmides and Tooby have mounted an impressive amount of evidence to support their claim that problems involving social contracts tap a specialization that is present in all human beings. The capacity would turn everyone into a Pinocchio, unable to hide their lies. The ability may be so trivial that cheaters might as well have long noses. The reason why subjects do better on the modified version of the Wason selection task is because they read the problem as a social contract, one involving a commitment. Finding the cheater is trivial, because we have evolved a mechanism to look for someone who takes the benefit without meeting the requirement associated with the initial commitment. In the case above, a person is a cheater if he drinks beer (benefit) and is under twenty-one (hasn't met the requirement). Putting these bits together, the reason why humans are better at solving this form of the Wason selection task is because our minds evolved a unique specialization to both understand social contracts and detect violations. Social contracts, whether stated over a beer or written in legalese, are commitments. They engage trust. Violating them engages distrust and a cascade of emotions designed to enhance vigilance and catalyze retribution.

Cosmides and Tooby carried their argument further. Not only is our capacity to detect cheaters universal, part of what evolution handed down to us in the context of negotiating over social contracts, but it is a specialized system of thought that operates unconsciously and automatically. When subjects read the social-contract version of the Wason task, it triggers, reflex-like, a mental search for the cheater. It's as if the social-contract problem played out like a "Where's Waldo?" scene, with the mind set to find the red-and-white striped shirt; with this search image cued up, Waldo pops out. The fact that we perceive these situations automatically doesn't mean that they are not susceptible to after-the-fact reasoning. We may have a rapid and automatic intuition about the situation, make a judgment, and then reassess. Our assessments may also change based on differences in context. But these effects do not alter the fact that there is an initial and often accurate judgment of whether someone has cheated.

Evolutionary relevance may be just one of a handful of contexts that can influence performance on the Wason selection task.[21] If other contexts also improve performance, then perhaps our facility with social-contract problems is merely one example of a more general facility with contextually relevant information. The cognitive anthropologist Dan Sperber and the psychologist Vittorio Girotto argue that people's performance greatly improves when there is some kind of payoff to finding the violation, and where the context's relevance depends on understanding the speaker's intent—what he or she wishes to convey. In such cases, the communicative message is transparent and easy to understand, and subjects appear to reason through the logical inferences without effort.

Let's return to the general conditional relationship of "if P, then Q." This is in some ways an ambiguous claim, because there are many different ways in which it can be represented. It logically follows from "if P, then Q" that "not-P or Q." In reasoning experiments, however, few subjects come up with this alternative way of representing the same conditional. To show how a simple change in wording makes one case easy and the other hard, consider the following two statements, which both stick to the same logical relationships but with different content: (a) is the straightforward "if P, then Q" case and (b) is the alternative representation "not-P or Q":

1a. If you sleep with my wife, I will kill you
1b. Don't sleep with my wife, or I will kill you

2a. If you are over twenty-one, then you can drink alcohol
2b. You are not over twenty-one, or you can drink alcohol

The first problem is trivially easy to understand. The second statement, 1b, seems to follow naturally and obviously from the first statement, 1a. In the second problem, the two statements seem utterly disconnected. What is it about the wording of problem 1 but not 2 that makes the difference? How do you get people to see that a conditional of the form "if P then Q" can be written in different ways? What Sperber and Girotto suggest is that by writing the case as a denial, people readily reason through the problem and search for the alternative representation of "not-P or Q." Moreover, when the conditional is written as a statement concerning moral rights, people interpret "if P then Q" as a claim about what is forbidden (i.e., "P and not-Q" is equivalent to cheating). Thus, Sperber and Girotto's point is that our capacity to detect violations in a conditional statement of the form "if P then Q" does not depend upon a specific ability to detect cheaters. Rather, by making the context relevant, the speaker's intent clear, and alternative representations of the conditional transparent, people perform exquisitely on many variants of the Wason task.

Cosmides, Tooby, and their colleagues have volleyed three specific responses to their opponents. First, they argue that studies showing enhanced performance on the Wason task by means of manipulating other, nonsocial contexts are irrelevant. They are irrelevant because the primary intuition is about an evolved adaptation to solve reasoning problems of social exchange and cheater detection. The possibility that there are other reasoning mechanisms for other problems is a separate issue. Further, it is possible that we evolved a specialized capacity for thinking about social contracts and detecting cheaters that was subsequently borrowed by the brain for other kinds of problem-solving, which emerged during our evolutionary history.

The second rebuttal comes from a series of studies involving important manipulations of the original Wason task. In an experiment by the evolutionary psychologists Gigerenzer and Hug, each subject read the following statement: "If an employee gets a pension, then that employee must have worked for the firm for at least ten years." The experimenter told half of the subjects that they should read the question as if they were employers (bosses), and the other half as if they were employees. Although each read the same question, the employees looked for cheating employers, and employers looked for cheating employees. Thus perspective alone flips the focus on particular kinds of violations. In an experiment by Cosmides and Tooby, subjects read the generic statement: "If you X, you must wear a piece of volcanic rock around your ankle." To figure out which card to turn over, you should look for people who are doing X and those who don't have a volcanic rock around their ankle. When X is some kind of reward, like going out at night to a party, people perform well on the task, turning over the violations. When X is not a reward, or is a punishment, people perform poorly. Together, these results seem to call into question Sperber and Girotto's line of criticism, as the only thing that changes subject's performance is the structure of the rewards and punishments relevant to some social contract.

The third response comes from a patient (RM) with damage to the circuitry connecting the base of the frontal lobes through to the amygdala.[22] As mentioned in chapter 4, this part of the frontal lobes is involved in inhibitory control and reward-processing, while the amygdala is significantly involved in emotional analyses. The cognitive neuroscientist Valerie Stone and her colleagues presented patient RM with two forms of the Wason task: a social-contract condition as before, and a precaution-rule condition. The social-contract condition always takes the generic form "If you take the benefit, then you meet the requirement." In contrast, the precaution rule takes the generic form "If you take a risky action, then you take a precautionary action first." An example of a precautionary rule is "If you jump off a cliff, you first attach to a bungee cord." Although both of these rules fall under the broader category of permissions,[23] we would obtain evidence for neural specialization by observing a loss of ability to solve one rule but not the other; in the neurosciences, this kind of distinction or dissociation is the telltale sign of two different circuits, each designed for a different function. Patient RM looks like normal subjects on precautions, but falls 40 percent below normal on social contracts. If reasoning is a domain-general, content-free system, then this kind of deficit can't be explained. The only way to account for patient RM's performance is by invoking different kinds of reasoning driven by different kinds of neural circuits. This is not a variant of the old adage "different strokes for different folks." Rather, it is an argument about the design of all human brains. Whether you are a hunter-gatherer living on the savanna or wear a suit on Wall Street, your brain is running different reasoning software for social contracts and precautions.

What about development? Paralleling our early competence for deceiving others, do we also show an early competence for detecting cheaters, individuals who fail to follow social conventions and moral rules? Early in life, parents and teachers bombard children with rules. Some are straight-forward: never hit your brother, comb your hair, don't pick your nose. Others are more complicated, involving conditionals that mirror those discussed above: if you eat your vegetables you can have dessert; if you clean up your room, I will take you to the movies; if you keep your seat belt on, we will go on an adventure. If Cosmides and Tooby are right that this domain of reasoning is part of an evolved specialization, then young children should show early competences in solving conditionals of the kind that mirror the Wason card-selection task. In contrast, if this form of reasoning is more domain-general, part of our general capacity to solve problems of induction, driven by exposure to the multitude of permis-sion rules, then the child's capacity should emerge gradually and show more substantial cross-cultural variation.

In a set of studies designed to directly extend work on adult com-petences, the developmental psychologists Nunez and Harris ran a child-friendly version of the Wason card-selection task with three- to four-year-olds.[24] Children listened to several stories, each with a similar format: a child wants to do some target action, but first she must com-ply with a parental request to do something else. Each story is therefore set up as a permission rule with a key conditional statement: if you take the benefit, you must satisfy a prior condition. Following the story, the ex-perimenter lays out four pictures, describes each one, and then asks the child to both identify the one representing a violation of the parental rule and explain why this is the case. For example: This is a story about Sam. One day, Sam wants to play outside. His mom says that if he goes outside to play, he must wear a hat. Here are four pictures of Sam. In this picture (pointing to top left), Sam is inside his house and is wear-ing a hat . . . (continue with description of each picture.) Point to the picture where Sam is being naughty. Why is Sam being naughty in this picture?

The task was simple for both three- and four-year-olds. Not only did they identify the naughty child in familiar cases, such as Sam wearing a hat to play outside, but unfamiliar cases as well, including a mother's request for her daughter to wear a helmet when painting indoors. Children at this age were also able to connect their knowledge of prescriptive permission rules to their understanding of others' mental states by distinguishing

between intentional and accidental violations of the rule. For example, Sam's mom asks him to wear a hat if he plays outside. In one picture, Sam is outside and takes off his hat while playing; in a second picture, the wind blows Sam's hat off while he is playing outside. Children recognize that only the first picture involves a violation. They understand that in moral evaluation what matters is not merely consequences, but the means by which they are attained. These data are accompanied by others, in which both physically and psychologically harmful actions, in either common or uncommon situations, are judged by young children as wrong, and these evaluations depend on the agent's intentions. For example, by about three years of age, children recognize that if an act causes harm, but the intention was good, then the act is judged less severely than when the intention was bad and designed to harm. Together, these observations indicate an early sensitivity to the underlying psychology of the agent, as opposed to the surface-level features of his behavior.

Contrary to both Piaget and Kohlberg, who explicitly denied early competences in the moral domain that were dependent upon a recognition of the agent's intent, these studies show that young children are equipped with the ability to identify cheaters, looking at the causal and intentional aspects of the event before weighing in with their moral verdict. Though limited in scope, these findings are in line with Cosmides and Tooby's predictions: Rather than a learned capacity, handed down from parental tutelage with permission rules, it appears that our ability to detect cheaters who violate social norms is one of nature's gifts.

COMPASSIONATE COOPERATION

The novelist Dorothy Sayers noted: "Envy is the great leveler: if it cannot level things up, it will level them down . . . rather than have anyone happier than itself, it will see us all miserable together."[25] Envy is a universal emotion, one tied to spite, and often a source of embarrassment and shame when recognized in oneself. Unlike its emotional sister, jealousy, envy has received far less critical attention. In a recently published series on the seven deadly sins, the essayist Joseph Epstein writes: "The origins of envy, like those of wisdom, are unknown, a mystery. People confident of their religion might say envy is owing to original sin, part of the baggage checked through on the way out of the Garden of Eden. The Bible is filled with stories of envy, some acted out, many subdued. Of the essence of envy is its clandestinity, its surreptitiousness." In a similar tone, the psychiatrist William Gaylin states: "Envy may indeed be a useless emotion. It seems to serve none of the purposes of other emotions. Unlike the emergency emotions of fear and rage, it does not serve survival; unlike pride and joy, it does not serve aspiration, achievement, or the quality of life; unlike guilt and shame, it does not serve conscience or community. It does not alert, liberate, or enrich us."[26]

The ethnographic literature on hunter-gatherers, together with studies in experimental economics and evolutionary psychology, suggest that Gaylin's diagnosis of envy is exactly backwards. Envy is useful, serving a key role in survival, motivating achievement, serving the conscience of self and other, and alerting us to inequities that, if fueled, can lead to escalated violence. Since envious people are a source of threat, addressing their concerns may be one way to avoid escalation and redress the imbalance. Seeing envy in this light does not deny Gaylin's perfectly correct conclusion that envy is a source of significant discontent and trouble. As Shakespeare expressed in *Henry VI*, "When Envy breeds unkind division, there comes the ruin, there begins confusion."

A first step in understanding the adaptive logic of envy—part of the Humean creature's toolkit—comes from seeing how it differs from jealousy. Whereas envy is strictly triggered by an inequity or disparity in the possession of valued resources, jealousy is triggered when one individual poses a threat, imagined or real, to an established relationship; we typically think of the relationship as romantic, but it need not be. Envy has therefore evolved in response to perceived inequities, capable of fueling competition in order to reestablish balance. In highly egalitarian hunter-gatherer societies, numerous mechanisms have evolved to maintain equity. Feeling envious and paying the costs to destroy someone else's reputation would be one way of alerting others to the start of a potentially volatile situation. As I discussed in chapter 2, a hunter-gatherer has violated a social norm if he returns from a hunt and boasts about his successes. A mild form of envy leads to gossip and often mockery as a relatively cheap mechanism to redress the inequity brought about by boasting. As Oscar Wilde mused, "Gossip is charming! History is merely gossip. But scandal is gossip made tedious by morality."[27]

In more contrived situations, such as the laboratories of experimental economists, subjects offered a raw deal, something substantially lower than a fifty-fifty split, act spitefully by rejecting the offer, incurring a personal loss but imposing an even larger one on their opponent. When individuals are winning a disproportionate amount of the resources in an economic game, losers are willing to spend a large amount of their own earnings to destroy the gains accrued by the winners. Envy may therefore act as a catalyst to reduce inequities. But unlike hunter-gatherer societies, where each person's reputation is well known throughout the group, most of us don't live in such fishbowl communities today. This demographic change may be partially responsible for the more nefarious consequences of envy, and the fact that it goes unchecked. Fueled by temptation, envy may level innocent individuals who have done no more than work hard for their earnings.

The action movie *Lara Croft: Tomb Raider* presents a tasty dilemma between the forces of good and evil; specifically, the emotional tug-of-war that arises in the context of a cooperative venture with helpers and cheaters. Croft—symbolizing good—recalls a conversation with her father about an ancient key that, if found and used during the planetary alignments, would unlock an extraordinary power capable of controlling time. Croft finds the first of three hidden pieces, which are then stolen by a secret group of elder statesmen—symbolizing evil. Each side knows that the other wants control, and each side has something that the other needs. Croft convinces the elders that they need her, that she has knowledge that will help locate the other pieces in time for the planetary alignment. A Faustian member of the elders offers Croft an opportunity to travel back in time and reunite with her dead father, something she desperately wants. They agree to cooperate. A commitment is made. The costs of defection are made explicit. Action! The elders have no intent to cooperate. Neither does Croft. Each side is using the other for selfish means. Croft's motivation is good. The elders' motivation is evil. Not surprisingly, good prevails, with Croft destroying the key and a few of the elders along the way.

While *Tomb Raider* is fiction, it captures a common source of conflict: commitment. This problem—a generic conflict between self-interest and cooperation—emerges in a variety of social arenas, including political relationships, investment deals, coordinated bank robberies, marriages, friendships, business collaborations, and even the kidnapper and kidnapped. Imagine a kidnapper who suddenly realizes, after weeks of detaining his victim, that there are personal costs to getting caught. He has a change of heart and contemplates letting the victim go. But he then realizes that the victim may go to the police. The victim assures the kidnapper that he will never mention the incident to anyone, including the police. It is in the victim's interest to be involuntarily committed to silence. Both realize, however, that once the victim is free, nothing binds him to secrecy. The kidnapper therefore concludes that he must kill the victim.

A commitment requires trust, which requires evaluative machinery that can detect cheaters. Detecting cheaters requires not only logical inference, as discussed earlier, but a method of reading emotions. In the last few years, a number of social scientists and evolutionary biologists, inspired by the Nobel laureate Thomas Schelling and Robert Frank, have stressed the importance of emotions in stabilizing cooperative relationships and anchoring commitment.[28] Emotions provide an involuntary mechanism for creating the equivalent of a binding contract. We can illuminate the force of this idea by reintroducing *Homo economicus*. Members of this self-interested species adhere to commitments and other social norms because of the ever-pending threat of punishment. Should they find themselves alone, or confident that no one would ever catch them in the act of norm violation, they would never feel guilt, shame, or embarrassment. As we learned in the last chapter, such individuals exist: aberrations of human nature called psychopaths.

Let's say that I have agreed to help a friend defend himself against the neighborhood bully who frequently goes around thumping innocent and weaker children on the block. One day, I notice that my friend is about to encounter the bully. I recall my commitment to help. From a selfish perspective, I may be tempted to renege, because of the costs associated with fighting. If selfishness is the winning psychology, then selection should favor an immunity to emotions that might compel one to act otherwise. In contrast, if emotions play a more powerful biasing role, and there are advantages to feeling good about cooperation and bad about defection, I should feel guilty about reneging. Feeling guilty should compel me to help my friend. Empathy toward him, and what he will feel like when thumped by the bully, should facilitate cooperation. Both sides of this conflict are real: selfish drives to defect and emotional leashes to stabilize cooperation.

To what extent are our emotions a safeguard against cheating, lying, defecting, and breaking commitments? Do feelings such as guilt, empathy, embarrassment, loyalty, envy, anger, and disgust provide a prophylactic against selfishness? Frank unambiguously claims that feelings commit us to act, providing the motivating force. It would be easy to leave a restaurant without tipping. But doing so might lead to feelings of guilt, shame, and embarrassment. These feelings may cause one's sense of honesty to decay, leading to further self-interested actions. Let's look at the psychology and neurobiology of emotions that are engaged in strategic actions involving material gains.

When we enter into a cooperative venture with another person, we use an assessment of trust to both launch and maintain the relationship. If there is information about past performance, this can be used to assess the odds of cooperation or defection. If you are playing an ultimatum game with someone who has rejected all offers under 30 percent of the initial pot, it is to your advantage to offer a larger proportion. If you lowball this player, the odds of rejection are high. What if there is no information about prior performance, but an image of a face?[29] Is there something familiar and trusting about that face? If the experimenter creates an image by morphing someone else's face into yours so that it looks more like you, you are more likely to trust them. Since selection favors altruism toward kin—nepotism—it will also favor mechanisms that enable kin recognition. The fact that people are more likely to trust those who look like them suggests that trust and kinship are correlated. We may feel more positive and willing to take risks with those who share genes in common with us.

Actions can follow from or lead to emotions. When we do something that, upon reflection, appears wrong, we may feel shame, guilt, disgust, or embarrassment. Sometimes, these emotions are instructive and can fuel change. Sometimes they can cause us to do the right thing. Consider guilt, an emotion we feel when we harm someone in a social setting that is characterized by mutual concern. Guilt is often triggered when we cheat and recognize the consequences of our act. But guilt may also play a stabilizing role, reversing an instability caused by deception. When people play repeated sessions of bargaining games such as the ultimatum game, those who admittedly feel guilty are more likely to cooperate in future rounds. Guilt is an emotion that jumps in for damage control, a prediction made almost thirty years ago by Robert Trivers.[30]

What's happening in the brain when we cooperate or defect? Neuroeconomics—a newly emerging field that fuses the technology of brain imaging with the theories and methods of classical experimental economics—has begun to provide some of the answers.[31]

The anthropologist James Rilling placed subjects in a scanner and watched their brain activation while they played repeated rounds of a prisoner's dilemma game. In this game, the payoff for defection is highest when the other person cooperates. This creates the temptation to defect, even though if both defect, both obtain lower payoffs than if both cooperate. Each subject played alternating rounds against an anonymous human actor secretly told to defect after three rounds of cooperation, and a computer playing tit-for-tat—a strategy that starts out nice and cooperative, then matches the opponent's move from then on, taking offense at defection, and responding in kind. When playing against another human, there was more activation in the striatum and orbitofrontal cortex than when playing against the other three opponents; as mentioned in the last chapter, both of these areas play a significant role in processing reward. Only mutual cooperation with a computer partner activated the orbitofrontal cortex. Reciprocating a partner's previous cooperation increased activation in areas involved in reward assessment as well as conflict resolution.

When reciprocity fails or the offer is unfair, imaging studies reveal significant activation of the anterior insula, an address of the brain known to play a role in negative emotions such as pain, distress, anger, and especially disgust. How interesting that cheaters might be considered disgusting. Equally interesting is the fact that when subjects engage in altruistic punishment of the kind described in chapter 2, paying a personal cost in order to impose a larger cost on someone else, the punisher experiences relief and satisfaction, evidenced by activation of the caudate nucleus, a key center for processing rewarding experiences.[32] When we punish, our brains secretly relish the process. Emotions are critically involved in our strategic decisions to cooperate and punish those who cheat.

In the final chapter of his 1988 book *Passions Within Reason*, Frank concludes that people "often do not behave as predicted by the self-interest model. We vote, we return lost wallets, we do not disconnect the catalytic converters on our cars, we donate bone marrow, we give money to charity, we bear costs in the name of fairness, we act selflessly in love relationships; some of us even risk our lives to save perfect strangers." All of this is true, but look at the proportion of any population enacting these altruistic, emotionally mediated actions, and the picture looks different.

Some people do risk their lives to save strangers. It happens in war, and in much more mundane arenas. But when someone jumps into a lake to save a drowning child, or steps in front of gunfire to save a general, these events make news not only because they are heroic but because they are rare. Opportunities to help others arise often, and are, more often than not, ignored.

Donations to charity present an equally grim picture in terms of emotions overriding the ruthlessness of the material world. Frank points to the fact that countries such as the United States contribute about $100 billion a year to charity. But a closer look at who contributes suggests a different picture of generosity. In several countries, most noticeably Britain, there is an inverse correlation between the level of donations and wealth: the richest give relatively little. In the United Kingdom, only about 30 percent of households contribute to charity, and a majority of these donations involve little cost or commitment. In the United States, almost 50 percent of all charitable donations go to the individual's church, a contribution that could be seen as selfish given that churchgoers get back personal returns on their gifts.

Bone-marrow donations show a similar trend. A 2003 survey by the Bone Marrow Donors Worldwide revealed a wide range in the number of people donating, with only 3 Austrians but 3 million Americans giving. Although I could not find any statistics detailing the number of people contacted and then rejected, it is clear that sites collecting bone marrow can't keep up; the same applies to organ donors. Although people feel good when they donate to charity or return a wallet, and feel bad when they bypass these opportunities, most people ignore flyers or announcements asking for charitable contributions, and many people walk right by a wallet on the street, or take the cash. In a comparative study of responses to finding another person's property, results showed that Japanese residents in Tokyo were far more likely to return property than American residents in New York City.[33] The Japanese legal system provides cash incentives for returning lost items, and punitive measures for keeping them.

I do not doubt that in the absence of morally relevant emotions, selfishness would run rampant. I also agree with Frank that the classical economic view of our species is false, and that emotions help explain why we are sometimes generous, and often cooperative. What I do doubt is Frank's optimism that emotions provide the ultimate safeguard, shepherding us through the temptation to acquire greater material wealth. We are a hybrid species, the fertile offspring of *Homo economicus* and *Homo reciprocans*. More often than not, we give in to selfish temptation. We are a lopsided hybrid. As Herman Melville astutely pointed out, "It is a very common error of some unscrupulously infidel-minded, selfish, unprincipled, or downright knavish men, to suppose that believing men, or benevolent-hearted men, or good men, do not know enough to be unscrupulously selfish, do not know enough to be unscrupulous knaves."[34]

NAVIGATING NORMS

Every society is founded on a set of norms—informal and often unstated expectations about how people ought to behave. All societies have at least two norms of altruistic behavior: Help people who can't help themselves and return favors to those who have given in the past. The first represents a norm of social responsibility, the second a norm of reciprocity. Reciprocity follows from a favor received, while responsibility starts from ground zero, with no expectation that the favor will ever be returned. Social scientists typically suggest that these norms are learned, instilled by personal and third-party observations. Helping others and returning favors brings praise and good feelings, while abstaining and reneging bring criticism and bad feelings. Together these experiences teach us the virtuous life.[35] Support for this idea comes from the observation that older children tend to provide more help than younger children in cases of both responsibility and reciprocity. For example, in one study, an experimenter read stories to children between the ages of five and ten. In each story, one child either did or did not help another, who was either in need (e.g., a child without any toys) or who had previously cooperated. Older children perceived helping in the responsibility case as more laudatory than in the reciprocity case, and were more likely than younger children to help those who had not helped in the past. With age, norms of responsibility appear to take precedence over norms of reciprocity. This developmental change is consistent with Kohlberg's framework. It suggests that young children stick to concrete rules of thumb, whereas older children, due to their experience, focus more on the maintenance of social order.

The fact that patterns of helping change over development is unambiguous. What is unclear is whether learning by experience is the only way to explain such change, and how experience molds the child. For example, perhaps the developing child's ability to reason in the moral domain changes because of abilities that have nothing specific to do with morality. As mentioned before, developing along with the child's moral capacity is also her ability to understand others' beliefs and desires, distinguish intentional from accidental actions, plan for the future, and recall detailed events from her past. None of these capacities is specific to the moral domain, but they are certainly recruited in the process of delivering moral judgments and attempting to justify them. Some of these abilities, such as the attribution of beliefs and desires, is on a maturational timetable that is quite independent of experience. Children don't learn to attribute mental states to others. It is a homegrown ability, more like seeing and hearing than working out the multiplication tables.

A different interpretation of the child's development is that our moral faculty is on a slowly maturing time course. In the studies mentioned above, children's judgments about the merits of someone's altruistic behavior were more consistent across age groups when compared with their own altruistic behavior in the same contexts of responsibility and reciprocity. Once again, competence in judging altruistic behavior shows one signature, whereas behaving altruistically—performance—shows another.

These results, together with the studies of early competence reviewed thus far, push for an alternative to the social science perspective that has classically driven work on moral development: We are equipped with a grammar of social norms, based on principles for deciding when altruism is permissible, obligatory, or forbidden. What experience does is fill in the particular details from the local culture, setting parameters, as opposed to the logical form of the norm and its general function. Building on the last two sections that focused explicitly on deception and the detection of cheaters, I now want to use the more general arena of social norms to bring back the Rawlsian creature and take a closer look at the grammar underlying these rules of conduct.

Studies of social rules have something of a split-personality disorder at present. One side argues that there are no principled differences between the different flavors of social rules. There is one, all-purpose inference generator that operates in the context of permission rules. The other side argues that the differences are real, with distinct psychological principles and parameters accounting for the taxonomy that includes moral, conventional, permission, precaution, and personal rules. The theoretical foundation for this view comes from evolutionary biology as well as the domain-specific perspective developed earlier. The logic of natural selection suggests that the mind is equipped with specialized reasoning abilities, designed to solve specific adaptive problems. Social exchange is one problem and precautions are another. The domain-specific perspective sees a more articulated set of computations, mediated by causal and intentional aspects of each event. If the analogy to language holds, complete with a moral grammar, then the second personality type must be normal, the first abnormal. In the last section, I provided some evidence in favor of this (psycho)analysis. Patient RM is an existence proof that part of the human brain can be damaged, selectively knocking out the capacity to reason about social contracts while preserving the capacity to reason about precautions. To place these findings in a broader context, let's return to two earlier distinctions: the very general difference between descriptive and prescriptive rules on the one hand, and the more specific difference between social conventions and moral rules on the other.[36]

The evolutionary perspective draws out key distinctions between different classes of rules. So, too, does the moral-reasoning literature that emerges from studies of development pioneered by Eliot Turiel and Judy Smetana, among others.[37] As mentioned in chapter 1, Turiel made the important point that among the variety of social rules or norms, some are culturally specified conventions, some moral, and some personal. When adults and children are confronted with these different norms, they give different functional justifications, show different patterns of permissibility judgments, reveal different intuitions with respect to the gravity of rule violation, and provide different verdicts in terms of the upper hand of an authority figure. Although there has not yet been extensive cross-cultural work on these distinctions, especially in terms of small-scale societies, the studies that have been carried out suggest that the different social rules are universally recognized, similar among boys and girls, and even consistent in cultures with seemingly different parental styles—in China and the United States. Social conventions are for group coordination, whereas moral rules are for issues of welfare or fairness. Social conventions are violable and may only apply to a select group of people, whereas moral rules are inviolable and universally applicable. Violations of moral rules are more serious than violations of social conventions. And authority figures can intervene to override social conventions, but even God cannot always override a moral rule.

The evolutionary psychologist Larry Fiddick ran a series of experiments to both highlight the limitations of the Wason selection task and provide further support for the idea that the human mind is equipped with a set of articulated principles for reasoning in the moral domain. These studies provide a key test of the Rawlsian creature's design specs.

One group of subjects responded to three different versions of a Wason task involving a precaution, social contract, and social convention; each centered on a fictitious drink called "tanka":

PRECAUTION: Tanka is a poisonous religious drink that could blind a person. However, the caffeine in coffee beans neutralizes the poison, so the chief of the tribe made the following rule: "If you drink tanka, then you must have a coffee bean in your mouth."

SOCIAL CONTRACT: Tanka is a desirable (nonpoisonous) drug made from a secret recipe passed down from mother to daughter. In order to acquire tanka, a man has to give his wife a gift, hence the chief of the tribe made the following rule: "If you drink tanka, then you must give your wife a gift."

SOCIAL CONVENTION: Tanka is a religious drink that people traditionally drank with a coffee bean in their mouth, hence the customary rule was: "If you drink tanka, then you must have a coffee bean in your mouth."

Following each scenario, subjects saw four cards and then selected among them on the basis of some criterion for rule violation. A second group of subjects read the same scenarios and rules, but instead of cards and questions concerning rule violation, they answered a questionnaire.

The goal was to assess whether these different social conditions trigger different judgments in terms of the importance of authority, consensus, universality, and whether they elicit different justifications. For each scenario, the questionnaire provided a set of judgments and justifications, and subjects responded to these by stating whether they agreed, disagreed, or were uncertain. For example, subjects read that members of the Mubata tribe should not drink tanka without a coffee bean, even if the chief and all Mubatas said that it was okay to drink tanka without a coffee bean. This statement taps the role of an authority figure (the chief) and consensus (all Mubatas). Next, subjects read a justification, such as "The chief made the rule for a social purpose."

Fiddick reported two central results: Subjects showed no difference in response to the three scenarios under the Wason selection task, but showed highly nuanced differences with the questionnaire. Overall, subjects perceived a difference between precautions and social contracts, but not between social contracts and social conventions. This provides support for the articulated taxonomy of social rules, and raises cautionary flags about celebrating the victories of a theory based on a single method.

In a second set of studies, borrowed from work by Paul Rozin on the moral emotions, each subject first read a scenario followed by a rule, then a statement indicating that a person had violated the rule. An experimenter then presented subjects with four photos, each revealing a person with a different facial expression—anger, disgust, fear, or happiness. The task: identify the rule violator's facial expression. If there are psychological differences between precautions, social contracts, and social conventions, then violations of these rules should result in different emotions, and thus different facial expressions of emotion. Here's an example:

You are an anthropologist studying the Jibaru tribe. The Jibaru hunt with blowguns and poison darts. The poison is a powerful neurotoxin obtained from a small tree frog and has been known to kill humans, too. In fact, several Jibaru have died preparing poisoned darts when the poison got onto their exposed skin. You had heard about this problem and brought a supply of rubber gloves for the Jibaru tribesmen to wear to avoid contact with the poison when making darts.

The tribal elders thought that using the gloves was a great idea and so they made the following rule: "If you make poison darts, then you must wear rubber gloves."

While you were studying the Jibaru, one of the tribeswomen caught a man breaking the rule.

If precautions protect others from hazards, then violations should be associated with fear. If social contracts maintain social cohesion or stability, then violations should be associated with anger. Results support these predictions. If a man makes poison darts but doesn't wear rubber gloves, fear is the appropriate response, not anger. If a man makes poison darts but doesn't share meat from his next hunt—breach of a social contract—then the appropriate response is anger, not fear.

To look more carefully at the causal-intentional aspects of these rules, Fiddick ran one final experiment, paralleling the studies by Harris and Nunez discussed earlier. One outcome of the first two experiments was that subjects judged social contracts as social rules, but precautions as nonsocial. Social contracts can be overturned by social consensus, but precautions cannot. If there is a hazard—dangerous chemical plant, faulty electrical system, flimsy bridge—then this is an objective fact about the world, and no town meeting can override this fact. Given this difference, the cause of the violation should matter for social contracts, but not for precautions.

Consider the following example from Fiddick's experiment: children can only take an advanced swimming class if they have paid $50 (social contract version) or if they are experienced swimmers (precaution version). Each story involved either a father sorting applicants or an elderly woman with Alzheimer's who was suffering from absentmindedness. The father intentionally allowed his son into the class either without paying the $50 or by failing to disclose his son's beginner status. In contrast, the elderly woman accidentally failed to check on the $50 payment or failed to note the child's beginner status. As predicted, subjects more readily detected the violation for social contracts in cases where it was intentional as opposed to accidental, but showed no difference in detection for precautions.

Taken together, Fiddick's experiments show that different principles underlie precautions, social contracts, and social conventions, and people are sensitive to them. People perceive differences among these rules in terms of function, universality, susceptibility to authority, and the seriousness of violation. Driving these principled differences are the core properties of our moral faculty: the causal-intentional aspects of action and the emotions they trigger—the Rawlsian and Humean creatures, respectively. Due to the details of Fiddick's experimental design, it is not possible to determine whether subjects had access to these principles, nor is it possible to assess whether these principles have been articulated at the right level of abstraction. In his study, an experimenter provided subjects with a relevant set of explanations for each scenario, as opposed to soliciting judgments about an action and then asking subjects to construe an explanation. The latter approach was adopted in the Web-based experiments on moral intuitions that I discussed in chapters 3 and 4, and provides one way of looking for a dissociation between people's judgments and justifications. Although people can undoubtedly reconstruct an explanation that is consistent with their judgments, my guess is that these explanations will turn out to be as insufficient and incoherent as those obtained from subjects reasoning about trolley problems or the nature of incest.

Concerning abstractness, my sense is that like the principle of double effect discussed in chapter 3, the principles of a social contract and a precaution rule are not at the right level of detail with respect to the computations that go into processing a complicated social event. They are general labels that cover up the ways in which subtle parametric changes have large effects on how we perceive permissible, as opposed to forbidden, violations. If our theories about this family of social rules is to have the same kind of explanatory power as theories in modern linguistics, we will need a more microscopic view. This will include manipulating the intentional aspects of the action, as well as the variety of options and consequences associated with action or inaction. In the context of a social contract, do people distinguish between intended and foreseen consequences? Is breaking a contract permissible if the intent is not to injure the partner but to provide greater help to a third party in need? In this case, injuring the partner is a foreseen side effect of the primary intention to effect a greater good. With answers to these questions, and others, we will begin the process of generating a grammar of social norms.

Characterizing the grammar of social norms has an added benefit: it may help inform age-old debates in philosophy, concerning moral objectivism.[38] If you are a moral objectivist, you think that if a moral judgment is true—that it captures what the world ought to be like—then it is true under all conditions. There is no room for caveats or exceptions that depend on certain conditions. There is no room for relativism. Now recall that a key distinction between social conventions and moral rules is that moral rules have the feel of universality—of being true under all conditions. Young children seem to understand that there is an objective truth to the claim that you can't gratuitously harm another person, and this claim holds for Americans and Africans, Jews and gentiles, boys and girls; further, there is no sufficient authority to trump this claim. Social conventions, on the other hand, have the feel of relativism, varying cross-culturally and trumpable by an authority. Young children also understand that moral labels like "bad" and "good" are objectively true characterizations, whereas "icky," "boring," and "yummy" are relative and subject to personal preference. Although it is not yet entirely clear how the objective-relative distinction is acquired, the capacity to make the distinction becomes available early, putting the child on a path that sees the moral arena through objective glasses. Understanding the details of the grammar that shifts rules from conventional to moral will thus feed directly into some of the central concerns of many moral philosophers, as well as legal scholars interested in connecting these principles to current and applied social problems.

GOING NATIVE

In 1932, the educational psychologist Helena Antipoff wrote that our sense of justice represents "an innate and instinctive moral manifestation, which in order to develop really requires neither preliminary experience nor socialization amongst other children. . . . We have an inclusive affective perception, an elementary moral 'structure' which the child seems to possess very easily and which enables him to grasp simultaneously evil and its cause, innocence and guilt. We may say that what we have here is an *affective perception of justice.*" This nativist view was immediately countered by Jean Piaget, who pointed out that because Antipoff based her conclusions on tests of three- to nine-year-old children, it is not possible to rule out the heavy hand of experience.[39] For Piaget, the child's understanding of morality is largely constructed out of her social experiences. Early on, moral judgments are defined by the sculpting effects of parental interactions. Particularly important at this stage is the child's sense that there are rules imposed by parents, and these are immutable, sacred, and unalterable. It is the child's feelings of respect that initiate her sense of obligation, or, as Piaget stated the case, "respect is the source of moral obligation and of the sense of duty: Every command coming from a respected person is the starting point of an obligatory rule. . . . Right is to obey the will of the adult. Wrong is to have a will of one's own." With time, and the construction of new relationships with others within and outside the family, the sacredness of parental rules washes away, replaced by a greater sense of independence. This final shift, again driven by experience, allows each child to evaluate what is just, fair, and expected from a social relationship.

What I hope the linguistic analogy makes clear is that Antipoff and Piaget were fighting the wrong battle. If only they had read Aristotle, who noted hundreds of years ago that "Neither by nature, then, nor contrary to nature do the virtues arise in us; rather we are adapted by nature to receive them, and are made perfect by habit." Ultimately, what we want to understand is how the mature state of moral knowledge is acquired. Piaget's position only serves to raise more mysteries: How do children move through the stages of moral development? Why do they move through in the particular way that they do—what constrains the pattern of change? Piaget, and those who have followed in this tradition, have failed to answer these questions. As Hume posited several hundred years ago, "Nature must furnish the materials, and give us some notion of moral distinctions."[40] It has to be the case that the human mind has certain innate capacities that enable us—but not chimpanzees, dolphins, or parrots—to see certain moral distinctions and appreciate their significance for our lives and the lives of others. Further, although the young child could presumably pick up any experience from the environment and add it to her repertoire of moral considerations, but doesn't, there must be some innate structure to guide which bits of experience are taken on as part of one's moral knowledge. If that is the appropriate diagnosis, then something about our DNA has enabled this psychological difference. And something about our DNA also enables us to acquire the unique signature of our local culture.

To understand what is at stake in this discussion, imagine three different phenotypes, or designs for the Rawlsian creature.[41] Each phenotype maps onto a different acquisition device, and therefore a different flavor of the nativist position. Let's call them "Weak," "Temperate," and "Staunch." The Weak Rawlsian is the default design: As a species, distinct from all others, it has the capacity to acquire morally relevant norms, but nature hasn't provided any of the relevant details. The Weak Rawlsian is endowed with a mechanism for learning about norms, but lacks both general principles as well as more specific ones concerning incest, reciprocity, and killing. The weakest of the Weak Rawlsians doesn't even grant that this acquisition device is specific to the moral domain. There is some general learning mechanism, perhaps married to a set of emotional biases, that turns social conventions into moral rules. The less weak version grants that something about this mechanism is specific to the moral domain, enabling it to immediately distinguish between conventional and moral transgressions.

The Temperate Rawlsian is probably closest to my own characterization, and the view adopted by most generative grammarians working on language. A Temperate Rawlsian is equipped with a suite of principles and parameters for building moral systems. These principles lack specific content, but operate over the causes and consequences of action. What gives these principles content is the local culture. Every newborn child could build a finite but large number of moral systems. When a child builds a particular moral system, it is because the local culture has set the parameters in a particular way. If you are born in Pakistan, your parameters are set in such a way that killing women who cheat on their husbands is not only permissible but obligatory, and the responsibility of family members. An American-born child faced with the same events perceives things differently due to differences in parameter setting: Killing an unfaithful woman is forbidden, except if you are from the South, where it once was (socially, if not legally) permissible, and the husband's responsibility. For the Temperate Rawlsians, culture affects their early development to set the parameters. Once set, culture has little impact. Here, acquiring a second moral system would be equivalent to acquiring a second language: slow, laborious, requiring rote memory, and hours of tutelage—something quite different from the effortless, fast, and almost reflexive acquisition of the first system.

The Staunch Rawlsian is equipped with specific moral principles about helping and harming, genetically built into the brain and unalterable by culture. For the Staunch Rawlsian, learning a second moral system late in life is equally slow and effortful.

To understand which phenotype best captures the evidence to date, we need to understand what counts in favor of or against each one of these characterizations. Typically, those arguing against a nativist position do so by attempting to undermine the aria of universality. They consider claims about universal incest taboos, hierarchical social structures, and prohibitions against harming, and find exceptions among the Bongo Bongo and others. For example, contrary to the Staunch Rawlsian, it is not the case that taboos against incest are universally held. In Graeco-Roman Egypt, reports indicate that up to 30 percent of urban marriages were between brothers and sisters. First-cousin marriages have been even higher and more omnipresent, across both time and place. Exceptions to rules involving harm are equally easy to find. Among some tribes, such as the Yanomamo of Venezuela, not only do members of one tribe systematically plan violent attacks on neighboring tribes, but such attacks are expected, relished, and possibly even obligatory, given the nature of their relationships. As the Hollywood extravaganza *Gladiator* made vivid to those unfamiliar with this history, the Romans adored the sight of two scantily clad men ripping each other to shreds. Harming another was not only permissible but deliciously entertaining, in some situations. Slayers of nativism spotlight these cases and claim victory.

These exceptions to universality do slay at least some potential claims of the Staunch Rawlsian. But I don't know anyone who holds this view of our moral faculty, and few hold to such an extreme position for any other faculty of the mind, including language. How do the Temperate and Weak Rawlsians fare against these kinds of attack?

Assume for the sake of argument that we have principles like the incest taboo and a prohibition against harm. The Temperate Rawlsian is equipped with these general principles, in addition to a set of parameters that are set by the local culture, early in development. It is the set of parameters that creates the potential for variation. "Potential" is the key word here, because it is at least plausible that every culture sets some parameters in the same way, leading to cross-cultural uniformity in how humans judge certain actions; the evidence from our Web studies suggest that certain forms of harm may be judged in this way. When our moral faculty evaluates incest, the computation runs over the nature of the sexual act (kissing, genital petting, intercourse), the degree of relatedness, and the costs and benefits of inbreeding. These are all potential parameters, and they may have evolved for adaptive purposes, designed to limit the potentially deleterious effects of particular mating patterns. Similarly, when our moral faculty evaluates an act that leads to harm, we evaluate the agent's intent, his goals, and the positive and negative consequences that ensue from his actions. These are all potential parameters, and they are most likely responsible for our shifting judgments in cases like the trolley problem.

The universal moral grammar is a theory about the suite of principles and parameters that enable humans to build moral systems. It is a toolkit for building a variety of different moral systems as distinct from one in particular. The grammar or set of principles is fixed, but the output is limitless within a range of logical possibilities. Cross-cultural variation is expected and does not count as evidence against the Temperate Rawlsian. When consistencies are found across cultures, this raises interesting questions about the acquisition device. Are certain parameters, possibly all, set to some default setting but open to alternatives as a function of current environmental constraints? For example, it appears to be a semantic universal that languages default to unmarked forms of certain verbs. For example, in English, "John *climbed* the mountain" means "John *climbed up* the mountain." I can force the marked case by saying that "John *climbed down* the mountain." Other default settings arise in our phonology, and there may well be cases in syntax. Do biases, if they exist in the moral domain, represent prior selective regimes in which certain settings have had greater success with respect to individual survival and reproductive success? Do moral systems work like language in the sense that choosing to set certain parameters influences subsequent settings? For example, once a language decides to set the subject-heading parameter and the word-order parameter, there are constraints on what can happen in terms of case agreement. When a culture decides that reciprocity is mandatory, or that all forms of incest are forbidden, how does this impact on other norms involving helping and sexual behavior? We don't yet have answers to these questions. By leaning on the linguistic analogy, however, we open the door to these questions, and wait for the relevant theoretical insights and observations.

Let's return to the slayers of nativism. During attacks on either the Temperate or Weak Rawlsian, the exceptions to universality are trotted out again, as is the possibility that a cultural-evolution story might provide a better explanation. Here's how Jesse Prinz puts the case: "If the universals could be culturally evolved, there is no pressure for them to be innate, and if the universals are not treated morally in all cultures, they may not qualify as an innate morality, even if they have an innate basis." The first part of this comment concerns what is learnable, the second concerns the nature of cross-cultural variation. Taking them in reverse order, the Temperate Rawlsian was not designed to produce a singular morality. Rather, it was designed to produce multiple moral systems. Finding counterexamples will not take this type down. Prinz, for example, trots out many examples of close relatives having sex, of individuals killing each other with glee, and of peaceful societies lacking dominance hierarchies. These are indeed interesting cases, but they are either irrelevant or insufficiently explained with respect to the nativist position. They may be irrelevant in the same way that it is irrelevant to cite Mother Teresa and Mahatma Gandhi as counterexamples to the Hobbesian characterization that we are all brutish, nasty, and short. The fact that some people are obsessively altruistic provides no counter to the fact that many are not, and most are nothing like these two saints. Ditto for incest. Even if the exceptions are relevant, they need to be explained.

Consider again incest, and the more general principle that might operate in guiding our sexual behavior. Presumably, when someone has looked hard enough, there will be a principle that takes the act of sexual behavior, examines the intentions and goals of the two relevant individuals, assesses their age and degree of relatedness, and delivers a judgment concerning the act's permissibility. How a given culture sets each of these parameters (or others) will most likely differ. But until we understand the sources of variation, as well as the extent of variation, we will be in no position to understand either the mature state of knowledge or its developmental history.

There is another problem with the use of counterexamples to slay the nativist. What we see people doing is unlikely to provide a clear view of their underlying moral competence. When a soldier jumps on a grenade to save his troop mates, we don't gain an understanding of how our moral faculty judges cases of rescue or self-sacrifice. Similarly, when we see a culture acting violently, we don't necessarily get their take on harming more generally. Prinz discusses the Gahuku Gama of Papua New Guinea who, like the Yanomamo, frequently go on raids and kill their neighbors. From these observations, he concludes: "They don't think it is morally wrong to harm members of other groups." But from these ethnographic observations, we learn only a limited amount about what they think is morally wrong. What we learn is that there are situations in which they kill others. As I pointed out before—perhaps once too often—the intuitive judgment we bring to bear on a particular case will often differ from what we actually do.

The first part of Prinz's comment concerns the possibility that universals are learned, passed down generation after generation through stories, religious teachings, and wise elders. We don't need to have an innately specified belief about gravity, because everywhere, throughout the world, we can readily learn about this physical principle by watching apples drop and people walking on the ground as opposed to floating in space. We also don't have an innate belief about the sun's location in the sky, because looking up a few times a day gives us the relevant input. What issues like this boil down to is a question about the relationship between a learning mechanism and the input it grabs on to. Consider the observation cited a couple of sections back, that three-year-olds can detect violations of permission rules, using the agent's intentions to navigate among the possible outcomes. The empirical question here is whether this capacity could be acquired through observation or teaching. If experience really drives the ability, then it should be possible to accelerate the timing of its expression by early training, giving children additional exposure to permission rules. If, on the other hand, the capacity to detect violations is part of our innate moral faculty, then children living in different cultures, with wildly different experiences in school and at home, should emerge with this capacity at around three years of age. One signature of an innate faculty is a narrow time window for expressing a skill that is relatively immune to differences in experience.

I have used the material in part II to provide a sketch of how we should think about the growth of our moral capacity. I am convinced that the observations tilt toward the Temperate Rawlsian design. We are endowed with a moral acquisition device. Infants are born with the building blocks for making sense of the causes and consequences of actions, and these early capacities grow and interface with others to generate moral judgments. Infants are also equipped with a suite of unconscious, automatic emotions that can reinforce the expression of some actions while blocking others. Together, these capacities enable children to build moral systems. Which system they build depends upon their local culture and how it sets the parameters that are part of the moral faculty.

PART III

Evolving Code

6

ROOTS OF RIGHT

---

*But it is curious to reflect that a thoughtful drone . . . with
a turn for ethical philosophy, must needs profess himself
an intuitive moralist of the purest water. He would point
out, with perfect justice, that the devotion of the workers
to a life of ceaseless toil for a mere subsistence wage, cannot
be accounted for either by enlightened selfishness, or by any
other sort of utilitarian motives, since these bees begin to
work, without experience or reflection, as they emerge from
the cell in which they are hatched.*

—THOMAS HUXLEY

CONSIDER THE FOLLOWING TWIST on the classic trolley
problems from chapter 3:
  John is on a footbridge over the trolley tracks and can see
that the trolley approaching the bridge is out of control. On
the track, there are five chimpanzees, and the banks are so steep that they
will not be able to get off the track in time. John knows that the only way
to stop an out-of-control trolley is to drop a very heavy weight into its
path. But the only available, sufficiently heavy weight is a large chimpanzee, sitting on the footbridge. John can shove this chimpanzee onto
the track in the path of the trolley, killing him; or he can refrain from doing this, letting the five chimpanzees die (see figure on p. 308).

My own intuition—not one that I comfortably accept—is that it is
permissible to push the large chimpanzee, even though in the parallel human case, it is not permissible—or, at least, less permissible—for Frank to
push the large person off the footbridge. American college students share
this intuition. What is the explanation or justification for the difference
between humans and chimpanzees? Why does the utilitarian outcome rule
for animals but not for humans? Logically, if it is impermissible to use one

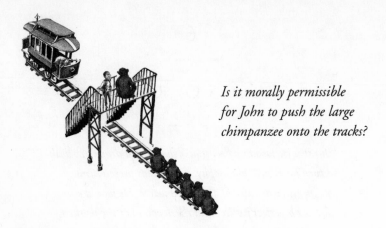

*Is it morally permissible
for John to push the large
chimpanzee onto the tracks?*

life as a means to saving many, this principle should apply with equal force to human adults, infants, brain-damaged patients, and animals. Although people contrasting these cases rarely come up with coherent explanations, many allude to distinctive differences between human and animal life, including our responsibilities to members of our own and another species. These types of explanation zero in on some of the central issues surrounding current debates over animal rights and welfare. When we make decisions about the treatment of animals, we often appeal to perceived differences between our mental wherewithal and theirs. We draw a line that sets us apart from them in terms of distinctive abilities, including language, consciousness, emotion, and a sense of the future. Mark Twain held this view, and believed it raised an important conclusion about our own moral faculty: "Whenever I look at the other animals and realize that whatever they do is blameless and they can't do wrong, I envy them the dignity of their estate, its purity and its loftiness, and recognize that the Moral Sense is a thoroughly disastrous thing."[1]

Critics of the drawing-the-line view respond with cases involving our own species, pointing out that even though a newborn baby is not as conscious as an adult chimpanzee, and is as linguistically challenged as its furry cousin, few would use a newborn baby to save five others. Pointing to psychological differences between us and them doesn't work. Perhaps the difference stems instead from our emotional attachment, built over millions of years, designed to guarantee the welfare of humans but not other species. When faced with the trolley case, our emotional attachment to humans is greater than our attachment to animals, and thus our judgments shift. If this view is correct, it brings us back to the role of the Humean creature in guiding our judgments. We might imagine, for example, that the weaker our connection with the particular target animal, the stronger our conviction that we can use one life to save many; we might even shift from a permissible judgment to an obligatory one, especially if the animals are endangered. Those who see all of life as sacred never draw the line, and thus hold to the logically defendable position that if it isn't permissible to push one human to save many, then it isn't permissible to push one caterpillar, canary, or chimpanzee to save many. Those who see differences between species draw the line, and allow the utilitarian principle to rule.

Discussions of animal welfare and rights are largely orthogonal to the central concerns of this final part, but they tie in nicely to this center: What is the mental wherewithal of animals such that it informs our interest in the evolution of the moral faculty? Here's Darwin's answer: "Any animal whatever, endowed with well-marked social instincts, the parental and filial affections being here included, would inevitably acquire a moral sense or conscience, as soon as its intellectual powers had become as well developed, or nearly as well developed, as in man." Darwin correctly suggests that animals with social instincts are the right sorts of animals for thinking about the origins and evolution of a moral sense. He is also correct in thinking that along the evolutionary path to our moral sense, nature must have added some extra accessories to the core, allowing individuals not only to care for others but to know why caring is the right thing to do, while harming is often the wrong thing to do. Darwin never provided a detailed depiction of what evolution added, nor why natural selection might have favored these extras. He did, however, leave open the possibility that minds "nearly as well developed" as ours could, in principle, have a moral sense, perhaps only in a rudimentary form, with selection acting to favor particularly moral groups. Jean-Jacques Rousseau was more explicit about the comparative issue, pinpointing a key difference between us and them, a uniquely human attribute:

"Every animal has ideas, since it has senses; it even combines those ideas in a certain degree; and it is only in degree that man differs, in this respect, from the brute . . . It is not, therefore, so much the understanding that constitutes the specific difference between the man and the brute, as the human quality of free agency. Nature lays her commands on every animal, and the brute obeys her voice. Man receives the same impulsion, but at the same time knows himself at liberty to acquiesce or resist."[2]

To Rousseau, humans have free will, animals don't. For Thomas Henry Huxley, Darwin's henchman, many of our good and evil attributes were gifts of evolution, but our capacity to squelch the bad and promote the good through a system of ethics was largely a human creation: "Laws and moral precepts are directed to the end of curbing the cosmic process and reminding the individual of his duty to the community, to the protection and influence of which he owes, if not existence itself, at least the life of something better than a brutal savage."[3]

With this comment and perspective, Huxley parted company with Darwin, suggesting that evolutionary theory, and the comparative method more specifically, would provide few insights into our moral psychology. As it turns out, Darwin was right, Huxley wrong, and, unfortunately, many evolutionary biologists have followed Huxley's authoritative voice.[4] Shifting from phylogenetic or historical concerns to adaptive function, Darwin first imagined a reproductive competition among individuals within a group that included "sympathetic and benevolent parents" on the one hand and "selfish and treacherous parents" on the other. Realizing that the brave men who risked their lives would perish, as opposed to the selfish cowards who stayed home, he concluded that natural selection would not increase the numbers of the virtuous.[5] In contrast, stepping from within group competition to between group competition painted a different picture: "A tribe including many members who, from possessing in a high degree the spirit of patriotism, fidelity, obedience, courage and sympathy, were always ready to aid one another, and to sacrifice themselves for the common good, would be victorious over most other tribes. . . . At all times throughout the world tribes have supplanted other tribes; and as morality is one important element in their success, the standard of morality and the number of well-endowed men will thus everywhere tend to rise and increase."

Darwin makes the assumption here that some sense of good will prevail over evil, creating a source of moral growth. He assumes that when one group supplants another, the group with the higher moral calling wins. But, as history reveals, Darwin's assumption is false, unless one is willing to grant a higher moral ground to Genghis Khan, Pol Pot, Adolf Hitler, Idi Amin, Efrain Montt, and Ratko Mladic—all leaders responsible for massive genocides guided by the call of ethnic cleansing. There is, however, one sense in which Darwin was correct. If we look to the positive effects of organizations such as the United Nations, we see the spread of particularly virtuous moral attitudes, including global decreases in slavery, subordination of minority groups, child abuse, capital punishment, and the harmful treatment of animals. It is thus possible for some groups to facilitate the spread of what many consider universal rights.[6]

We can answer problems of adaptive function in at least two ways. The classic approach is to document how specific behaviors contribute to an individual's survival and reproduction. Consider, once again, the problem of altruistic behavior. For Darwin, being nice to someone else at a personal cost made little sense in light of the logic of natural selection. Not only are Mother Teresa and Mahatma Gandhi aberrations, but so, too, are those who leave tips at restaurants, take care of other individuals' offspring, and contribute to charities. These acts reduce each person's potential for self-promotion. If the Darwinian theory is right, selection should wipe out those silly enough to lower their reproductive worth— and ultimate genetic posterity—by investing in others. As the evolutionary biologists William Hamilton, George Williams, and Robert Trivers argued, we resolve this paradox if we think about behavior at the level of the gene. What appears to be genuinely altruistic, and good for the group, is actually the covert operation of selfish genes. We act nicely to kin because our genetic posterity is wrapped up in theirs. What is good for them is good for our genes. When we lack kinship, we act nicely if we have some guarantee of a reciprocated return. This isn't an act of kindness. Reciprocity is an act of self-interest, because it is driven by the expectation of a fair return: food for food, backrub for backrub, babysitting for babysitting.

From the gene's-eye view, the way to think about the evolution of moral behavior is to think selfishly. Instead of asking "How can I help you?" ask "How can my helping you help me?" In the simplest case, you would compare two strategies, moral versus immoral, and tot up the number of babies for each. If the moral strategy wins, both in terms of reproductive output and in terms of its immunity to immoral invaders, then selection will favor the moralist and eliminate the immoralist. Life isn't that simple, but the logic of the argument is.

The second approach is to look at the source of an object's design features. Calling upon Reverend Paley's *Natural Theology*, Richard Dawkins argued that chance alone can explain neither the complicated and functionally precise design of a watch nor that of a living creature. While Paley appealed to God to account for complexity, Dawkins appealed to Darwin. While God has a vision, natural selection is blind. Natural selection builds organisms with complex design features based on a nonrandom but directionless process. Poorly designed variants are eliminated, well-designed ones favored. When we see an organism or organ with complex design, we see the handiwork of natural selection, a tinkerer that fine-tunes the raw material to the adaptive problem at hand. This argument applies with equal force to an animal's overall body shape as well as to its eyes, brain, and mind.

In the last chapter, I discussed Cosmides and Tooby's use of the design stance, to argue for an evolved cheater detector. As they suggest, a central problem for our Pleio-Pleistocene ancestors was to cooperate in the service of social exchange. When individuals engage in this kind of exchange, they implicitly or explicitly set up a social contract. Given that individuals can break social contracts by taking the benefit without paying the cost, selection will favor those who can detect such cheaters. Reverse-engineering the problem, we should find the psychological machinery required to identify cases of cheating, and Cosmides and Tooby's work suggests that we are so endowed. It is this kind of evidence that fits the logic of the adaptive design stance.

The controversy surrounding work on cheater detection ties into the second half of the "Why did it evolve?" question. By assuming that the cheater-detection system evolved as an adaptation for life among Pleio-Pleistocene hunter-gatherers, Cosmides and Tooby imply that this is a uniquely human adaptation. This is certainly one plausible story, but in the absence of observations of other animals, it remains uncontested. Proclamations about human uniqueness lie within the realm of speculation without studies of other animals. And, as I discuss later, there are numerous cases of deception in animals, and several cases where cheaters are detected.

In this chapter, I examine which parts of the moral faculty, if any, evolved prior to the emergence of our own species. I use reciprocity as a centerpiece for discussion, both because of its prominence in our own species and because it naturally raises questions about the psychological prerequisites. To initiate and maintain a reciprocally stable relationship, individuals must recognize each other, recall what was given to whom, how much, when, and with what costs. Individuals must also recognize whether the resource was given intentionally or as an accidental by-product of an otherwise selfish goal, and whether the exchange of resources was done contingently. Like other social interactions, this form of cooperation relies upon many other abilities, including the establishment of expectations, emotional responses to actions that satisfy or violate these expectations, the capacity to acquire, follow, and enforce rules, and a sense of responsibility for the health of the relationship. And in humans, at least, these social relations often depend upon the development of a rich sense of self, empathic concern for others, and the ability to generate predictions about others' states of mind without any direct experience of their behavior. When we generate moral judgments about another's action, we make use of many of these capacities, even though we are often unaware of the underlying process. Perhaps these are the bits of psychology that Darwin had in mind when he considered the development of our intellectual powers. Perhaps these are the bits that, if nearly as well developed as in our own species, would give some animals a primitive moral sense, a capacity that we might feel satisfied in calling an evolutionary precursor.

DARWINIAN NODES OF ACTION

When the nineteenth-century physiologist Ivan Pavlov taught his dogs an association between hearing a bell and receiving food, he taught them to expect food once they heard the bell. What we don't know, however, is what exactly these dogs expected, because Pavlov never explored whether they would have been satisfied by the appearance of any old food or whether they expected a particular kind of dog chow, and thus felt ripped off—cheated—when it was some other kind. The question then is what, specifically, do animals expect and think about prior to the occurrence of a predicted event? Whatever answer we give will not directly resolve questions of moral significance. However, because expectations are formed in the social sphere as well, it is important to understand the nature of expectation more broadly. If animals set up expectations and detect violations, then they should be capable of judging when an individual does something socially right or wrong, and, possibly, morally right or wrong.

In the 1920s, the psychologist Eduard Tinkelpaugh set out to determine whether rhesus macaques and chimpanzees create specific expectations about the kind of food shown and then hidden within a container. In one set of experiments, he concealed different kinds of food in one of two containers while a subject watched. He then placed a screen in front of the containers, hiding both from the subject's view, waited a short period of time, and then removed the screen so that the subject could search for the hidden food. Sometimes the content matched what was concealed and sometimes it did not. If they saw Tinkelpaugh hide a banana and then found the banana, they cooed with delight. If they saw Tinkelpaugh hide a banana and then found lettuce instead, they were either furious or puzzled.

We don't fully understand what it is like to have a primate experience of expectation satisfaction and dissatisfaction. But Tinkelpaugh's experiments have been repeated several times, including studies that reveal the neural code underlying matching expectation and detecting an error.[7] They show, without doubt, that the primate brain has evolved to set up expectations, anticipating outcomes that matter in terms of survival.

Here, I return to the theme set out in chapter 4 for human infants, and ask whether nonhuman animals set up expectations about actions and events, using the causes and consequences to detect violations. With apologies to Jim Watson and Francis Crick, I refer to these primitive detectors as DNA, for Darwinian Nodes of Action.

The first and most basic principle of action focuses on the capacity for self-propelled motion. This is a starting point for discriminating between animate and inanimate objects:

PRINCIPLE 1: *If an object moves on its own, it is an animal or part of one.*

In the natural world, objects that move on their own are animals, and those that can't are either dead animals, plants, or inanimate objects. When animals see these kinds of objects, what kinds of expectations do they form about their movements? Do they expect all animals to move where they please? Do they expect all inanimate objects to stay put unless contacted by some other object?

In a series of studies carried out with my students, we presented wild rhesus monkeys and captive tamarins with a two-chambered box, separated by a partition with a hole at the bottom.[8] In every condition, an experimenter placed one object into one chamber, covered the box with a screen for a few seconds, removed the screen, and revealed the object inside the same chamber or the opposite one. When these monkeys saw an apple placed into one side, or a ball rolled in, they looked longer when the objects appeared in the opposite chamber than when they appeared in the same chamber. These are inanimate objects. They have no capacity to move on their own. Rhesus monkeys and tamarins are therefore surprised when a stationary apple or a human-propelled ball appear to move, on their own, to a different location. They showed the same pattern of looking when the experimenter placed a clay object with eyes in the center of the chamber and then, by means of magnets, invisibly caused it to move within the chamber. Thus, even though this object moved on its own from a stationary starting point—the definition of self-propelled—these monkeys were surprised to see it move to the adjacent chamber. However, when the experimenter placed a live animal—tree frog, mouse, hermit crab—into one chamber, both rhesus monkeys and tamarins looked as long when these animals appeared in the opposite chamber as when they appeared in the starting chamber. In the mind of a rhesus monkey or tamarin, therefore, living things hold a privileged position: Unlike nonliving things, animals can move where and when they want, or they can stay put. Though self-propelled motion may provide a relevant cue, it isn't enough. When it comes to predicting an object's potential for trading places, these monkeys look for cues to animacy, hints that the thing they are looking at is alive, breathing, and capable of moving elsewhere.

Results from these experiments with monkeys lead to a potential difference with human infants. For our own species, a self-propelled object appears to provide sufficient cues to predicting an object's goals. The object in question can be as simple as a ball or a two-dimensional disk on a screen. A second experiment, however, suggests that monkeys may understand a corollary of principle 1: An inanimate object can only move if contacted by another object.

To further explore principle 1, the cognitive scientist Laurie Santos presented tamarins with one red and one blue train on a track. She then concealed the red train with a screen and launched the blue train. In one event, the blue train moved behind the screen, and, soon thereafter, the red train emerged from the other side of the screen. In the second event, the blue train only partially disappeared behind the screen, and soon thereafter the red train emerged. Thus, in the first event but not the second, the blue train made contact with the stationary red train. The second event is physically impossible, since the red train has no capacity to move on its own, and the blue train never made contact. Tamarins detected this impossibility, looking longer at the second than the first event. These results suggest that tamarins are equipped with the corollary to principle 1: Inanimate objects can't cause others to move without making contact.

Principle 2 builds on principle 1 by making goals an explicit part of the event:

> PRINCIPLE 2: *If an object moves in a particular direction toward another object or location in space, the targeted direction picks out the object's goal.*

To check whether this principle is part of the mind's code, we can present an incongruous event, at least from the perspective of a normal human adult: an individual moves toward or attends to an object or location, and then heads off in a different direction or picks up a different object. Would, for example, an animal be surprised to see one individual dash over to join another and then lie down and fall asleep? Would an animal be surprised to see another look toward a coconut but then reach for the banana? Woodward's experiments, described in chapter 4, addressed these exact questions. Babies watched as an experimenter looked at one of two objects on a stage and then reached either for this object or the other one. Infants looked longer when the experimenter reached for the unattended object. So, too, did cotton-top tamarins.[9]

It looks like we share principle 2 with at least one other animal. It is a principle of action with far-ranging moral implications, including our ability to detect rather perverse actions. When a mean-spirited parent teases her child by offering a toy that she can never reach, we perceive this as a moral infraction—as morally wrong. We recognize the perversity by recognizing the child's goal—grabbing the toy. Without a capacity to recognize goals and goal-directed behavior, we wouldn't have a category of morally perverse teasing. Animals, such as tamarins, have some of the requisite psychological machinery, even if they don't attribute moral perversity to teasing. Chimpanzees, however, apparently do make such attributions, and I will provide the evidence in a moment.

In 1984, while I was watching vervet monkeys in Kenya, I noticed a vervet infant who seemed to be irritated by something on its left thigh. It kept picking at this one spot. The infant's mother was some distance away, on the other side. All of a sudden, this infant leaped up in the air, bounding forward. Given the irritation, I assumed that something had poked or pinched the infant, causing her to leap up and forward. The infant's mother immediately dashed in to see what was wrong. But what did this vervet mother think? Did she assume that something pricked her infant, causing her to jump? Or was she puzzled at her child's apparent attempt to leap up and over an invisible barrier? Did she think that her child was acting irrationally? Principle 3 addresses this exact issue:

PRINCIPLE 3: *If an object moves flexibly, changing directions in response to environmentally relevant objects or events, then it is rational.*

Gergely and Csibra provided the key test of this principle with human infants, and the developmental psychologist Claudia Uller provided a replication with infant chimpanzees.[10] Each chimpanzee sat in front of a television and watched as a small square moved toward and over a barrier, and then settled next to a large circle. Watching multiple reruns of this show, they then watched two new shows, each with the barrier removed. In one show, the square moved forward a bit, then arced up and down and then straight over to the circle; this mimicked the original trajectory, but from a human perspective—both adult and infant—it appears bizarre and irrational. In the second show, the square moved straight across to the circle—a perfectly rational action.

Chimpanzee infants looked longer at the irrational square, implying that they expected rational action from a geometric figure faced with a new environment. Principle 3 appears to be an evolutionarily ancient piece of the psychology of action—part of primate DNA.

When chimpanzees in certain parts of Africa engage in a grooming bout, one animal initiates the interaction by raising its arm. The partner, if interested in grooming, responds in kind, and then both lock hands in what is called hand-clasp grooming. Here, the timing of the initial arm-raising, followed by the mirrored action, sets up a contingent response. It sets up a social interaction. This is the core aspect of principle 4:

PRINCIPLE 4: *If one object's action is followed closely in time by a second object's action, the second object's action is perceived as a socially contingent response.*

Grooming is one form of cooperation, seen among a wide range of animals. Grooming bouts can be carried out sequentially, with large or small gaps between bouts, or at the same time. Other forms of cooperation involve similar sequential exchanges, including babysitting, alarm calling, and food sharing. To maintain these cooperative exchanges, animals must have some sense of contingency. They must have something like principle 4, even though no study to date has actually carried out an explicit experiment.

Where contingency arises for some animals is in games of cooperation, especially those that involve either two animals working together to achieve some common goal or some form of reciprocity. In a reciprocity game with tamarins, an experimenter trained one animal to play a unilateral altruist strategy, always giving food to its opponent when the opportunity arose. The experimenter trained a second animal to play a unilateral defector strategy, never giving food to its opponent. If contingency matters, then when the untrained animals offer food to their trained opponents, food comes back from the unilateral altruist but never from the unilateral defector. All of the untrained animals cooperated with the altruist, which paid off handsomely, but not the defector. These observations, together with other experiments, provide evidence that contingency can play a central role in either stabilizing or breaking apart social relationships in animals.[11]

By their nature, all social animals have the skills to pick out the cooperators and cheaters, the kind and ferocious ones, the dominants and subordinates. Many animals form coalitions with trustworthy partners to gang up and defeat those higher up in the pecking order. Among monkeys and apes, when a dominant male moves near, meek subordinates spread their lips, baring their teeth in a display of submission that typically provides them with a protective shield—a passport against random acts of violence. From these observations, however, we don't gain the requisite insights into how these skills are acquired and how they are represented in the mind. We need to understand whether there is a key principle of action that determines how animals judge particular social interactions, assigning some to the category of helping and others to the category of harming. Are animals guided by principle 5?

> PRINCIPLE 5: *If an object is self-propelled, goal-directed, and flexibly responsive to environmental constraints, then the object has the potential to cause harm or comfort to other like-minded objects.*

The most relevant experiments are ones by David Premack and the psychologist Josep Call, both focusing on an actor's goals and the relationship between actions and the personal nature of their consequences.[12]

Premack recruited his star chimpanzee, Sarah, for this complicated task. Following years of experiments, Sarah had trainers that she liked and ones that she disliked. Premack selected one of each for this experiment. For every test, Sarah first watched a videotape of a trainer attempting to grab food that was just out of reach. Next, an experimenter handed her an envelope with three photographs. One showed a picture of the trainer using a proper action to solve the problem; for example, the trainer picked up a long stick to rake in the food. One showed a picture of the trainer using an improper action to solve the problem, such as picking up a short stick that didn't quite reach the food. And the final picture showed the trainer using a proper action but an irrelevant one to solve the problem; for example, the actor stood on a chair, a proper response to food hanging from the ceiling but an improper response to food placed out of reach on the ground, only accessible with a long stick. Would Sarah pick different actions depending upon whether she was watching the likable or unlikable trainer? If she was like us, she would want the likable trainer to succeed and the unlikable trainer to fail. If she was like us, she should pick the proper action for the likable trainer and the improper or irrelevant action for the unlikable trainer. If she was unlike us, she might just pick what happened next in the sequence, regardless of which trainer she was watching.

Sarah acted the way we would. In every condition, she picked the proper action for the likable trainer and either the improper or irrelevant action for the unlikable actor. These results suggest that chimpanzees recognize their own goal states, and can also represent the goals of others. And they can marry these representations of others with an assessment of their own emotions to choose actions that benefit some and potentially harm others. This capacity is central to morality, as it leads to the strategic use of cooperation with those whom we like and rejection of those whom we dislike.

In Call's studies, chimpanzees paired up with a human experimenter who controlled access to food—a highly desired grape. In some situations, the human experimenter cooperated, giving the chimpanzee a grape, and in other cases they didn't. At stake, however, was whether the chimpanzees would distinguish between actions that on the surface were similar but that differed in terms of the experimenter's underlying intentions or goals. Consider teasing versus clumsiness. In the teasing condition, Call held out a grape, moved it toward an opening in the partition, and then as soon as the chimpanzee reached for the grape, pulled it back. Clumsiness, in contrast, involved the same actions, except that Call accidentally dropped the grape each time he moved it toward the opening. For both interactions, Call moved the grape toward the chimpanzee, and the chimpanzee never received the grape. If chimpanzees only cared about getting food—if they were merely consequentialists—then from their perspective, an experimenter who teased them would be no different from an experimenter who was clumsy. The consequence would be the same: no grape. If chimpanzees cared about why they did or didn't get food—if they cared about the means—then these interactions were different. Call would be morally perverse—going back to principle 2—in the teasing condition, but merely annoying in the clumsy case.

Chimpanzees see the difference between these two conditions. In response to teasing, and in contrast to clumsiness, they leave the testing arena earlier and show greater signs of frustration—banging on the window, aggressively calling. Whether they perceive the teaser as morally perverse is anyone's guess at present. So, too, is the question of whether they would generate the same attributions to an inanimate object performing the same actions, paralleling the studies of infants watching geometric shapes move on a television monitor.

Premack and Call's studies suggest that chimpanzees may have access to principle 5. Minimally, they appear to read beyond the surface features of action to the intentions and goals of the actor, using these as a foundation for distinguishing between those who help and those who harm. And, presumably, this is part of their psychological design, because selection favors capacities that ultimately feed self-interest, even if it is in the context of cooperating with others.

In contrast to the wealth of information on the human child's developing concepts about living and nonliving things, we know relatively little about the animal equivalent. This makes the evaluation of some of these principles less than satisfying when it comes to the moral domain, as we would ultimately like to understand how moral judgments shift as a function of the individual's understanding of life and death. As mentioned in chapter 4, the human child's understanding of death is a relatively late development. Do animals have anything like a concept of death? Is it like the young child's, anchored in facts about breathing and moving? Or is it richer, more theoretically informed, tapping notions of growth and reproduction? Unfortunately, we only have anecdotal observations to go on.

Some animals, such as the ants that Ed Wilson has described, clearly don't. When an ant dies, it is dragged out of the colony and deposited. But dead ants secrete oleic acid that, when placed on living ants, causes them to be deposited in the ant cemetery as well. For an ant, dead = oleic acid. For other species, the story is richer, but nonetheless unclear. Studies of elephants, monkeys, and apes suggest that individuals, especially mothers, go into a state of mourning upon losing their offspring. These observations indicate that the loss of a group mate causes a change in others' behavior, and, we presume, their emotional states. But it tells us little about their understanding of death, whether they have any expectations about this individual's future, whether they will ever return, or carry on in some altered state somewhere else. Yet without an understanding of their understanding of the life cycle, the connection between principles of action and moral significance remains tenuous.

WHO AM I?

While I was observing rhesus monkeys on Cayo Santiago, a BBC film crew paid me a visit. They were shooting a documentary on the emotions of animals and wanted to get some footage of rhesus monkey social life. They also wanted to set up a large mirror, to see what these monkeys would do. I warned them in advance that some of the more rambunctious juveniles and adult males may break it into small pieces. A large adult male soon kicked the mirror kung fu–style, shattering it into smithereens. End of film sequence.

As a group of rhesus looked on, we cleaned up whatever pieces we could find. Later that day, with the film crew gone, we saw four adult females walking around on three limbs, using the fourth to carry a small piece of mirror, periodically stopping to take a good, long look.[13] What were they seeing? What were they thinking? And why only the females? Given the size of the mirror, they couldn't possibly think that they were carrying someone else? And if not someone else, than whom other than "me"? Were the females self-absorbed beauty queens, trying to look their best for the macho boys?

A cottage industry of animal research has developed around the use of mirrors to understand an animal's sense of self.[14] Charles Darwin initiated this approach with his studies of captive orangutans. But in 1970, more than one hundred years after Darwin, the comparative psychologist Gordon Gallup developed a more refined and informative method. Gallup provided chimpanzees with access to a full standing mirror and watched their behavior. Like the orangutans that Darwin had tested, the chimpanzees looked and made facial expressions at their mirror image, and also looked behind the mirror, as if they were trying to locate the individual inside, staring back. These behaviors did not lend themselves to a clean diagnosis. Gallup then took a further step. He anaesthetized each chimpanzee and, while they were unconscious, placed an odorless red-dye mark on one eyebrow and on one ear. Once they were conscious again, Gallup placed the mirror in front of them and watched. Immediately, the chimpanzees looked in the mirror and touched the dye-marked areas. This behavior can be interpreted in two ways. One, the chimpanzees figure out that when they move, the mirror image moves as well, in perfect synchrony. They conclude: "That's me." Two, they see the mirror image as another chimpanzee with red marks and wonder if they have the same. In both cases, the behavior reveals something to the staring individual about themselves. The second explanation seems unlikely, given the fact that once chimpanzees recognize the dye marks and touch them, they then proceed to use the mirror to look at previously unseen parts of their body. The mirror has become a tool.

A slightly different kind of experiment by the comparative psychologist Emil Menzel enriches our understanding of self-recognition in animals. Menzel wanted to understand whether chimpanzees and rhesus monkeys could use a video monitor of their arm to find a concealed target location. Rhesus monkeys never made it out of the initial training phase of the experiment, so there is nothing to report. Chimpanzees, in contrast, were not only able to use the video-monitor projection of their arm to find a concealed target, but were also able to reposition their arm when the image was inverted, and stopped reaching altogether when the monitor revealed a previously filmed version of their arm—in other words, when the real-time dynamics of their own arm moving stopped, they stopped moving as well. From the chimpanzee's behavior, we infer that it was thinking: "That's my arm on TV."

Following Gallup's lead, several researchers wondered if their animals were also equipped with this ability, this sense of self, or whether they were as clueless as rhesus. One by one, as if Noah were administering some standardized test for admission onto the ark, experimenters marked parrots, pigeons, crows, elephants, dolphins, tamarins, macaques, baboons, orangutans, gorillas, and bonobos, showed them a mirror, and watched their response. And, one by one, most of these animals failed to touch the marked areas and failed to use the mirror to explore previously unseen private body parts. With the exception of dolphins, those that passed were close evolutionary relatives to the chimpanzees—orangutans and bonobos. Only one gorilla showed any evidence of mirror recognition— the language-trained and human-reared Koko, certainly not your average specimen.

Some researchers claim that chimpanzees, bonobos, and orangutans are special, while others claim that they are no more special than dolphins and gorillas, who also appear to pass this test. At the heart of this debate, however, are two uncontroversial points. First, not all animals will show evidence that they recognize their image in the mirror. Species differences could arise either because some animals lack this particular sense of self or because they are not particularly sensitive to changes in the visual domain, which would lead to detection of the dye marks. Instead, they may show greater capacities in other sensory modalities, such as hearing, smelling, or touching. For example, in a wide variety of species, especially songbirds, individuals respond differently to their own song played back from a speaker as opposed to the song of a familiar neighbor or an unfamiliar stranger; and in some songbirds, once an individual has acquired its own species-specific song, there are neurons that will only fire when the bird hears its own song. This suggests that, at the neural level, individual songbirds recognize their own song.

Second, the mirror test says nothing at all about what the individual thinks when it recognizes its reflection. We don't know if these individuals are appalled by their appearances, indifferent, or narcissistically mesmerized. We don't know what they know, how they feel about such knowledge—if they feel anything—and what they can do with it, assuming they can raise it to some level of awareness. One relevant piece of evidence comes from a set of experiments asking whether rhesus monkeys know when they are ignorant. To set up the problem, consider the movie *Memento*, a thriller that explores the nature of human memory. Although the actual story line is left intentionally vague, what is clear to everyone in the audience is that the protagonist can not remember any recent events. To aid recall, he tattoos key events onto his body, and posts sticky notes and photographs all over his room. He effectively offloads what would be stored in memory into an external videotape of his recent past. This trick works because the protagonist knows what he doesn't know. He is aware of his deficit and this allows him to counteract the problem.

The cognitive neuroscientist Robert Hampton[15] ran a series of experiments with rhesus monkeys designed to test whether they are aware of what they don't know. In one task, he presented subjects with a sample image, turned it off, and then offered a choice between a discrimination test or a pass. The test included four images, one of which was the same as in the sample. Hampton rewarded subjects for touching the matching image and punished them with a long lights-out period for picking any of the other, incorrect images. This is a standard matching-to-sample test, used in countless studies of nonhuman primates. Hampton's insightful twist on this standard was the pass option. On some proportion of trials, he gave subjects the option of passing up the opportunity to take the test trial and on the remaining proportion he forced them to respond. The idea was to give them the option of passing on the test when they were uncertain, perhaps because they had forgotten the details of the sample image. The key finding was beautifully simple: When Hampton forced rhesus to take the test, they did far worse than when they were in control of which test trials to take and which to pass over. Rhesus appear to recognize when they have forgotten, seeing ignorance as a deterrent to performance. This is one of the few clear pieces of evidence that animals know what they know, and can use this knowledge to aid action.

With studies like Hampton's, we can begin to see how to connect the different strands that constitute the animal's sense of self, especially the connection with their emotions and beliefs. Animals with these pieces in play would feel guilty about their own actions or expect guilt in another, recognize the difference between their own and another's beliefs, and use this knowledge to guide action and the judgment of another's actions. As the American philosopher Herbert Mead noted, organisms may only be able to build a sense of self by recognizing the harmonious resonance be-tween their own behavior and its mirrored reflection in another's behavior: "Any gesture by which the individual can himself be affected as others are affected, and which therefore tends to call out in him a response as it would call out in another, will serve as a mechanism for the construction of self."[16]

CROCODILE TEARS

In 2002, the Takara Corporation in Japan released Bowlingual, a digital device that translates dog barks, growls, and squeals into Japanese or En-glish. The press release described the device as an "animal emotion analysis system," designed to "fulfill the realization of real communications between humans and animals." The device is rather simple, taking only three steps to deliver a translation. Step one: record the dog's vocalizations. Step two: analyze its acoustic morphology. Step three: convert the acoustic signal into one of six categories corresponding to different emotional states. If the analysis detects frustration in the dog's voice, perhaps because Rover wants to go outside while owner Bob is couched up watching the Super-bowl, Bowlingual spits out phrases such as "I've had enough of this!" or "You're ticking me off!" If the analysis detects sadness, Bowlingual throws back "I'm bored" or "I'm sad."

The device was an immediate success. *Time* magazine dubbed it one of the best inventions of 2002, and the spoofy scientific magazine *Annals of Improbable Research* awarded Bowlingual its Ig Nobel Peace Prize for inspiring harmony between species. I imagine that for some pet owners, Bowlingual's decoding takes all the pleasure out of living with another species. As the American political commentator Andy Rooney once said, "If dogs could talk, it would take all the fun out of owning one."

But for those who have bought into Bowlingual, there is some finan-cial investment in the company's promise of emotional decoding. But is this what you get for your money? Is there an acoustic signature of emo-tional frustration that maps onto behavioral frustration? Can sadness be plucked from the waveform? Many biologists, myself included, have spent significant parts of our careers trying to decode what animals are saying, and none of us feels as comfortable as the Takara Corporation in labeling each acoustic signal with a descriptive label that is as clear as the ones Bowlingual generates. Perhaps scientists are too cautious, or perhaps they have struggled to find coherent explanations of what animals feel when they communicate. The staff at Takara Corporation has a different mission, presumably driven by money as opposed to accuracy. Within the first few months of launching their product, they had sold 30,000 within Japan, with a price tag of $220 per unit. Sales skyrocketed to 300,000 by March of 2003, with comparable sales on the international market.

But what do we learn about animal emotions from the work behind Bowlingual? Is it a cute gimmick or something more? Ever since Darwin, it has been clear that animals have emotions. Who could doubt that a growling dog is angry, a purring cat content, or a screaming monkey afraid? Controversy arises, however, in assessing whether the words we use to describe these emotions actually reflect the animal's experience, whether there are emotions that other animals experience but we don't— and vice versa—and whether animals make decisions that are not only fueled by the emotions but reliant upon them.[17] Here, I use this controversy to think about how our current understanding of animal emotions contributes to our understanding of their cooperative and competitive acts, behaviors that are guided by principles essential to the healthy functioning of any social system.

Consider fear, an emotional state that is apparently experienced by many animals, presumably because of its adaptive role in avoiding predators and competitors.[18] The logic of emotions, like the logic of our conceptual knowledge and systems for learning, may also be domain-specific. Fear of snakes is different from fear of heights or impending pain. As the social psychologist Susan Mineka has demonstrated, humans and other animals are equipped with a kind of mental readiness to respond with fear to snakes. If a group of rhesus monkeys with no snake experience watches an experienced group express fear toward the snake, the observers will readily absorb this fear, responding with alarm the next time they confront the snake. In contrast, if a naïve group of rhesus watches other rhesus show fear toward a bed of flowers, the fear doesn't spread; the next time they confront a bed of flowers, there is no fear at all. It would take a lot more to convince the primate mind that flowers count. And even if they could be convinced, the process of associating flowers with fear would be different from the spontaneous fear that emerges in the context of seeing a snake. This kind of fear, also exhibited in humans with or without extensive experience with snakes, is different from anxiety.[19]

In contrast with monkeys, apes, and humans, rats do not have a characteristic facial expression for fear. They do, however, have both a freezing and a withdrawal response to things that they find threatening. Rats, monkeys, apes, and humans all show a cascade of hormonal and neural changes when frightened. For example, when frightened by an aversive event—a loud sound or visual cue previously associated with a physically painful experience—all mammals show activation of the amygdala.

Due to the overlap in physiological and behavioral responses, many argue that rats, monkeys, apes and humans experience fear. Others disagree, arguing instead that the actual experience is different, even if there are parallels in behavior and some aspects of the physiology. For example, the developmental psychologist Jerome Kagan argues that "One good reason for distinguishing between the state following a painful shock in rats and in humans is that the latter have a much larger frontal lobe. When humans hear a tone that had been associated with electric shock, the frontal lobes are activated and the person quickly acquires control of the biological signs of fear after only two exposures to the tone. That phenomenon could not occur in rats."[20] Although Kagan may be right, his comment concerning species differences depends on two untested assumptions: The size of the frontal lobe is crucial for the experience of fear, and the speed with which we acquire an association between tone and fear matters. The fact that fear activates the frontal lobes in humans, and not in rats, is interesting in terms of what areas of the brain are involved. But from a description of brain areas, it is impermissible to jump over and assume that the processing and experience are different. Rats may process the situation in a different part of the brain, but then experience the emotion in the same way we do. The issue of speed runs into a different problem. Although we may form the association between tone and fear faster than a rat or monkey, once acquired, each species may experience fear in precisely the same way. What is different is the learning mechanism that facilitates making the association, and this may, in fact, be due to our larger frontal lobes. But this interpretation shifts the argument from species differences in emotion to species differences in learning.

Kagan is absolutely correct in pointing out that from the rat's behavior we must not leap into an inference concerning its subjective experiences—its feelings. When I say that I have a fear of heights, you certainly can't understand my experience if you don't have a fear of heights, and even if you do, you can't understand exactly what it is like for *me* to feel such fear. However, because members of our species share a common neural and physiological substrate, some aspects of our experience will be shared. Consequently, when you say that you are afraid of heights, I have a general understanding of what you mean. I also know, because I am a native speaker of your language, that when you say "fear," it refers to a distinctive kind of emotion. In the case of animals, we simply don't have access to all this information, nor is it necessarily reasonable to make the same kinds of assumptions. The same concerns hold for human infants. When an animal or human infant freezes, presents an increase in heart rate and the stress hormone cortisol, and then heads in the opposite direction from the apparent triggering event, we reasonably call these the signatures of fear; these are, after all, the same kinds of responses that human adults often make when they are afraid. These signatures indicate that some part of the brain has made an evaluative judgment about the situation that causes fleeing or fighting. But what we don't know is what fear feels like to each individual when they are in the throes of the experience. Let's put this hard question to the side, and consider instead how the perception of an event—imagined, anticipated, or real—triggers an emotion, and, on occasion, a follow-up action.

In socially living animals, either in the wild or in captivity, emotions undercut much of their daily life. Animals engage in political strategizing, attempting to climb the social hierarchy or avoid dropping any further within it. Climbing requires motivation, risk-taking, and aggression, while maintaining the current status quo requires sending signals of submission and fear to others higher up in the hierarchy. Mothers, and sometimes fathers, must contend with weaning their offspring—an often trying experience, as the infants' capacity to pester, torment, and manipulate are unmatched. To cooperate, individuals have to muster motivation and trust. Fights will happen, perhaps mediated by feelings of revenge and a thirst for retaliation. But staying angry at someone that you have to live with isn't productive. Making peace is better.

Some of the most revealing work in this area comes from the detailed observations by the biologist Frans de Waal. Beginning with his classic book *Chimpanzee Politics*, de Waal has helped show the complexities of primate social life, highlighting the role that emotions may play in fueling competition and stabilizing cooperation in the service of preserving peace.

Following aggressive conflict, many nonhuman primates—and some nonprimate species, including dolphins, goats, and hyenas—attempt to reconcile their differences by engaging in a variety of peace offerings, ranging from hugs, kisses, and testicle-touching to grooming and the exchange of food.[21] Conflict is associated with stress, reconciliation with the reduction of stress. Researchers measure stress in animals by watching their behavior and recording physiological markers, including heart rate and levels of blood cortisol. Though stress serves an adaptive function, placing individuals in a ready state for action, prolonged stress compromises the immune system and can lead to selective neural death and, ultimately, early mortality. Among rhesus monkeys and baboons, heart rate and cortisol levels skyrocket following aggressive conflict, and remain above normal resting levels for several minutes. But when conflict is followed by a peace offering, heart rate and cortisol levels drop, as do accompanying behavioral correlates of stress. Though we don't know whether the experience of stress in monkeys, apes, and humans is the same, there are many behavioral and physiological parallels, including convergent changes following reconciliation.

The broad distribution of reconciliation among mammals is accompanied by important differences between species in how, when, and how often they do it. This tells an interesting story about the biology of reconciliation, especially its development and plasticity. Some species, such as the despotic rhesus monkey, rarely use reconciliation as a response to postconflict stress and ambiguity. Rhesus are much more likely to redirect aggression: If rhesus A beats up rhesus B, B is more likely to go and pound rhesus C than to hug rhesus A. In contrast, the egalitarian and closely related stump-tailed macaque is more likely to hug than fight. To determine whether these differences are part of each species' innate repertoire, and unlikely to change even in a different environment, de Waal and his colleagues carried out an experiment involving some baby swapping. Rather than have rhesus grow up in their native environment, these youngsters were transported at an early age to a stump-tailed macaque colony. Would these young rhesus carry the flag of their despotic heritage or bend at the will of an egalitarian society? They bent. Rhesus monkeys reconciled their differences using stump-tail gestures. When they returned home to their native environment, these rhesus monkeys preserved their peacenik style, using reconciliatory gestures to manage conflict. Bottom line: genes enable certain species to reconcile their differences, but details of the local society guide whether they reconcile, how often, in what contexts, and with what techniques.

The work on reconciliation shows that emotions play a central role in the maintenance and guidance of certain social norms, even if the more immediate goal is to reduce stress and violence. If we had a simple method to evaluate primate judgments, we might say that the Humean creature fuels its judgment concerning what constitutes a permissible or possibly even obligatory situation for reconciliation. If a chimpanzee watched a film of two individuals fighting, and then saw a follow-up sequence in which they did or did not reconcile, what would be the more surprising case? What would they expect? What counts as a violation or social transgression? Although emotions play some role here, we are also left with the same dilemma that confronted our account of human judgment. To evaluate the interaction, chimpanzees must also recognize it as a case of aggression, assess whether harm was intended as a direct or indirect consequence, evaluate the time elapsed postconflict, and consider the local society's expectation with respect to the form of reconciliation. Given this calculation, carried out without emotional input, a chimpanzee might judge whether reconciliation is permissible or obligatory. The Rawlsian creature is back. Unfortunately, few researchers in this field have looked at reconciliation with respect to this kind of appraisal mechanism,[22] leaving the door open to at least two different accounts: Both emotional and action analyses drive their expectations, or emotions follow from the analysis of action. Whichever way this turns out, there is one obvious conclusion: In species with reconciliation, as well as other dyadic or even triadic social relationships, there are principles of action in play that generate expectations about how animals ought to behave.

At present, there are two competing explanations for why reconciliation evolved as a form of conflict resolution in animals. One possibility is that selection favored reconciliation because of its role in preserving long-term, valuable social relationships. A second is that selection favored reconciliation because it enables individuals to send benign signals of intent, designed to reestablish cooperative alliances for short-term resource gains. Both explanations put a premium on the value of the relationship, either for its own sake or for the immediate resources it affords. Here, then, is a way of marrying the Humean creature with values, and some measure of utility. We can ask how much are such relationships worth? Do animals feel that social relationships are part of their natural-born rights? How hard are they willing to work for them? Is depriving an animal of a social relationship a violation of an implicit moral code?

A way to get at these vexing questions about what really matters to animals comes from an unexpected source: a series of experiments explicitly designed to address questions of animal welfare and rights. In the 1980s, the ethologist Marianne Dawkins and her students developed a brilliant line of experiments based on a simple economics model.[23] The work starts from the premise that for the near future, our species will keep other species in captivity so that we can eat them or use them for some biomedical purpose. Some readers will vehemently disagree with this policy, but the fact remains that many humans enjoy eating animals and, for a variety of human ailments, research on animals provides the only current hope for a remedy. Given that we are going to keep animals in captivity, the only humane thing to do is to treat them with respect, and give them what they need. We can figure out what they need by studying what they do and what they have in their native environments. Finally, we can use what we learn from these observations to create an economy in which animals can work—pay—for what they want, and thus, presumably, for what they need; we call this a closed economy, because there are only a set number of products that an individual can purchase.

In one of the first studies to adopt this approach, Dawkins explored what domestic chickens need. The experiment was motivated by a decision from the British government stating that, due to rising costs, chickens could no longer be supplied with wood shavings on the bottom of their cages. Dawkins argued that chickens need such shavings because it allows them to carry out their species-typical scratching behavior. Dawkins placed a hen on one side of a two-chambered box, separated by a door. The only difference between the two chambers was that one had wood shavings on the floor and the other was bare. Hens placed on the side with shavings stayed put, while hens placed on the bare side immediately moved over to the side with shavings. Next, Dawkins made it more difficult for hens to move from one side to the other by increasing the tension on the door's spring. Although the costs of moving increased dramatically, hens placed on the bare side rammed into the door, eventually making their way to the wood shavings. Chickens not only want wood shavings, they need them.

A similarly designed study examined what mink want, in order to evaluate their housing conditions in fur farms. Each mink started in a standard cage, but with an opportunity to upgrade by choosing different commodities, each placed behind a different door. Behind door 1, mink found a larger cage; behind door 2, a second nest site; door 3, a raised platform; door 4, a tunnel; door 5, some toys; and, behind door 6, a water-filled pool.

Mink consistently opened door 6, content with the opportunity to bask in the water. And, like Dawkins's chickens, mink paid the high price of admission to water by ramming through the heavy, spring-loaded door. Most significant, from the perspective of welfare and our understanding of the mink's emotions and values, mink denied access to water pools were physiologically stressed, almost to the level of mink denied access to food. If their evolved right to live with and in water is taken away, these animals are continuously stressed. Continuously stressed animals develop compromised immune systems. Animals with compromised immune systems are more susceptible to disease, and therefore more likely to die prematurely. That seems unfair and wrong.

What do mink want? Water pools. Why? Because in nature, mink spend a considerable amount of time in the water. Water is a necessary commodity.

From Dawkins's initial insight, designed to infuse objectivity into the often-subjective debates about animal welfare, we gain a new understanding of how animal emotions connect with animal values. We learn what animals need, what they will fight for, and how selection molds a relationship between the value of a commodity and their motivation to work for it.

Crocodiles don't shed tears, and elephants don't weep. No animal expresses its sorrow by turning on the eye faucets. This is a uniquely human expression. But underlying this human specialization are a heart and mind that share many commonalities with other animals. And in this sense, the Humean creature has an ancient evolutionary heritage. That it has such a legacy does not imply a static psychological system that is no different today than it was when we diverged some 6–7 million years ago from a chimpanzee-like ancestor. How we experience emotions must, in some way, differ from how animals experience emotions. But so, too, must chimpanzee emotions differ from elephant emotions, which must differ from crocodile emotions, which must differ from ant emotions. The main point here is that whatever emotions animals have, they are involved in individual action and the evaluation of others' actions.

NATURAL TELEPATHY

In the 1960s, the computer scientist John Conway developed a program called Life. Though built from a few simple rules, it provided an elegant example of how chaos can morph into order. The game is played on a grid. Each cell has eight neighboring cells, and each cell is either alive or dead. Only three rules bring this static grid to life:

1. If a cell has one or no living neighbors, it dies of loneliness
2. If a cell has four or more neighbors, it dies of overcrowding
3. Whenever an empty square has exactly three living neighbors, a new cell is born

From a few randomly filled-in grids, we move quickly into a series of organized clusters of life, as some cells die and others are born.

The standard game of Life involves extremely simple creatures, perhaps mindless, guided by three rules. These creatures have no social relationships.

What happens if we input social relationships into the game of life? Imagine a game involving a fictional species with two distinctive types. Let's call them B and M, for Behaviorist and Mentalist, respectively. These two types look the same on the outside, but are different on the inside. Bs make decisions about social interactions and relationships using only their prior experiences. By accumulating data, they spend more time with some than others. They use simple statistics to classify the population into friend or foe. Their prior associations define what they do to and with other group members. Ms make use of experience to guide their interactions, but go one step further. They make inferences about what is unobservable: the beliefs, desires, and intentions of other group members. They are mind readers, using information about what other individuals can or cannot see, or what they do or do not know, to predict what they will do next. Ms make predictions about behavior in the absence of having experienced behavioral interactions with others. Where someone is looking represents a proxy for what that individual knows. What they can't see, they can't know, assuming that the senses of hearing, touching, or smelling are out of commission. Ms can use their knowledge of what others know to teach and to deceive. This ability to infer what can't be seen means the Ms are better behavior readers because they go deeper into what behavior implies about believing and knowing.

Now imagine a simulation on the grid of Life. Here, we are looking for not only a shift from chaos to order, but an insight into who will win out and why. If both Bs and Ms reproduce, who will make more babies, winning the Darwinian footrace that is measured in terms of genetic prosperity? If one wins the reproductive competition, then there is room for selection to operate, favoring one and weeding out the other. Selection will favor the best design given the environmental circumstances. Ms are faster and more insightful than the Bs, and they are up to the challenge of both a novel habitat and completely novel social interactions. Bs sit around and wait for more data. They rely on highly familiar cues for deciding the next move. As a result, Bs make silly mistakes, failing to distinguish two actions that look the same but differ, because one was done intentionally and the other accidentally. The chaotic population of Bs and Ms will end up as an orderly grid of Ms. Individuals that can predict what is going to happen before it happens are like good chess players: They are several steps ahead and thus can manipulate their opponents by seeing where others will fail or succeed. In the Darwinian competition of life, Ms live, Bs die.

Up until a few years ago, most essays on human evolution concluded that we are the only Ms; all other animals are Bs. While we are uniquely mind readers, everyone else is a mere behavior reader. In my book *Wild Minds* (2000), I echoed the consensus view that animals fail to make inferences about others' ". . . beliefs, desires and intentions—they lack a theory of mind." I followed this comment up, however, with a more cautionary note, based partially on wishful thinking and partially on an insider's knowledge of new experiments by a young graduate student: "We must be cautious about this conclusion, however, given the relatively thin set of findings, weak methods, and incomplete sampling of species and individuals within a species." Here I want to capture the current state of play in a rapidly changing field, including what we know and how it bears on the central ideas that David Premack set in play about twenty-five years ago.[24] Does any animal, other than the human animal, move beyond behavior and into the minds of other individuals? If so, what kinds of psychological states can animals read, using this information to predict behavior before it happens?

Two sets of experiments, one on macaques and the other on chimpanzees, dominated the comparative landscape up until the end of the millennium.[25] Both led to the same conclusion: Animals, even chimpanzees, are strict Behaviorists! Dorothy Cheney and Robert Seyfarth showed that macaque mothers expressed the same level of alarm when their offspring could see an oncoming predator as when they could not. In the context of predation, ignorance is not bliss. But macaque mothers acted as if there was no difference. They failed to distinguish between an ignorant and knowledgeable infant. They also failed to take into account what infants could see and, therefore, what they would know. And the same story plays out in studies of other monkey species: In baboons living on the savannas of Botswana, for example, mothers don't call back to their distressed offspring, even though this would provide the babies with explicit information that the mothers are aware of their plight.

The anthropologist Daniel Povinelli presented comparable findings based on a series of studies of chimpanzees. In the general setup, a chimpanzee entered a test room and, for each condition, had an opportunity to beg for food from one of two experimenters. In each condition, one experimenter could see the begging chimpanzee and the other could not. For example, one experimenter faced the chimpanzee while the other turned his back; one experimenter looked off to the side while the other looked straight ahead; one had a blindfold on his eyes while the other had a blindfold on his mouth; and one had a bucket on his head while the other held the bucket to the side of his head. With the possible exception of one person turned around while the other faced forward, the chimpanzee's begging behavior was random, even with massive amounts of training.[26] They were as likely to beg from someone who could see as from someone who could not. Like the macaques, these chimpanzees were as likely to make a request from an ignorant experimenter as a knowledgeable one. Macaques and chimpanzees are mind-blind.

There are at least two reasons why these findings seemed paradoxical at the time.[27] First, there was a mountain of anecdotal evidence from wild and captive monkeys and apes showing that they are sensitive to where someone is looking; their sensitivity shows up in what biologists describe as tactical deception, the strategic manipulation of another's access to information for some self-serving benefit. For example, low-ranking animals sneak copulations or pinch a piece of food when the dominant alpha isn't looking. Though everyone acknowledges the need for caution when interpreting these single-observation cases, piling them up amounts to an impressive set of observations, raising the possibility that nonhuman primates deceive by taking into account what others can see and potentially know. Other work on plovers and jays suggested that these birds consider where someone is looking when they engage in concealment behavior, hiding their nest in the case of plovers and hiding a food stash in the case of jays. Second, several studies showed that monkeys and apes attend to where others are looking, and can use this information to pick out what someone is looking at. For example, if a chimpanzee enters a test room and sees a human experimenter staring up at the ceiling, he will immediately look up to the same area; seeing nothing at all on the ceiling, he will then glance back at the experimenter to recheck the direction of gaze and then look up again. Given the combination of anecdotal evidence on deception and experimental work on reading visual perspective, there was a growing tension in the field between the believers and the nonbelievers. This tension helped set the stage for Brian Hare—the unnamed graduate student from a few pages back.[28]

Hare's insight was simple. Chimpanzees in the wild compete more often than they cooperate. Their competitive skills have evolved to handle other chimpanzees who have similar interests in limited resources, including food and potential mates. Povinelli's experiments, in contrast, involved cooperation, and, in particular, cooperation across the species' divide: between a chimpanzee and a human. Might chimpanzees recognize the relationship between seeing and knowing if they had to compete with each other for access to food?

Hare's experiments involved a competitive task between two chimpanzees of different rank. Each separate experimental setup or condition explored the same question, but from different angles: Would these two competitors use information about seeing to make inferences about knowing, and then use this information to guide the next competitive move? Each condition imposed different constraints on what either the subordinate, dominant, or both could see. For each condition, the subordinate sat in one room, the dominant in an adjacent one, and a test room between them. When the dominant and subordinate had the same visual access to the available food in the test room, the subordinate stayed put and the dominant ran out and grabbed it all. But when the subordinate could see hidden food that the dominant could not, the subordinate headed straight for it. For example, in one condition, Hare set out two opaque barriers in the center test room. While the subordinate watched, and the dominant looked away, he concealed one banana on the subordinate's side of the barrier. Although subordinate chimpanzees typically avoid conflict over food when dominants are nearby, in this condition, they beelined to the hidden piece of food, taking advantage of their exclusive visual access. These results, together with several other conditions, show that chimpanzees can use seeing to outcompete others. They imply that chimpanzees can use seeing as a proxy for knowing.

These results are interesting on another level. The patterns observed do not reflect individual personalities, but rather, the relative ranks of each individual in the pairing. In some contests, an individual held the dominant position, and in other contests the subordinate position. Their behavior changed as a function of their current ranking. For example, in a condition in which Hare placed one banana in the open and one hidden behind an opaque screen, individuals changed their strategies depending upon their relative rank: When subordinate, they first moved to the hidden banana and then to the visible piece, whereas when they played dominant, they moved to the visible banana first and then to the hidden piece. What determines how an individual competes for food is not how his opponent behaves, but what his opponent can see and therefore know about the current arena of competition.

Hare's results opened the floodgates to further studies of chimpanzees and other species—including monkeys, apes, jays, and ravens—pushing the logic of the initial experiments, especially the use of natural, untrained behavior.[29] For example, studies of captive chimpanzees and wild rhesus monkeys explored the relationship between seeing and knowing by using Povinelli's original design but with one critical change: Instead of subjects cooperating with a human experimenter, they competed. Consider the rhesus results carried out by the psychologists Jonathan Flombaum and Laurie Santos as they more directly parallel Povinelli's design, and also push the evidence further back in evolutionary time to a species that diverged some 30 million years ago from the branch that ultimately became human. Two experimenters approached a lone rhesus monkey on the island of Cayo Santiago, separated by a few feet, and then each placed a white platform with a grape on top next to his feet. For each condition, one experimenter kept an eye on the subject, while the other either looked away or couldn't see due to an opaque barrier. For all conditions, rhesus monkeys selectively snuck food away from the experimenter who couldn't see them.

Animals as distantly related as birds and primates use seeing as a proxy for knowing. These animals have evolved the ability to go deep, reading minds to predict behavior.

These new results on mind reading are only the beginning. There are controversies here and there, as well as further pieces of the story to map out. We need to understand in what ways mind reading in humans and other animals are similar and different; both similarities and differences bear on the extent to which animals can recruit an understanding of beliefs, desires, and intentions to make judgments of moral importance.

One way in which humans and other animals may differ is in the extent to which they deploy their mind-reading capacities across different contexts. Across several studies of chimpanzees, results show that individuals successfully use information about what another individual knows and intends to guide competitive interactions, while failing to use the same information to guide cooperative interactions.[30] From a human perspective, these results are puzzling. If I know that you are ignorant about the location of a hidden piece of food, I can both outcompete you because of your ignorance or guide you to the right location in order to facilitate cooperation. The context is irrelevant, because our capacity to mind-read is more general and abstract. How shall we interpret the results on chimpanzees?

Several students of animal behavior have noted that selection appears to have favored highly context-specific adaptations, designed to solve a small range of problems. This has led to the idea that animals have laser-beam intelligence while we have a floodlight of brilliance. One explanation of the chimpanzee results is that their capacity for mind reading is different from ours, able to use seeing as a proxy for knowing when in the heat of competition, but not in other contexts. This specialization is akin to the honeybee's famous dance-language. When first described by the ethologist and Nobel laureate Karl von Frisch, it was described as a language, because the dance was symbolic, providing detailed information about the distance, direction, and location of food displaced in time and space—all characteristics of words, and the more general capacity to refer to objects and events in the world. As it turned out, however, the honeybees' capacity lacked generality. It was remarkably specific, restricted to food and nothing else besides food. Although it is conceivable that honeybees have little else to talk about, further work by von Frisch and other students of bee biology have noted the rich complexity of their social lives. Bees have much to talk about, but don't, at least not with the referential precision of their foraging dance. Their communication system is an example of laser-beam intelligence. The social psychology of the chimpanzee may be another example.

There is an alternative explanation for the chimpanzee results, one that takes us back to chapter 4, cheater detection, and the Wason card-selection task.[31] To recap, here is the argument that Cosmides and Tooby

used to both motivate and interpret their results. Humans have been selected to solve problems involving social contracts as these are the kinds
of problems that we evolved to solve in our hunter-gatherer past. In contrast, we did not evolve to solve abstract, socially detached problems of
logic. Proof comes from human performance on the standard Wason
logic test and Cosmides and Tooby social-contract version. We draw the
correct inferences when the logic is translated into the language of a social
contract, but not when it is in a more pure, unadulterated form. There is
a context effect that plays on our ability to draw logical inferences. From
these results, we do not conclude that humans are like honeybees, with a
laser-beam intelligence for generating logical inferences. Rather, we conclude that context can sometimes uncover masked abilities. A similar explanation is possible for the chimpanzee results. It is only in the context of
competitive interactions that we can unmask what lies behind the chimpanzees' eyes. Given the chimpanzees' prowess for cooperation in the
wild, my guess is that it is only a matter of time before someone reveals
comparable cases of mind reading in this context as well.

If we put together all of the results on mind reading in animals, the
conclusion seems clear: We are not uniquely in possession of this capacity.
Premack's early intuitions about the chimpanzee's theory of mind were
right. How far does this capacity reach in animals? Do animals recognize
that others can have false beliefs? Do animals recognize the difference between accidents, mistakes, and informed choices? At this point, it is too
early to say. In the absence of such information, however, we can't say
how rich or impoverished the animal mind is with respect to judgments
of others' social actions. We can't say whether violations of social norms
are judged on the basis of consequences or the causes that drive them.
There is an urgent need to know more about what animals know about
each other.

WEIGHTING WAITING

Many birds and rodents stash food in secure places for weeks or even
months, and then use their razor-sharp memories to return to these hiding
places for a feast. Many spiders, fish, and cats sit for long periods of time,

quietly watching a parade of prey before pouncing on an inattentive individual. A wide variety of primate species spend considerable time peeling,
stripping, or cracking into highly protected fruits before reaping the rewards
of their efforts with a delicious slice of flesh. And most animals face the
general problem of whether to stay with the current patch of food or move
on to greener or fruitier or meatier pastures. For each of these cases, individuals must fend off the temptation to feed an immediate desire, waiting
for a more profitable but delayed return. They must delay gratification. It
looks like evolution equipped animals with a healthy dose of self-control.

Foraging problems, such as those mentioned above, involve decision-
making. Assume, as is standard in the field of animal behavior, that natural
selection has designed animals to maximize foraging returns, converting
energy into babies. In absolute terms, a small amount of food is worth less
than a large amount of food; ditto for low- and high-quality food items.
Where things get interesting is when the small or low-quality food item is
available immediately whereas the large or high-quality food item is available at some point in the future. For example, imagine that a leopard sees
a small, lame, juvenile gazelle only a few feet away, but a large, fat, and
healthy adult female one hundred feet away. If the leopard attacks the juvenile, it will succeed and feed immediately. If it passes up this opportunity and hunts the bigger adult, it will take more time and energy but the
returns will be greater. The central problem is how time influences this
choice process. Waiting for a larger or more valuable food item is risky:
The probability that food will be available in the future decreases over
time, as other competitors may jump in and snatch it away, or the vagaries
of climatic events may damage it. We want to understand the kinds of calculations animals make as they look at the trade-offs of taking something
immediately as opposed to waiting. How far does the value of a food
packet sink with time? Are there some trade-offs that no animal would
ever contemplate, thinking that no matter how tasty a piece of food might
be now, it would never have the same value as the life of an offspring? If
animals show limited self-control, acting impulsively in the face of temptation, then they will break down when called upon to follow social norms.
They will succumb to self-interest in the face of helping another. The
short-term gains to self outweigh the potentially long-term but delayed
gains from cooperation and being nice to others.

The relationship between value and time falls under the general topic of temporal discounting: the longer the delay to accessing the resource, the lower its value. There is a vast literature on discounting in rats and pigeons, and a smaller set of studies in less traditional laboratory animals, such as starlings, jays, tamarins, marmosets, and macaques.[32] Paralleling studies in humans, the central question is: How does the value of an item or action change as a function of time? Economists tend to think of the relationship between value and time as an exponential curve: The subjective value of a reward some time in the future decreases at a constant rate. This decay is therefore a measure of risk, of potentially losing everything by waiting for the larger reward. In contrast, students of human psychology and animal behavior tend to think of this relationship as a hyperbolic curve. Like the exponential model, there is a trade-off between subjective value and time, but with two distinctive differences: value is inversely proportional to time delay, and preference reversals arise when the time delay to both rewards stretches out into the future. Preference reversals are real in humans, a fact that annoys economists with a bent toward rational choice, but delights psychologists interested in the basis of subjective preferences. The exponential model can't explain why a human who prefers $10 today over $11 tomorrow would flip this preference when offered $10 in thirty days and $11 in thirty-one days. Since the difference in delay is the same and the monetary rewards are the same, the preference should be the same. The hyperbolic model, in contrast, predicts context effects such that rewards dispensed in the future have an inherently different subjective feel than rewards delivered in the immediate present. Humans flip-flop their preferences depending on time. The hyperbolic model predicts this pattern.

Give pigeons the choice between one and ten food pellets. They consistently pick ten; so will every other animal. Now, make pigeons work for their food. If they peck the left button, they immediately get one pellet, whereas if they peck the right button, they get ten pellets later. If "later" is much more than a few seconds, pigeons will consistently peck the left button for one pellet. They can't resist. The value of one piece of food drops precipitously after a short wait. Their impulsivity persists as long as there is a good-sized difference between the small and the large and there is some waiting period for the large and little or none for the small. In species as different as pigeons, rats, tamarins, and macaques, the ability to wait for a larger reward is on the order of seconds. Humans given a similar task will wait for hours and even days. No contest. When it comes to patience, we are the paragon of animals.[33]

In some sense, preference for the smaller immediate quantity is irrational. If selection favors long-term gains, because these impact most on survival and reproduction, then animals should wait it out. As the behavioral ecologist Alex Kacelnik rightly points out, when a pattern of behavior is observed in a wide variety of species, and when the consequences of this behavior appear to go against the ultimate goal of maximizing genetic fitness, it is high time for evolutionarily minded scientists to figure out why. The experimental economist Ernst Fehr offers this explanation:

. . . throughout evolutionary history, future rewards have been uncertain. An animal foraging for food may be interrupted, or, in the case of reproductive opportunities, die before it is successful. For humans, the promise of future rewards may be broken. And if the risk faced by a person varies over time, he or she applies various discounts to future events and so behaves inconsistently.[34]

This account implies that animals, humans included, are nonoptimal, failing to maximize their potential intake because they are chained to the ghost of uncertainty. But sometimes what appears to be a maladaptive solution may represent an appropriate solution under different circumstances. Consider, for example, the typical laboratory task offered to pigeons and rats, and what the naïve animal must learn. At first, the individual wanders aimlessly around his cage, doing nothing much at all. Eventually, it stumbles onto a lever, pecks or presses it, and something happens either immediately or with some delay. Since the causal force of an action is greatest with short delays, contacting the lever associated with no or little delay is immediately most effective. Consequently, there will be a bias in the learning phase to make contact with the lever associated with the small immediate reward. And this bias maps onto the natural foraging behavior of most animals in most feeding contexts. In nature, foraging decisions almost never entail an action followed by passive waiting; in those cases where it occurs, such as the food storing of birds and rodents, there is an entire period devoted to storing and then a long follow-up period devoted to waiting prior to retrieval. We can therefore explain the bias to grab the more immediate small reward by the fact that there is a more transparent relationship between grabbing and getting a reward. Learning to wait for some abstract period in the absence of doing anything is unnatural—at odds with the biologically engineered machinery for learning.

To circumvent some of these problems, some students of animal behavior have followed Kacelnik's lead, using nonstandard laboratory animals to explore decision-making under more realistic conditions. In particular, whereas the traditional laboratory experiments on discounting give animals choices between two options and explore how waiting impacts upon choice, the more realistic tasks translate waiting into a behavioral measure of expenditure, using insights from subjects' native environments to establish appropriate experimental conditions. When animals forage, rarely do they just sit still and wait for food to arrive on a silver platter; the exceptions are the sit-and-wait predators. Most species walk, run, fly, scratch, peel, and pry in order to eat. Foraging animals *behave*. In one of Kacelnik's studies, he gave starlings a choice between walking or flying for a reward; each activity was associated with a particular energetic cost and return rate for food, with flying associated with higher costs but better returns. Starlings followed a hyperbolic pattern that maximized intake per unit of time. In Kacelnik's terms, starlings deployed a rational strategy, given the constraints.

Taking into account a species ecology can also reveal how selection molds different patience functions. The behavioral ecologist Jeff Stevens compared the discounting behavior of two closely related New World monkeys—cotton-top tamarins and common marmosets. Both species are cooperative breeders with one dominant breeding pair and their offspring who often stick around to help rear the next generation. Both species have similar brain-size-to-body-size ratios, group sizes, and life spans. And both species live in the upper rain-forest canopies, foraging for fruits, insects, and tree sap. There are, however, two key differences: feeding specializations and territory size. Tamarins specialize in insects, whereas marmosets specialize in sap, and tamarins have significantly larger territories. These differences generate two interesting predictions. Given the tamarins' preference for insects, they should be more impulsive or impatient than marmosets, who specialize in sap. When insects are about, there is no time to wait. Foragers must attack whatever they see, immediately. In contrast, sap feeding requires patience. The forager must scratch at the surface of the tree until it starts the flow of sap, and then sit and wait for it to ooze out; leaving the area and then coming back isn't an option, because another individual can readily profit from the original forager's efforts to break through. The differences in territory size lead to a different prediction. Given the larger size of tamarin territories, they should be willing to travel greater distances for food than marmosets. Thus, if we imagine distance as a proxy to time and effort, marmosets should devalue distant rewards more steeply than tamarins. Marmosets should settle for a small piece of food that is nearby over a large piece of food far away, whereas tamarins should be willing to travel the extra mile.

When these species worked against the clock, tamarins were impulsive and marmosets patient: Tamarins waited about half as long for the larger reward as the marmosets did. When these species worked against the tape measure, tamarins traveled significantly longer for the larger reward. Together, these results show that in our attempt to understand the evolution of patience, we must not ignore the essential role that a species' ecology plays in shaping their minds. What appears to be irrational may actually be a perfectly rational and adaptive solution under realistic constraints.

Animals are capable of extreme patience in highly specialized contexts: stashing food in birds and rodents, sit-and-wait-predators waiting for prey, and in some primates when extracting food. But in parallel with our discussion of mind reading, this is likely another example of laser-beam intelligence, a unique specialization locked into one or a few contexts, with no evidence of flexibility. What we have yet to explore, however, is whether the impatience animals show in the context of foraging extends to social situations involving violence and cooperation, problems that hook us back to morality.

DOMESTICATING VIOLENCE

Dominance hierarchies, unwritten rules of territoriality, and property ownership work well, most of the time, to control aggression. Physical aggression, harassment, and withholding resources also function in the service of unwritten rules of punishment. These policing mechanisms are, however, weak, bound to a narrow range of contexts, and rarely if ever used in the service of moderating cooperative relationships among animals. If a lion lags behind in a context requiring cooperation, there are no costs to the laggard. If a capuchin monkey fails to help a group member acquire food, it is not beaten for its apathy. If a dolphin fails to join in on an alliance, it is not chased out to another ocean or excluded from further alliances. In the social domain, there is always someone breaking through the lines of cooperation, defecting when it pays and the costs are small. Often it is the strong over the weak, and the smart and savvy over the dolts. But the weak and dim fight back, fueling an arms race of competition.

Among animals, killing is relatively rare. Animals threaten and fight one another, but rarely attack to kill. Our own species counts as an exception, but not the only exception. The lack of killing raises two interesting questions concerning the nature of violence in animals: What stops and starts it? Are there principles of harm that guide violence in animals, paralleling some of the principles uncovered for humans? To answer the second question, we need some answers to the first, focused on what controls the impulse to fight and sometimes kill others, which we will discuss further in chapter 7.

The ethologist and Nobel laureate Konrad Lorenz suggested that the aggressive instinct is often controlled or suppressed by the submissive gestures of other individuals—a point I raised in discussing James Blair's theory of morality as viewed through the eyes of a psychopath. A snarling dog is likely to go no further if it sees its victim look away with its tail between its legs. It may refrain from an aggressive attack, because once the submissive signal is launched, there is no additional benefit from pushing further. Some authors have argued that submissive gestures work by tapping the aggressor's compassion or empathy. Empathy—feeling what another is feeling—can operate at a strictly physiological level, without any awareness. The snarling dog may stop because it feels what the subordinate feels following an attack, and this suppresses any further aggression. Empathy can also operate with awareness. Perhaps the snarling dog imagines what it would be like to be in the subordinate's place, and this turns off his aggression. At this point, there is no evidence that dogs imagine what it is like to be another dog, but there is also no evidence to rule out this possibility.

One context associated with both aggression and conflict concerns emigration and immigration into a new group. For emigrants and residents, there are impulses that are likely to push in one direction or another: to leave or stay, to fight or flee. In socially living mammals, including most of the nonhuman primates, a tension arises in the life of a young male when he reaches reproductive maturity. He can either stay in his own natal group or leave to join another. While living with his natal group, he will have antagonistic relationships with his neighbors. But once he decides to leave, he never looks back. There are, however, costs associated with leaving. An attempt to find a suitable group with mates and an opportunity to climb up the social hierarchy will undoubtedly involve at least one good fight. From the resident's perspective, seeing a foreigner elicits curiosity, but it may also elicit aggression and a bit of fear, especially in cases where the immigrant males go on an infanticidal rampage, killing all of the new infants in the group.

Studies of wild and captive monkeys show that the hormone serotonin plays a role in these social contexts, as it does in parallel human contexts.[35] Animals with low levels of serotonin are more impulsive, emigrating from their natal group at an earlier age and approaching threatening intruders more quickly than individuals with high levels of serotonin. Adolescent males in general have lower levels of serotonin and are more impulsive, with the same pattern holding for subordinate as compared with dominant males. One can even show that serotonin causally influences social impulsivity, as opposed to being merely correlated with it. Treating animals such as vervet monkeys with the drug fluoxetine—Prozac in common parlance—decreases the uptake of serotonin, thereby increasing the levels of serotonin. Vervets with higher serotonin levels are less likely to approach a threatening intruder.

An important link to the work on serotonin and impulsivity are studies of aggression and the hormone testosterone. As Dave Barry has often mused, especially in his *Guide to Guys*, much of the chest-puffing machismo of men is due to testosterone poisoning. Fortunately, serotonin and testosterone are engaged in a physiological ballet. Testosterone motivates aggression, while serotonin regulates the level or intensity of aggression. If testosterone levels are high, then the odds of a fight are high as well. Serotonin may then act to reduce the chances of a fight by diminishing the tendency to strike out at the slightest provocation. When serotonin levels are low, impulsivity is high, and the brain relinquishes control of aggression.

In a study of wild rhesus monkeys living on an island off the coast of South Carolina,[36] young males with high levels of testosterone frequently threatened other males, though they did not necessarily suffer any injuries. Individuals with low levels of serotonin, however, had not only more fights, but more severe injuries than individuals with high levels of serotonin. Young males with low levels of serotonin were also more likely to take leaps across large gaps in the canopy, a dangerous move that suggests risk-taking in contexts other than aggressive ones. Testosterone is trouble, as the biologist Robert Sapolsky[37] has pointed out, and what makes guys act macho—and stupid—as Dave Barry points out. Fortunately, for some animals at least, serotonin saves the day, turning knee-jerk, impulsive aggression into more controlled and calculated attacks when fighting is necessary.

Although there is ample evidence that natural selection has played a role in the design of aggressive impulses, little is known about how such selection works on the brain, how rapidly it can alter brain structure and chemistry, and the extent to which it, as opposed to other factors, has contributed to each species' aggressive profile. But there is a different way into this problem: artificial selection by means of domestication.

Anyone familiar with dog breeds will attest that there is a continuum of types running from the let-me-at-your-jugular pit bull to the please-rub-my-belly Labrador. Breeders have created this variation. For domestication to work, however, animals must lose both their fear of humans and their tendencies to be aggressive to each other. But in creating differences between breeds, and by selecting against aggressive impulses, the selective process has resulted in a series of unexpected characteristics that provide a window into the mechanisms of control.[38] Looking across domesticated animals as a group, including dogs, cats, and many farm animals, not only has there been a general reduction in aggression relative to the wild type—compare dogs with wolves, or cats with lions—but there has been an overall decrease in brain and canine size, along with an increase in what appear to be unrelated bits of anatomy, such as lop ears and coats of

fur with distinctive white splotches. All of these changes suggest that domestication leads to a shift back to juvenile qualities, or what biologists refer to as "paedomorphosis."

The most detailed study of the domestication process comes from
work on the silver fox by the biologist Dmitry Belyaev. His goal was to explore the process of domestication by selecting for tameness. The tech-
nique was simple: approach a wild fox and note the distance at which it
runs away. Define tameness as approach distance. Take those foxes with
the shortest approach distances and breed them. Take the next generation
of offspring and repeat this process. After forty years and thirty generations of artificial selection, Belyaev had produced a population of tame
foxes, with newborn kits as friendly as newborn puppies. Further, and
paralleling all other cases of domesticated animals, this new generation of
foxes looked different, exhibiting a white patch of fur on the head, a
curled tail, lop ears, and a significant reduction in skull size relative to the
wild type. At a more microscopic level of the brain, the domesticated
foxes also showed a higher level of serotonin. Recall that higher serotonin
levels are associated with greater control over impulsivity and, thus, lower
levels of knee-jerk aggression. As Belyaev reported, although they had
only selected for tameness, they ended up with more than they expected:
a fox with a different appearance, brain, temperament, and social savviness that comes from hanging out with humans.

The punch line, one that we can derive from hindsight, is that when
humans select for a particular trait, there are always unanticipated consequences because of hidden relationships or correlations between traits.
Further, although the focus and intensity of artificial selection may be
different from natural selection, it is clear that selection can rapidly transform the brain of a mammal as complex as the silver fox, leading to dramatic changes in behavior. Selection can change the dynamics of the arms
race, favoring either impulsivity or control.

Can we be certain that Belyaev selected for tameness, and only this
characteristic? Although he used approach distance to characterize each
generation, it is possible that he inadvertently selected for something else.
For example, perhaps those foxes that allow humans to approach closest
have higher levels of serotonin. In breeding these individuals, selection is
operating on serotonin levels. Alternatively, perhaps those individuals

with the shortest approach distances are the ones that maintain eye contact, and are thus more socially skilled and attentive. Breeding these individuals would select for differences in attention or social cognition. The
point of these challenges is not to undermine the results but to question
their cause. Tameness is simply a description of behavior. In selecting for
an outcome that we describe as tame, we don't necessarily capture the psychology that enables such behavior. A wild fox that doesn't run away from
a human may do so for a variety of reasons. The fox experiments show
that artificial selection can change impulsivity—over a short period of
time—but they don't show how the process occurred. In terms of our
moral faculty, these studies show that intense selection can rapidly change
the temperament and social savviness of a complex vertebrate. This sets
up a significant challenge to those who believe that the human mind was
largely sculpted in the Pleio-Pleistocene period of evolution, and kept relatively mummified since. Though it is possible that we have held on to
many of our hunter-gatherer thoughts and emotions, as these were surely
good tricks for survival, the story of the silver fox opens the possibility of
significant and rapid changes in brain evolution.

TEMPTED BY THE TRUTH
OF ANOTHER

When is it permissible for one animal to harm another? The discussion
thus far suggests that animals harm others during predation, while attacking members of a neighboring group, while beating up a lower-ranking
group member, during an infanticidal run, and while redirecting aggression as a mechanism to reduce postconflict tension. Paralleling our discussion of human violence, there is no single deontological principle guiding
animal violence that dictates, plain and simple, that harming another is
forbidden. Nor is there a principle that states that harm is permissible
whenever it feels right. We explain variation in the expression of harm by
appealing to principles and parameters that are grounded in action, and
especially the causes and consequences of different actions. But in addition to the Rawlsian contribution, there is also a Humean component.
Let's return to an earlier example to see how this might work.

When an aggressor reconciles with its victim, there is some sense in which this interaction looks like a sympathetic or perhaps empathetic response. In chapter 4, I discussed some of the work on human empathy, inspired by Hoffman's pioneering research, and mapped out more recently in terms of development and neural correlates by Nancy Eisenberg, Andrew Meltzoff, and Tania Singer. For some, empathy entails more than feeling the same way as someone else. It entails knowing or being aware of what it is like to be someone else. In its simplest formulation, empathy grows out of a mirror neuron-like system, where my perception of an event is mirrored by my enactment of the very same event. Once in place, however, this form of empathy is transformed—either in evolution or in development—by the acquisition of mind-reading skills. With this new capacity, individuals can think about what someone else feels, imagine how they would feel in the same situation, work out what would make them feel better, and from this deduce how to make the other person feel better.

Do animals have anything like the first or second form of empathy? In my discussion of empathy in humans, I mentioned the interesting observation that people who are more empathetic are more susceptible to yawning. Yawning is generally contagious. But it is really contagious if you have a big heart, unable to turn off your compassion for others. Based on this correlation between yawning and empathy, the psychologist James Anderson wondered whether other animals might also be susceptible to contagious yawning.[39] Captive chimpanzees watched videos of other chimpanzees yawning and doing other things. Though inconsistent across individuals, some individuals consistently yawned back. We can't say that the yawners are empathetic while the non-yawners are not. What we can say is that given the observation that contagious yawning is a signature of empathy in humans, it is possible that the same holds true for chimpanzees and other species. This possibility, as well as other observations of caring in animals, sets up a more specific look for empathy.

In nature, rats forage in the company of other rats and often learn from them. In the laboratory, naïve animals learn what to eat either by following knowledgeable individuals or by smelling their breath. Although rats are social eaters, they do not naturally forfeit the opportunity to eat so that someone else might have a chance.[40] To examine whether one rat might forfeit the opportunity to eat because of the benefits to another, an experimenter taught a rat to press a lever for food. The experimenter then introduced a second rat into an adjacent cage and changed the wiring of the apparatus. Now, when the rat with access to the lever pressed it, he delivered a strong shock to his neighbor. This shock had not only a direct effect on the recipient, but an indirect effect on the actor rat in control of the levers. The actor actually stopped pressing for a while and thereby forfeited access to food. In so doing, the actor incurred the cost associated with hunger while relieving the recipient of pain. This is altruism, at least in the biological sense: cost to actor, benefit to recipient. It suggests that rats can control their immediate desire for food to block an action that would cause pain to another. This looks like empathy or compassion, but simpler explanations abound. Seeing another rat in pain might be aversive. When something is aversive, animals tend to stop what they are doing. Alternatively, when the rat pressing the lever sees the other in pain, he may stop for fear of retribution.

Although these results are open to various interpretations, they provide a parallel with the discounting experiments on pigeons described earlier: At some level, the actor rat must control the temptation to eat immediately. These studies differ from the discounting experiments in that the control problem is not between some food now versus more later. The choice is between some food now versus none later because pressing causes pain to another rat. Although rats initially curtail their pressing, ultimately they go back to pressing. This makes sense, given that a failure to relax control over pressing would lead to starvation. Even though it may be wrong to shock another, and even though the rat is directly responsible for the shock, self-interest carries the moment.

In a follow-up study, an experimenter taught a group of rats to press a lever to lower a suspended block of Styrofoam to the ground; if the subject failed to press the lever, the experimenter delivered a shock. Once the rats learned to press the lever, the experimenter eliminated the shock and thereby eliminated lever pressing; in the absence of either punishment or reward, motivation to press disappears. For half of the rats, the study continued with a Styrofoam block suspended by a harness and the lever available for pressing. For the other rats, the experimenter replaced the

Styrofoam with a live rat suspended by a harness, a stressful position that leads to wriggling and squealing. Rats confronted with a suspended Styrofoam block do nothing at all. Rats looking at a suspended rat immediately start pressing the lever. Although the experimenter had no intention of shocking these rats for apathy, nor rewarding them with food for pressing, they nonetheless pressed the lever and thereby lowered their compatriots, relieving them of the stress associated with suspension. This is altruism. The actor rat incurs the cost of pressing and thereby benefits the suspended individual by lowering him to safety.

What do these results tell us about the evolution of altruism and morality more generally? Perhaps seeing another in distress triggers in the actors an emotional response that blocks off the desire for more food. In many of us, seeing an elderly person struggle to open a door or carry a bag triggers an almost reflexive and sympathetic response that results in our attempt to help, as opposed to resuming lunch or a conversation. There is no control problem, because there are no alternative choices. Seeing another rat in pain or distress is sufficient to cause a sympathetic response. Alternatively, perhaps seeing another in distress is aversive. When rats experience something that is aversive, they do what they can to stop it. Pressing the lever isn't altruistic at all.

Each of these studies looks at what rats do in situations where they can help. They leave open what rats might perceive if they watched others, some acting altruistically, others selfishly. Would they prefer to interact with altruists? Would they reject the selfish individuals from joining their group? There are no answers to these questions. Until we have them, we can't distinguish between an animal's judgments or perceptions of action and their decision to act. Studies of nonhuman primates are no better off, but do move deeper into the nature of the phenomenon.

An experimenter trained a rhesus monkey to pull one of two chains in order to obtain its daily ration of food. Subjects readily complied and fed themselves. Next, the experimenter introduced another rhesus monkey into the adjacent cage and, in parallel with the rat studies, hooked up one of the chains to a machine that would deliver a shock to the newly introduced neighbor. Mirroring the rats' behavior, rhesus also stopped pulling the chains. But unlike rats, most of the rhesus showed far greater restraint, far greater inhibitory control. Some individuals stopped pulling for five to twelve days, functionally starving themselves. The extent to which rhesus refrained from pulling was related to two important factors: experience with shock and identity of the shockee. Individuals refrained from pulling for longer periods of time if they had the experience of being shocked, if they were paired with a familiar group member as opposed to an unfamiliar member of another group, and if they were paired with another rhesus monkey, as distinct from a rabbit.

The rhesus experiments are open to the same alternative explanations as are the rat experiments. Though rhesus may feel compassion or empathy toward another in pain, they may also see the expression of pain as aversive. Seeing another in pain is aversive. Seeing a familiar cage mate in pain is more aversive than seeing an unfamiliar rhesus. Seeing a rabbit in pain is irrelevant. Rhesus may also think that all bad deeds are punished, and thus expect retaliation if they continue to eat, thereby shocking their neighbor. But even if rhesus know that pulling leads to pain, there is no reason to conclude that they stop pulling in order to alleviate another's pain. They may stop because it is distracting, or because they expect shock themselves. Although these experiments, and those on rats, do not yield clear interpretations, they raise the possibility—discussed in the next chapter—that recognition of another's emotional state may trigger an inhibitory response. As the psychologists Stephanie Preston and Frans de Waal[41] have discussed, this could happen in a completely unconscious way, recruiting circuitry in the brain that has been designed to unify how individuals act with how they perceive others acting.

In this chapter, I have pressed on the possibility that some of the core capacities underlying our moral faculty are present in nonhuman animals. We have seen that animals experience emotions that motivate morally relevant actions, including helping and harming others, as well as reconciling differences in the service of achieving some modicum of peace. We have also seen that animals are endowed with several, if not all of the core principles of action that underlie the human infants' initial state, and that these principles ultimately hook into a capacity for mind reading and some capacity for self-reflection. Differences between humans and other animals emerge as well. Birds and mammals are remarkably impulsive, exhibiting little control in the face of temptation; their discounting curves are steep, creating problems when it comes to delaying gratification in the context of helping another at a personal cost. Perhaps the most intriguing difference is that whereas individual species exhibit some subset of these capacities, only humans appear to have evolved a complete set.

7

FIRST PRINCIPLES

We can imagine a society in which no one could survive
as a social being because it does not correspond to
biologically determined perceptions and human social
needs. For historical reasons, existing societies might have
such properties, leading to various forms of pathology.
—NOAM CHOMSKY[1]

THROUGHOUT HISTORY, and in all the world's cultures, various groups have articulated various versions of the Golden Rule. Sometimes it has been stated with a positive angle, sometimes a negative one. The general principle has, however, always been the same:[2]

BUDDHISM: "Hurt not others in ways that you yourself would find hurtful."

CONFUCIANISM: "Surely it is a maxim of loving kindness: Do not unto others what you would not have them do unto you."

TAOISM: "Regard your neighbor's gain as your own gain and your neighbor's loss as your own loss."

JUDAISM: "What is hateful to you, do not to your fellow men. That is the entire Law; all the rest is commentary."

CHRISTIANITY: "All things whatsoever ye would that men should do to you, do ye even so to them; for this is the Law and the Prophets."

ISLAM: "No one of you is a believer until he desires for his brother
that which he desires for himself."

One interpretation of this sample is that when humans live in social
groups, the Golden Rule emerges as an obligatory outcome, an explicit
imposition that is handed down from on high. Religions make it explicit,
because individuals tend to forget the second half of the Golden Rule,
taking from others without giving back. Evolutionary biologists have developed a rich theoretical framework to explain the selfish instincts that
drag the Golden Rule down. Here, I wish to use these ideas to explore the
battle between the heavyweight self-interest champion and the cooperative contender. The goal is to use the adaptationist's lens to zoom in on
some of the most ancient principles that guide helping and harming in the
animal kingdom, and tie these back to our characterization of the mature
state in humans.

To determine what is special about the moral faculty, we need to run
two critical tests. First, we must determine whether any of the mechanisms
that support our moral faculty are shared with other animals. We take all
of the components that enable our moral faculty to operate and we subtract the components that we share with other animals. Those components
left over are unique to humans. Second, we take those components that
are unique to humans and then ask whether they are unique to the moral
domain or shared with other domains of knowledge. Here again, we run
a subtraction operation. We take those components of the moral faculty
that are unique to humans and subtract the ones that are operative in other
domains of knowledge. We must be prepared for an outcome in which either none of the components is unique to humans or those that are
unique are also used by other domains of knowledge. It might still be possible though that some of the components are unique to morality. Similarly, it is possible that all of the components are unique to humans, but
none of them are unique to the moral faculty. Part II already showed that
there are properties of the mind that are involved in our moral judgments,
which are not uniquely involved in morality, including basic aspects of action perception, theory of mind, and some emotions. What appears to be
unique to the moral faculty is how we implement these shared capacities
to create judgments of permissible, obligatory, and forbidden actions. To

tackle these issues, we begin with evolutionary theory and the comparative method that Darwin conceived.

Richard Dawkins opened the preface to *The Selfish Gene* with these words: "We are survival machines—robot vehicles blindly programmed to preserve the selfish molecules known as genes." Like many other readers, I found this sentence haunting, especially given the follow-up claim: "This is a truth which still fills me with astonishment." After a series of beautifully crafted and defended arguments, Dawkins ends the book with a sentence that should have provided comfort but instead has left me puzzled for almost twenty-five years: "We, alone on earth, can rebel against the tyranny of the selfish replicators." In the opening sentence, Dawkins clearly did not mean "we" in the restricted sense of "we humans." Rather, he meant "we" in the broader sense of "we living organisms"; Dawkins's metaphor of the selfish gene is not tied to a particular species, certainly not humans. But the final sentence, the final breath of ink, refers to "we humans." Why did Dawkins believe that we alone were capable of overcoming our selfish nature? Why wasn't he willing to consider the possibility that other organisms might lead a rebellion and tell their genes to take a hike? And what gave him confidence that we could rebel? What would constitute a stable rebellion against human nature? What emotions and principled conceptions of action might we—but no other animal—recruit to fight against the militia of genes and psychological states that they help construct?

Prior to the publication of *The Selfish Gene*, the dominant explanation for altruism in humans and other animals was that it had evolved to serve the greater good of the group. Selection favors niceness because it benefits the group.[3] But being nice entails a personal cost, and herein lies the paradox. For example, an individual giving an alarm call potentially attracts the predator's attention, increasing its own chances of dying but simultaneously benefiting others who may escape detection. An individual giving food to another sacrifices the personal benefit of eating the food, while benefiting another who may starve to death without it. What defends against selfishness? Why give an alarm call or give up food if someone else can do the work for you? The theory of group selection suggests that over evolutionary time, a sacrifice for the group is of far greater benefit to the species than selfish behavior. But in a group consisting of team players, there will always be an egoist looking out for number one.

The Selfish Gene heralded a new solution to the problem of altruism, shifting the focus from individuals and groups to genes. Individuals give alarm calls not to protect their group but to protect their genes. Individuals promote the replication of genes linked to alarm calling, either directly by saving their own skin or indirectly by warning their blood relatives. Females control the number of eggs laid or babies born not to aid in population regulation and save the group from extinction, but rather to optimize the number of offspring that survive and reproduce. Unlike the Golden Rule, which transcends the biological relationship between individuals, Hamilton's Rule—named in honor of its creator, the late evolutionary biologist William D. Hamilton—explicitly targets genetic relatives. For Hamilton, the rule reads: Do unto others to the degree to which they share your genes.[4] With this simple formulation in mind, the mystery of altruism vanishes. I am willing to incur a personal cost if it benefits individuals who share my genes. From the gene's-eye view, I should sacrifice myself to save two brothers, four grandchildren, or eight first cousins.

The selfish gene view does not deny the possibility that selection can operate at other levels, including individuals, groups, and even species. It is certainly possible, as the evolutionary biologist David Sloan Wilson has argued, that a group of cooperative altruists will outcompete a group of selfish cheaters, a point that is consistent with Darwin's early intuition. Inter-group differences provide variation for selection at the level of groups, and, as some have argued, may account for the extraordinary forms of cooperation observed among humans but no other animals; I return to this possibility at the end of the chapter and in the epilogue. For now, I focus on the relationship between adaptive design and psychological constraints.[5] I begin with the clearest case of unbridled care, and a candidate starting point for the evolution of our moral concern for others: the bond between parent and offspring. It is from this context that we will see the unfolding of a set of principles guiding cooperative behavior.[6]

CUDDLERS AND KILLERS

If you are not a biologist, your vision of parenting is colored by your own experience as a mammal, which may include giving birth, witnessing someone else doing the same, or taking a trip to the zoo, spending time with the other, furry lactaters. Your vision is further colored by the documentaries that portray chimpanzee mothers playing with their cute innocent offspring, baboon mothers carrying their young to safety, wild dogs licking their pups as they each suck on a nipple, and large hulking elephant mothers gently nudging their little ones to move forward and keep up with the herd's movements. Even if you branch out to birds, chances are that your image of parenting is of a clutch of warm, cozy, featherless nestlings, tucked under their mother or father, waiting for the next food delivery. There is a sense in which we conceive of parenting as obligatory in the animal kingdom, as part of what all social species ought to do as part of their day jobs. Abandonment is a no-no in Mother Nature's eyes.

This cooperative view of parenting pleases us because it provides an echo of our past, resonating with what we think we ought to do. And I would be negligent if I argued that animals lack these nurturing tendencies. But as with our own species, the evolutionary biology of parenting is more complicated and interesting, and provides a way into thinking about the principles and parameters of helping others.[7] Here, as in previous sections, we want to acknowledge that parents are typically motivated to care for their young; a selective regime that made parenting purely optional would certainly be a genetic dead end. But we also want to ask about the conditions under which parents do otherwise. We want to understand the parameters that provide caveats to the general principle of caring for the young. As we will see, these parameters include conditions in which parents harm their young, either directly or indirectly by allowing others to do the dirty work. Some animals, in some conditions, are no different than some humans in some conditions: infanticide, siblicide, and even suicide are all options, supported by none other than Mother Nature.[8]

Consider the following news clip:

COUPLE FOUND GUILTY, SENTENCED TO LIFE
IN PRISON FOR OVERPRODUCING

Jake and Sylvia Darner started having children as soon as they graduated from high school. Before their 40th birthdays, they had conceived 15 children. By the time they had reached their 45th birthdays, only 2 of the 15 were alive. These statistics caught the attention of local social workers, who went to visit the Darners. The house was a shambles, the cupboards were empty, and the two children were barely dressed. Since the Darners started having children, they had little money to support them. In fact, as legal counsel uncovered, they never had any intention of supporting them. As Jake stated in court, "We knew we couldn't support all of these children. We always wanted to produce as many as we could, hoping that at least some would survive." Yesterday, after the full hearing, Judge Klingston sentenced the Darners to life in prison for intentionally overproducing children.

Most of us presumably find this passage, made up as it is, shocking, because we can't imagine parents playing roulette with their children. Roulette shouldn't be part of the psychology of creating a family. We ought to first think ahead, estimating income for the coming years and the costs of raising one child. We should then determine what counts as a reasonable number of children, and have that many. Wouldn't it be nice if we were all so rational and the environment so predictably cooperative? But we are neither perfectly rational nor supremely capable of running the economic calculus necessary to predict an unpredictable environment. In many human societies, and in countless animal species, individuals do overproduce—generating more offspring than they can support given the resources at hand. Neither humans nor animals need to plan on losing some of their young. All that is needed is the cold, blind force of natural selection. Selection will favor individuals that overproduce and turn out more offspring than individuals who exhibit supreme control and planning, but produce fewer offspring. For example, among fur seals living in the Galápagos, many females give birth while they have a one- to two-year-old pup that is still nursing. On the face of it, this seems like a reasonable thing to do as long as she can simultaneously or sequentially feed both pups. Unfortunately, she usually can't. In less than a few months, these newborn pups die of starvation. In several bird species, such as white pelicans and black eagles, there are typically two eggs per clutch, laid sequentially, often with a significant time lag between the first and second. By the time the second egg hatches, the first nestling is relatively well developed. Before the second has had much time to enjoy the care and warmth of the nest, the first launches an all-out attack in what biologists call "obligate siblicide." In one observation of black eagles, the firstborn delivered over fifteen hundred blows with its beak to the secondborn, while the parents stood by and watched their youngest die. The second-born rarely has a chance, and the parents don't appear to care.

These examples of overproduction, and the consequences that ensue, may seem either cruel or stupid, depending upon one's perspective. In fact, neither indictment is correct. The logic of natural selection works the same way: Fur seal mothers who quickly produce a second pup after the last one, and happen to hit upon a bonanza year for resources, score in the reproductive game of life relative to conservative females who wait until each pup is fully weaned. The principle reads something like this: Take care of your children in bonanza years, but abandon the youngest in poor years. The principle includes a parameter that entails letting the young die. In fur seals, and many other species, parents let the young die of starvation if there are limited resources. And fur seal mothers most likely don't feel guilty or depressed about the abandonment. There is no reason for selection to favor such a psychology, and every reason for it to be selected against. In siblicidal birds, the same logic applies. Produce a second egg with the hope that it will be a good year. If it is not, let the firstborn take care of the dirty work, removing the burden of feeding a second mouth. In both cases, the logic of parental care allows for harming; in the case of pelicans and eagles, harming is obligatory. In other species, parents and even siblings help by intervening on behalf of younger and weaker offspring. What the animal kingdom reveals is that the nurturing environment of parental care can also be an arena for violence. Where there is cooperation, there is competition, and sometimes it is lethal. Some of the mysteries are solved by working out the parameters that, when switched on or off, guide the principles of harming within the nurturing environment of parental care.

The number of offspring produced has dramatic implications for family dynamics. This is where the logic of Trivers's parent-offspring conflict comes in, especially his intuitions about parental investment. Following on the heels of his insights into cooperation, and especially his uncovering of the more sinister side of the Golden Rule, he similarly turned things upside down when he proposed that we look at the dark side of the parent-offspring relationship. We need to look at the problem as a repeated game involving intensely competitive players with competing interests. We can best understand the logic of the game and its dynamics by looking at the starting conditions:

CONDITION 1. From the earliest stages of fetal development up to and often beyond weaning, offspring play an active role in sequestering resources from their parents. The young are not passive receptacles for their parents' donations. They are active players in the game. They are competitors. Recall from chapter 5, and our discussion of genomic imprinting, that even the human fetus has been engineered to manipulate its mother's resources, often taking more than she had planned on giving. Animal engineering can include physical devices that enable competition among siblings before they see the light of day. In the sand tiger shark, a pregnant female may produce twenty thousand eggs. Tucked within, safe and sound, each egg rapidly develops teeth and the capacity to swim about freely. With teeth and mobility comes cannibalism; some of the mother's eggs are even generated late in the pregnancy in order to feed the older and more developed ones, a form of uterine fast food. By the time she is ready to give birth, the winning cannibal baby emerges, having eaten all of its siblings. As often happens in science, this somewhat shocking observation was first discovered by a biologist who, following a dissection of an adult female shark, noted, "When I first put my hand through a slit in the oviduct I received the impression that I had been bitten. What I had encountered instead was an exceedingly active embryo which dashed about open-mouthed inside the oviduct."[9]

CONDITION 2. By using the logic of Hamilton's rule, focused on genes as opposed to individuals or groups, we see that each offspring is related to itself by 1.0, and to each of its parents by 0.5. This same offspring, however, is only related to its full siblings by 0.5. Parents are related to all of their offspring by 0.5, assuming that both mother and father are the same throughout. These different values set up an asymmetry of interests, and, from the offspring's perspective, entitlement to making certain demands. Asymmetries of interest set up conflict, both between parent and offspring and between siblings. If fatherhood from season to season is uncertain, as is typically the case in all polygynous species, then siblings will be less than full siblings related by 0.5, and selection for greediness will be even higher. Although Hamilton's rule predicts greater acts of kindness among kin than non-kin, it also predicts greater acts of kindness among kin that are closely related. The same logic holds for acts of violence. The degree of genetic relatedness is a parameter that dictates both when helping and harming are favored.

Adopting the broad implications of Hamilton's insights, we can understand how selection molds psychological and physical tricks that enable offspring to manipulate their parents and to enable parents to distinguish truth from spin. In a wide variety of birds and mammals, the young emit begging cries; as mentioned in chapter 5, human infants are no different, except perhaps for their ability to marry the acoustics to tear ducts. When the young beg for food, parents must estimate whether their requests are honest, truthful indicators of nutritional needs. Several studies now show that properties of the beggars' cries reveal how healthy they are. Parents use this information to guide their investments, sometimes giving more to those in need, and at other times, backing off when the requests are uncalibrated. Among vervet monkeys, all mothers respond to their offspring's cries when they are young; these cries seem to be closely matched to need. Within a few weeks, most individuals start crying at much higher rates, and, in many cases, these cries are gratuitous and unnecessary, designed to draw the mother back in because weaning is on the horizon. Mothers tend to recognize the mismatch, and begin to decrease their responsiveness. Most infants see what is happening and back off, returning to a level of crying that is consistent with their needs. A few either don't see the problem or don't care, and keep pushing, crying at higher and higher rates. Like the village response to the boy who cried wolf, mothers continue to ignore these cry babies, and most die before they reach their first birthdays.

CONDITION 3. If we keep our eyes on the genes, as opposed to the body or the individual or the group, then selection for genetic fitness can happen in several ways. Thus, a central outcome of Hamiltonian thinking is that we shift from looking at how many babies each individual generates to how many related gene copies each individual helps pass on. We shift from a notion of individual or direct genetic fitness to a more inclusive sense of fitness, which includes the number of babies produced plus the help this individual allocates to others who share genes in common.[10] Thinking in this way helps resolve some initially paradoxical observations in animals. For example, in several bird species, and some primates, reproductively capable individuals forgo the opportunity to breed. If optimizing genetic output is the evolutionary goal, celibacy is the ultimate maladaptive dead end. What happens in these seemingly bizarre cases is that individuals forgo reproduction in order to help their parents with the next generation of offspring. Helping their siblings helps their genetic success. Similarly, there is the puzzle of why human females, and perhaps a few other species, such as chimpanzees and short-finned pilot whales, live on for so many years after menopause; for example, like humans, pilot whale females stop reproducing when they are between thirty-five and forty years old, but then live until their sixties, with some evidence for continued lactation until their early fifties. One explanation, originally put forward for humans but potentially applicable to other species as well, is that grandmothers shift from investing in their own offspring to investing in grandchildren and other closely related kin. This logic leads to the intriguing possibility that grandparents provide an insurance policy for our genetic prosperity, helping both themselves and their kin produce more and healthier children.[11]

CONDITION 4. Most sexually reproducing species with parental behavior have more than one offspring in a lifetime. Sometimes
the parenting is restricted to pregnancy, as when a female sun-
fish waits for her 300 million eggs to be fully fertilized, and
then launches them en masse into the unpredictable ocean currents. Sometimes it is more extensive, as in the Australian brush
turkey. Males build a large mound of leaf litter. Once it has
rotted, females pay a visit, carry out a home inspection, and, if
satisfied, mate and then deposit their eggs inside the mound.
With the exception of a bit more tidying up of the mound by
the male, this ends their period of parenting. When the chick
hatches, there is no one around to help; born into a mound of
litter, the brush turkey chick climbs out and begins life, alone.
At the opposite extremes are the many birds and mammals that
tend to their young for many months and even years after they
are born, giving them food, transport, defense against predators
and competitors. Given that a parent's resources for investment
are limited, allocations to one offspring potentially reduce what
is available to another. For parents, spreading the wealth is a
guiding principle. For their offspring, taking the wealth is a
more appropriate principle.

These initial conditions dictate how the game transpires. Parents want
offspring to survive and reproduce. Offspring want to survive and reproduce. Offspring want to optimize their individual chances. Parents want to
optimize their lifetime reproductive output, a forward-looking strategy that
attempts to estimate the number of potential offspring in a life well lived.
Because they are genetically related to their siblings, by either a half or some
amount less depending upon parentage, they also want their siblings to
survive, as long as it doesn't cost them too much. What Trivers's parent-
offspring conflict model therefore generates is a way to think about the dynamics of family conflict. It is a game in the game theory sense discussed in
parts I and II, because there are different strategies and the payoffs to one
strategy depend upon what others do. It differs from these games because
the time scale extends from one or a few repeated games played within
an individual's lifetime to many games played over the course of a species'

lifetime. Although each reproductive round constitutes a one-shot game, the consequences ramify over the course of multiple rounds.

Using these model games, theoretical biologists have mapped out the conditions in which infanticide or siblicide should be obligatory as opposed to facultative, and when parents should invest more as opposed to bailing and trying again the next season. These conditions, written out in the form of mathematical rules, are the beginning steps of characterizing the principles and parameters of helping and harming within the context of family dynamics.

Although parental care is only one small corner of the space that covers human and animal social relationships, this brief sketch of the issues highlights two central points in my account of the moral faculty. One, we can only understand the principles of harming and helping by looking at the parameters that, when switched, determine when particular actions are operative and when they are not. We tend to think of human parenting as obligatory, with abandonment and infanticide as not only forbidden but morally abhorrent. From a descriptive perspective, this is not necessarily the case. If we take into account the conditions under which humans evolved, and consider some of the parameters mentioned above, not only are abandonment and infanticide operative in human societies, but they are expected under certain conditions. Neither infanticide nor siblicide are necessarily aberrations, even though they can be triggered in humans and animals by extreme conditions. In all animals, they are in part the consequence of unconscious principles that guide when individuals harm other family members. We, of course, can consciously support these unconscious processes, or fight against them.

The second point, also on a descriptive level, is that we share with other animals some of the same principles and parameters in the context of parental care. Many animals take care of their young, doing what they can to guarantee a long and successful life as breeders. But taking care of their young is not obligatory. There are no rules stipulating that all parents must take care of their young and that siblings must never fight. There are exceptions both within and between species, dictated by environment, typical family size, and patterns of growth. What is not yet known is whether individuals ever witness violations of expected patterns of parental behavior or sibling interactions, judge these as counter to species-specific norms, and respond with appropriate measures of intervention. By "violation" I mean a pattern of behavior that is inconsistent with current principles and parameters of parental care for the species, at this moment, in this particular habitat. For example, in siblicidal species, if resources are plentiful and the firstborn still kills the secondborn, are parents indignant but passive nonetheless? Or do they rise to the occasion and intervene on behalf of their youngest? Biologists have already played around with some of the relevant manipulations, taking away the firstborn, adding more food, and beefing up the secondborn. In some cases, as in the siblicidal egrets, changing these factors effectively turns on and off the severity of the attacks. Thus far, however, none of these manipulations alters parental intervention. Parents either don't worry about violations, don't see them as violations, or don't have the flexibility to change the evolved bias favoring indifference. In raising these possibilities, my hope is to stimulate biologists to carry out manipulations of this kind so that we may better understand how animals judge violations of principles dictating harming and helping in the context of parental care, as well as for other social relationships.

PROPERTY RIGHTS

In the spring, in grassy fields throughout North America, male redwinged blackbirds set up a space in which they first sing to announce their arrival and then continue singing to attract potential mates. The stomatopod, a beautifully colored marine shrimp living in shallow reefs, defends a small cavity that is both a source of safety against predators and a playground for enticing willing mates and fighting off competitors. Due to the stomatopod's biology, individuals continue to grow throughout life and thus are frequently in search of new cavities that better fit their size. Hermit crabs have a much more defined space that they call their own—the shell on their backs. Like stomatopods, they, too, grow throughout life and are often in the market for a new port-o-home. These are all cases in which property is purchased in the currency of space or a physical object that is functionally equivalent to space. Sometimes, animal property takes the form of one or more other individuals. In harem societies, such as gorillas, hamadryas baboons, and some ungulates (gazelles, horses, deer), one male monopolizes access to a group of females. They are *his* females, as defined by exclusive mating rights, a willingness to let them feed in the same space, and strong motivation to defend them against predators and competitors. Needless to say, females are not passive, letting the males have their way. When things are bad—a deadbeat dad, lousy defender, or fertility dud—they leave.

All of these cases raise issues linked to property rights. When we speak of property rights, we refer to an individual's or group's control or jurisdiction over an object. This broad definition is useful, especially in today's world, as it allows less tangible turf, such as intellectual property, to count as a commodity that can be protected and is worth protecting. Making the concept explicit and legally enforceable fuels competition, creating a clear delineation between haves and have-nots.

Do animals have a concept of property and, if so, what is it? What does ownership entail? How is it enforced? What are the principles and parameters that guide property ownership? How do individuals or groups handle violations?

Animal societies don't have written rules of conduct. They do have unwritten rules that function in the regulation of dominance relationships, sexual behavior, and the defense of space. These unwritten rules set up expectations about patterns of interaction, about likely outcomes, regularities, and bankable resources.

Territorial ownership is established by delineating boundaries. Sometimes this is done by placing "no trespassing" signs along the perimeter, in the form of urine or feces. Sometimes it is announced by calling, as is the case for redwinged blackbirds and many other species. For most territorial species, there is a precedence effect. If an individual without a territory flies or walks into a space and the owner is present, the intruder will back off almost immediately, without a challenge. There is an unwritten rule that states: Space occupied by another individual constitutes their uncontested property. Challenges do arise, and often they lead to escalated chases or attacks. Territory owners usually have the home-court advantage when things get nasty, though, sometimes, repeated attempts to garner a piece of land leads the owner to relent just to get the intruder off its back.[12]

For animals like stomatopods and hermit crabs, other factors determine when it is permissible to challenge ownership. Differences in size and fighting weaponry—what biologists call "resource-holding potential"—tend to determine both the state of play and the outcome of the challenge. Hermit crabs, for example, are often in the market for a new shell. Hermit crabs do not typically hand off the shell from their backs, so those who are shopping must evaluate whether the current owner is well suited to his or her home. Shell size doesn't provide an accurate indicator of hermit crab size. To assess fit, hermit crabs tap on each other's shells with their claws. The sound produced by this tap provides information about fit. If the animal fits well, the resonance properties of the shell are different than when the individual is small and there is a lot of air space. If a big individual taps and then hears the signature of a small hermit crab, he will begin to push and ultimately oust the little one from his home; the smaller hermit crab doesn't put up much fuss, because his own tapping has generated information about the size of his opponent. Here, then, an animal's fighting potential plays a role in guiding when ownership challenges are expected and when they are not.

Questions of ownership are different when the property can move or be moved. This is true of food. In some lovely studies with macaques, the biologists Hans Kummer and Marina Cords presented individuals with a raisin-filled tube that was either fixed to a wall or freely movable.[13] If the subordinate gained access first, dominants snatched the tube away if they were fixed or if the subordinate relinquished physical control. However, if the subordinate held the tube close to its chest, then the dominant sent an approving nod of ownership.

Apparently, macaques have a rule of ownership that is established by proximity. This rule effectively controls the dominant's behavior and provides subordinates with a sense of peace when they do gain access to food.

In many nonhuman primates, exclusive relationships are formed between males and females. These relationships, characteristic of monogamous and harem societies, represent attempts to restrict sexual interactions to the pair or to the one male and his group of females. Members of each of these groups see the others as property, in some sense. Among humans, Western cultures tend to think of harems as a collection of wives under single male control. In Arabic cultures, however, where harems appear to have originated, the word stands for "forbidden" or "secluded." These two interpretations place emphasis on a restricted resource, which is borne out by attendant eunuchs, as opposed to male servants who might provide competition. Hamadryas baboons also live in harem societies and represent an interesting case where an unwritten rule establishes control and restricted access to a valuable resource—property. A hamadryas harem consists of one adult male, several adult females, and their offspring. Like the regulation of territorial space, the rule for harems is: If you have one, everyone else respects it. Neither other males with harems nor young males without ever mount a challenge, attempting to take adult females away. Given this respect, how do harems form? Typically, adult males watch out for budding young females and herd them into their harem by trying to show off their leadership qualities. Under captive conditions, you can watch such dynamics unfold and also observe the respect that emerges between males. If you put male A into an enclosure with an unknown female C, he will immediately begin the herding process, grooming her and staying close by. If a male B watches this process and then moves into the enclosure, he will stay away from the honeymooners. But now reverse the situation and let B herd female D while A watches. When A moves in, he, too, respects B and D's honeymoon. These unwritten rules help maintain societal conventions in the face of individuals tempted to cheat. And in the case of hamadryas baboons, these rules maintain property rights—the entitlement an adult male has when he establishes a harem.

Paralleling the line of questioning for parental care, we can ask how individuals might respond to an experimentally imposed violation. How does hamadryas male A respond to B if B moves in on female C while A watches? If B never saw A consorting with C, he may be ignorant, as opposed to some macho stud intentionally trying to muscle in on someone else's girl. A might want to regain his turf, but should be less aggressive toward B than in a situation where B watches A and C consort and then nonetheless tries to take over when alone. Here, we are not only considering what counts as a violation of the social norm, but whether animals take into account the causes of a violation. We are back to the Rawlsian creature and questions concerning the capacity to process the causal and intentional aspects of an event. Though we don't have answers for this kind of experiment, or others concerning the defense of space or sexual partners, we do have partial answers in a different context: the power struggles that animals experience in social groups characterized by dominance hierarchies. These interactions present one of the premier contexts for looking at the kinds of principles that have evolved to guide access to and control of resources.

Dominance relationships are about power, and they are prevalent in the animal kingdom.[14] They differ in terms of their rigidity and the principles that determine winning and losing fights. Among some species, such as the eusocial insects (bees, wasps, ants) and the naked mole rat, individuals are born with a rank, and there is little hope for change. In these species, low-ranking animals might dream of an overthrow, but that rarely happens. In other species, there is less rigidity and greater opportunities for individuals to change their dominance status. Among several mammalian species, especially those characterized by a polygynous mating system, size largely determines rank. The elephant seal provides a classic illustration. The harem master is the biggest male, a gargantuan individual who dwarfs all the females, most of the males, and, as a result, controls over 90 percent of the matings. After one or more seasons as harem master, and several fights and copulations under his belt, another male comes along and deposes number one. Size rules as the only relevant factor. In many primate species, in contrast, an individual's dominance status is determined by size, age, sex, the mother's rank, and the availability of coalition partners. In most but not all primates, adults outrank juveniles and males outrank females. Among baboons, when a male reaches reproductive maturity, he often leaves his natal group. Upon arrival into his new group, he typically assumes the position of top baboon, the alpha male. In contrast, when rhesus monkey males leave their natal group, they drop to the bottom of the heap in their new group, even if they were the top rhesus in the natal group. With many Old World monkeys, such as the macaques, if your mother is high ranking, so are you, even if you are small in size.

Rank establishes who wins the footrace to resources, including food, resting spots, and mating opportunities. Low-ranking animals are typically obedient, respecting the authority that comes with high rank. When they break with tradition, attempting to overthrow a higher-ranking animal, they often recruit the help of another individual. These coalitions operate with political savvy, as illustrated by de Waal's description of chimpanzee behavior.[15] To the unsophisticated observer, the interaction looks like two chimpanzees beating up a third. To the skilled observer, the coalitionary attack represents the culmination of a carefully plotted over-throw. One case involved the recently deposed alpha male Yeroen, the newly crowned alpha Luit, and the beta male Nikkie. Soon after Luit's as-cendancy, Nikkie formed a coalition with Yeroen to take over the alpha status. Without Yeroen, Nikkie's rise to power would have been impossi-ble. Further, Nikkie needed Yeroen's coalition support once he attained alpha status. Since most animals can't commandeer coalitionary support, they offer goods likely to encourage such support. Nikkie offered first dibs on mating, which Yeroen accepted, mating twice as often as Nikkie. Once Nikkie's tenure was secure, he resumed his mating advantage while also maintaining his alliance with Yeroen.

Coalitions represent an added layer of complexity to our understand-ing of the principles guiding dominance interactions. Although we can assign a rank to each individual in most socially living animals, for those that use alliances in battle, their partners provide them with an added de-gree of leverage. Predicting the outcomes of fights is therefore more diffi-cult, as it depends on the conditions in which the coalition is engaged. Regardless of the outcome, these interactions illustrate that there are prin-ciples guiding the control of resources within groups. These principles de-termine when it is appropriate to respect power and when it is appropriate to challenge it, seeking help in the service of inflicting harm.

In some species, an individual's skills or savvy may be so great that reputation may override dominance rank, giving the individual extra leverage in the competitive arena. To illustrate, consider a stunning exper-iment by the biologist Eduard Stammbach.[16] In each of several groups of long-tailed macaques, Stammbach removed the lowest-ranking individual and provided him with special training on a popcorn dispenser; popcorn was a delicacy in this particular captive colony. Each low-ranking animal learned, on its own and away from its group, to press a series of levers in combination to deliver the popcorn. Once these specialists learned the trick, Stammbach placed each one back into its group along with the popcorn machine. When the "ready" light lit up on the dispenser, the spe-cialist marched over, pressed the levers, and watched the popcorn fall into the bowl. But before the specialist could obtain his reward, a more domi-nant individual snatched the popcorn away. This sequence—subordinate works the machine and dominant eats the popcorn—occurred repeatedly until the specialists went on strike. The light on the dispenser glowed, dominants looked to the specialists, and the specialists looked away. What happened next, however, was remarkable. In the absence of any training by Stammbach, the dominant individuals stopped threatening the specialists away from the dispenser and started grooming them more often. Soon thereafter, the specialists returned to the dispenser, pressed the levers, and now sat and ate popcorn with the dominants. Although the specialists never gained in dominance rank, like the court jester they enjoyed "quality time" with the royalty because of their skills. What makes this outcome particularly interesting is the fact that the dominants did not try coercion as a means of achieving access to food. Stammbach's work shows that animals can assign reputations and use these attributions in the service of getting food. Instead of harassment, a strategy that works in many primate species, these animals opted for niceness.

Socially living animals tend to know a great deal about their own status and the status of others in their group; this capacity appears to be most exquisite among the primates.[17] When a subordinate steps out of line, attempting to garner resources typically earmarked for more dominant animals, they do so as covertly as possible, often using the dominant's line of sight to work out what they can get away with. If they step out of line, and the dominant catches them, a chase typically ensues, often escalating into an all-out fight. To prevent such violations, and to keep subordinates in line, dominants sometimes attack for what appears to be no good reason. As Joan Silk put it, in many primate species, dominant individuals "practice random acts of aggression and senseless acts of intimidation." On the face of it, these might look like violations, actions that go against the principles guiding harm. However, these acts fit well with the logic of animal contests in which individuals are relatively well matched for size, encounter each other frequently, recall past encounters, and incur significant costs when they fight. Here, the dominant's best strategy is to occasionally, at random, launch an all-out attack so that the subordinate has little hope of retaliating. Launching such attacks will keep subordinates in check, preserve them in a relatively high and costly state of stress, and maintain the stability of the social hierarchy. In addition to these attacks, however, are also signals that dominants send to subordinates to indicate that they are most definitely not in an attacking mood. These are also important parts of the social system, as they enable dominants to form coalitions with subordinates, to handle their babies, use them as babysitters, and engage from time to time in friendly interactions.

Cheney, Seyfarth, and Silk used the fact that some vocalizations are only given by individuals of a certain dominance rank to ask whether baboons recognize anomalous interactions—situations in which a subordinate appears to step out of line.[18] Among baboons living in the Okavango delta of Botswana, high-ranking mothers often approach lower-ranking mothers and grunt. Grunts are often signals sent by dominant animals toward subordinates, and sometimes the other way around. Following a grunt, subordinate mothers sometimes respond with a submissive fear bark. Dominant animals never give these fear barks to subordinate animals. Of significance, grunts are given without triggering barks, barks are given in the absence of grunts, and dominant females can continue to grunt well after the subordinate female has ceased barking. The only fixed vocal pattern occurs when one animal grunts and another barks. Here, the grunt always comes from the dominant animal while the bark always comes from the subordinate.

Taking advantage of these signals of dominance, Cheney and colleagues played back either anomalous sequences or consistent ones, and used the logic of the looking time method to probe their expectations. Baboons looked longer at the anomalous sequence than the consistent one—they detected a transgression.

Showing that baboons detect this violation is a first step. It demonstrates that they are sensitive to particular principles of dominance and how they are expressed behaviorally. What is needed next is some measure of how listeners judge such violations, and whether it changes the dynamics of their social relationships. What we need is a situation in which individuals are given the opportunity to either redress the imbalance by punishing the violator or take advantage of this situation to shift the rules of the game.

Once we step outside the sphere of parental care, we find other principles and parameters guiding patterns of harming and helping others. These principles guide when individuals compete for resources, when they are expected to defend their property with little or no contest, and

ANOMALOUS PLAYBACK SEQUENCE

1. DOMINANT "GRUNT"
2. SUBORDINATE "FEAR BARK"

RANK-3

RANK-2

CONSISTENT PLAYBACK SEQUENCE

1. DOMINANT "GRUNT"
2. DOMINANT "GRUNT"
3. SUBORDINATE "FEAR BARK"

RANK-1

RANK-3

RANK-2

when they should expect to be harmed by a group member because they have violated a local principle. My discussion has only scratched the surface of these interactions and the theoretical models designed to explain them. In general, we have little understanding of the appraisal side of animal contests, the side that aligns with the Rawlsian creature. We don't know what constitutes an infringement of property rights and we don't know what psychological resources animals have to evaluate these situations. We don't know whether animals take into account others' intentions in deciding whether an act that causes harm in another is punishable. But the best starting position is to recognize that these questions exist and are worth addressing.

IT TAKES TWO

Among Shakespeare's historical plays, *King Lear* stands out for its intricately crafted plot, filled with acts of cooperation, deception, familial strife, status-striving, investment strategies, parental favoritism, and sibling rivalry. The story begins with the somewhat pathetic Lear asking his three daughters—Goneril, Regan, and Cordelia—to articulate the depth of their love for him. This request amounts to a green flag for competition among full siblings, an attempt to inspire rivalry in the service of commandeering a substantial inheritance. Goneril and Regan play the game, each attempting to one-up the other. Cordelia, Lear's favorite, sees through her sisters' transparent flattery and decides to speak the truth, telling her father that she loves him as a daughter should, but no more. Lear is outraged and removes Cordelia's title and inheritance, allocating his wealth to Goneril and Regan. Goneril and Regan, greedy and power-hungry as they are, think their father is foolish and unworthy of his title, and plot to remove him from power. They cooperate, each realizing that they are more powerful together than alone. As the rest of the play unfolds, we witness the breakdown of Goneril and Regan's alliance at the hands of selfishness, and the rebuilding of Lear's love and respect for his favorite child, Cordelia. In this story, we see the power of cooperation, the adaptive significance of coalitions, the dissolution of kinship, the competition for resources, and the challenges to maintaining stable cooperative alliances among kin.

Some animals, notably lions, hyenas, dolphins, and many of the monkeys and apes, form coalitions to outcompete individuals for access to food, space, and sex. These alliances, built among kin and among nonkin, require coordination, commitment, and cooperation. Though cooperation is present throughout the animal kingdom, it takes on various forms based on different principles, mediated by psychological capacities and adaptive goals. To get to the point, although humans and other animals share many of these forms of cooperation, humans stand out in two distinctive ways. We are the only animal that cooperates on a large scale with genetically unrelated individuals and that consistently shows stable reciprocity, exchanging within the same market currencies or different ones. I next turn to some of the principles underlying our shared capacities for cooperation, using the remaining sections of this chapter to characterize the differences, explain why they arise, and why we have uniquely evolved particular forms of helping.

As Hamilton's rule makes clear, cooperating with kin is no longer a dilemma for evolutionary biologists. As the illustration below indicates, *kin cooperation* evolves and remains stable because the altruist's costs of

A. Kin cooperation [relatedness A–B > 0]

A incurs cost,
B receives benefit

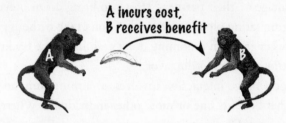

B. Byproduct mutualism [relatedness A–B ≥ 0]

A alone or A+B incur cost,
A+B receive benefit

C. Reciprocity [relatedness A–B = 0]

A incurs cost,
B receives benefit

T

T+1

B incurs cost,
A receives benefit

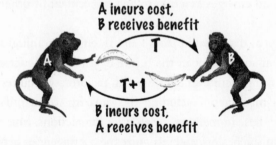

giving are outweighed by the benefits to the recipient who carries the same genes. The fact that this solves the problem of kin-directed helping doesn't mean that kin will always be nice to each other—think King Lear and the siblicidal egrets and eagles. Kin-directed cooperation is, nonetheless, common among animals, and, for many socially living species, represents the dominant pattern of helping. The puzzle comes from interactions between genetically unrelated individuals. As Darwin stated in the *Origin of Species* (p. 228), "Natural selection cannot possibly produce any modification in any one species exclusively for the good of another species." Though Darwin's challenge focused on interactions between species, we can apply it with equal force to individuals of the same species. We need to explain why, for example, lions out on a cooperative hunt don't cheat, taking advantage of their partners' efforts to bring down prey; why vampire bats regurgitating blood to each other don't take without giving back; why primates engaged in grooming don't cash in on the backrub and fur cleaning without returning the favor.[19]

Every cooperative interaction involves, at a minimum, an act by one individual that benefits one or more other individuals. Where things get interesting from a Darwinian perspective is when the act is costly, and kinship provides no explanatory help. *By-product mutualism* arises when the outcome of an act benefits both participants. A good example is cooperative hunting, where each individual, A and B, has a selfish interest in hunting success, and needs at least one other individual's help to maximize the odds. Hunting itself is costly to both, but both benefit if they bring down a prey item together. In these situations, it is possible that both the costs and benefits are differentially shared, perhaps one individual incurring greater costs and reaping greater benefits. The key point is that cooperation emerges as a by-product or accident of otherwise selfish interests.

Reciprocity, as discussed in parts I and II, entails an initial act of altruism followed at some point in the future by a reciprocal act of altruism. Although seemingly quite straightforward, reciprocity requires substantial psychological machinery, including the capacity to quantify costs and benefits, store these in memory, recall prior interactions, time the returns, detect and punish cheaters, and recognize the contingencies between giving and receiving.

For all forms of cooperation, we ask the same question: Do cooperators do better working with one or more other individuals than they would do on their own? Consider two examples: cooperative hunting in lions and alliance formation in dolphins.[20] In both cases, we want to know which parameters, when turned on, favor cooperation and which select against it. To collect the relevant observations, the biologists Craig Packer and Anne Pusey dedicated countless hours on the Serengeti plains of Tanzania to watching lions hunt, counting the numbers of hunters per chase, their success rates, prey capture size, and amount of food per individual. In study areas where lion groups cooperate, success rate increases with group size, with groups doing better than solitary hunters.

However, within this general pattern that favors cooperation, there are different strategies during the hunt. In particular, lions of the Serengeti adopt one of three strategies prior to an opportunity to hunt: They *refrain* from hunting altogether, *conform* to the general pattern of hunting exhibited by others, or *pursue* an active hunting role that varies relative to others in the group, sometimes leading the chase, sometimes hanging back. What determines a lion's strategy is, to a large extent, the size of the target prey and the particular composition of animals who do or do not choose to join in on the hunt. When prey are small, like warthogs, solitary hunting is common, and most members of the pride refrain. When prey are large, like buffalo, cooperation is not only more common but necessary, as large prey are dangerous and a greater challenge to take down. Refraining from a buffalo hunt is a form of cheating, as is pursuing but taking the least active role. What is puzzling about lion cooperation is that there seem to be virtually no costs to cheating, and no benefits to being a great and active hunter. In this sense, though there are principles that guide when cooperation arises in lions and is maximally beneficial in terms of yield, there are no detectable consequences for being a savanna potato, hanging back while others do the work, and then reaping the rewards of their efforts.

Cooperative group-hunting in lions represents a case of mutualism, because those involved mutually benefit, even if there are asymmetries in the upfront costs and the ultimate returns. Lions also cooperate in territorial defense against intruders, and, often, brothers join forces to take over a pride, commit a few cases of infanticide to prime female reproduction, and then settle into their groups. Dolphin alliances arise within large fission-fusion communities, in which overall community size can reach hundreds, but on a day-to-day basis, individuals interact within the context of small, ephemeral groups. Like Packer and Pusey, the biologist Richard Connor has spent countless hours at sea, observing bottlenose dolphins living in Shark Bay, Australia. Given the parallels between dolphin and chimpanzee social organizations, Connor expected to find evidence of coalitions, and he did. Early on in the project, he observed the emergence and maintenance of male coalitions, typically two or three individuals joining forces to outcompete loners or other coalitions for access to potential mates. Paralleling the cooperative hunting work in lions,

Connor showed that coalitions had much higher success in guarding females than did individuals on their own. Moreover, he showed that success rate increased with increasing coalition size, culminating in an extraordinary superalliance of fourteen males, each of whom had five to eleven alliance partners within this larger group; like many primate groups, this shows that dolphins also have a hierarchical social organization with different levels of relationships. Although Connor has yet to observe different strategies within his alliances, as seen in lions, there is a division of labor in the context of hunting. When dolphins hunt cooperatively, one individual exclusively takes on the role of "driver," herding prey toward a group of individuals that take on the role of a "barrier." Given that many have remarked on the apparent uniqueness of a division of labor in human evolution, examples like these should not only humble us but shift our focus to the conditions in which individuals assume similar or different roles in a cooperative relationship.

In cases where animals form alliances, several questions arise:[21] How do individuals pick their partner or partners? Do animals take into account not only their partners' loyalty, work ethic, and physical power, but also their skills? To form an alliance, do individuals require direct experience with their potential partners or can they use their observations to decide who would cooperate and who would defect? As mentioned earlier, the evolutionary biologist Richard Alexander noted long ago that one way in which animals might engage in stable cooperation is by collecting data on who cooperates and who cheats, selectively cooperating with the former and ostracizing or ignoring the latter. This would count as a form of indirect reciprocity, and may carry less of a psychological load than direct reciprocity—our next topic. Observations may also provide individuals with information about who lags in an alliance and who leads. How many laggards can an alliance tolerate while maintaining its effectiveness and yield per individual? In theoretical models and observations of lions, cheetah, and dolphins, alliance size is typically three, especially when the target of the alliance is a sexually receptive female. What is unclear is how the composition of these threesomes influences their success and stability. Recent work by Connor suggests that in dolphins, the most stable coalitions consist of individuals who synchronize their activities.

FOOD TRAITORS

Naguib Mahfouz, the 1988 Nobel laureate in literature, noted: "Food offers a better explanation of human behavior than sex." Yet even with all the food in the world, sexless lives are genetic dead ends. Trading food for sex is an ancient habit, as is exchanging food during times of duress. Food was most likely the first commodity used by humans in trade, and especially reciprocal exchanges. Of the many interesting observations in nature of animals exchanging resources—including vervet monkeys, who help each other in a fight and then return the favor with a groom-fest, and antelope, who take turns grooming the inaccessible parts of an-other's body—food is the most common currency. Focusing on food exchange therefore offers the best opportunity to look for reciprocity across a broad array of species.

We owe the initial theoretical arguments for reciprocity to Robert Trivers, who, using the logic of Hamilton's rule, turned the Golden Rule into a selfish strategy. Recall from parts I and II that for reciprocal altru-ism to evolve, individuals must satisfy three conditions:

1—small costs to giving and large benefits to receiving
2—a delay between the initial and reciprocated act of giving
3—multiple opportunities for interacting, with giving contingent upon receiving

What distinguishes reciprocity from mutualism is the time delay—condition 2. The hurdle is to surmount the delay, a period in which recip-ients might never return to pay their dues. Condition 3 helps constrain those who might take off: Reciprocity should only begin when the initial donor has reason to expect further opportunities to interact with recipi-ents. Given conditions 2 and 3, we would expect to find reciprocity in highly social, long-lived species, where individuals have multiple oppor-tunities not only to observe others but to interact with them, either pro-moting the reciprocal relationship or ending it by punishing those who renege.

Vampire bats have relatively large brains—for bats of their size, that is. They can live for almost twenty years, spending much of their time in large, stable social groups where there are multiple opportunities to interact with the same individuals. Individuals have distinctive voices, thereby enabling individual recognition even in the darkness of a hollow tree—their home. Therefore, individuals can recognize their social partners and interact frequently with them. A vampire bat's survival depends critically on the consumption of blood. If an individual goes for more than sixty hours without a blood meal, it dies. On any given day, therefore, an individual must either obtain its own meal or convince someone else to regurgitate some of the undigested blood. These attributes make vampire bats ideal subjects for studies of reciprocal altruism.

The biologist Gerry Wilkinson observed over one hundred regurgitations among vampire bats. Because blood is valuable, giving it up represents a cost. Regurgitating is altruistic. Why do it? Of the cases observed, nearly 80 percent were between mother and infant. These were not examined in any detail, because there's no puzzle: Regurgitating to your offspring makes sense, since you share half of your genes with them; there is no expectation of reciprocation here, and Hamilton's rule explains the kin bias. Of the remaining regurgitations among more distantly related individuals, about half were between grandparent and grandchild; these, too, can be explained by an appeal to kinship and maximizing genetic self-interest. Reciprocity isn't required.

It seems that regurgitation among vampire bats is largely motivated by kinship, with an extremely small proportion of cases among genetically unrelated bats. Nonetheless, given that some regurgitations were delivered to non-kin, these cases require some explanation. There are two possibilities: Either some bats made mistakes, failing to recognize their kin and thus accidentally giving blood to non-kin, or they purposefully gave blood to non-kin with the expectation that they would receive blood back in the future.

To better understand what motivates regurgitations among non-kin, and to clarify whether giving is contingent upon receiving, Wilkinson conducted a laboratory experiment with eight unrelated vampire bats. Over many days, he removed one bat from the colony before feeding while providing the other bats with a two-hour-long bloodfest. He then returned the now-starving bat to the group of blood-stuffed bats. The pattern of blood-sharing was clear: Individuals regurgitated blood to those who had regurgitated to them in the past. Although the number of players and the number of reciprocal regurgitations were small, these experiments provide evidence of a contingency: Bats give blood unto those who have given to them in the past.

Although this is beautiful biology and fascinating behavior, there are at least two reasons for expressing caution in accepting the vampire-bat case as evidence of reciprocal altruism. One: the number of naturally observed cases is small and could be explained as errors of recognition, as distinct from reciprocation among non-kin. Though regurgitations are given to unrelated animals, these are infrequent, and there is no evidence that individuals recognize the recipients as non-kin as opposed to kin; Wilkinson didn't conduct any tests to show that bats recognize their kin and, if so, to what degree of relatedness. The consequence of contingent regurgitation may benefit non-kin, but the payoffs and mechanisms may have evolved for kin, occurring among non-kin as an accidental by-product with insufficient fitness consequences for selection to operate. Two: even if we accept these few cases, it is not at all clear whether reciprocal altruism among non-kin plays a significant or trivial role in individual survival. The fact that individuals need blood to survive is clear. Whether or not they depend upon reciprocation with non-kin to survive is a different issue.

A second way to test for reciprocal altruism in animals comes from work on captive blue jays trained to peck keys in one of two classic economic games.[22] Although the task—pecking keys—is highly artificial, jays in the wild are cooperative breeders, meaning that a large extended family is responsible for jointly bringing up the young. As such, their biology might predispose them to cooperate even under unusual situations. The biologists Kevin Clements and Dave Stephens set up a "prisoner's dilemma," in which the payoffs for mutual cooperation were higher than for mutual defection but lower than for one player defecting while the other cooperated. Trivers originally used the prisoner's dilemma to show why cooperative games are open to defection. As illustrated, the dilemma arises because the best payoff for both is to cooperate, but on any given turn, the best individual payoff is always defection. In the second game, called "mutualism," there is no dilemma because the best payoff for both individuals, alone and together, is to cooperate; there is no temptation to defect.

Every game involved two jays, each with access to a "cooperate" and a "defect" key. One jay started off, pecking either the "cooperate" or the "defect" key. Immediately after the first jay pecked, the second jay had an opportunity to peck, but with no information about his partner's choice until the food reward emerged; the experimenter made the food payoff depend upon the jay's choice, indicated below by the relative size of each circle within the two-by-two table. When the jays played a prisoner's dilemma game, they rapidly defected. No cooperation. In contrast, when the jays switched to a game of mutualism, they not only cooperated but maintained this pattern over many days. That jays switch strategies as a function of the game played shows that their responses are contingent upon the payoffs associated with each game.

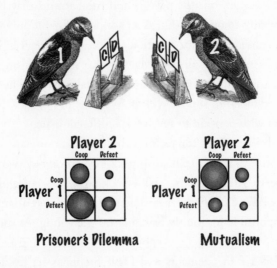

To determine if other conditions might enable cooperation among jays in the prisoner's dilemma, Stephens and his colleagues ran a second experiment, this time targeting a central aspect of reciprocal altruism: The temptation to take an immediate benefit outweighs the benefits of waiting for a larger payoff. As discussed in chapters 4 and 6, several studies of animals and humans reveal that waiting for a payoff devalues the item's worth. A small payoff now is better than a large payoff later. Humans and animals discount future rewards. In studies of human cooperation,

shorter delays to the large reward lead to higher rates of cooperation—
that is, less defection. In the original jay work, pecking brought an immediate payoff of some amount. In this second study, Stephens and
colleagues delayed the payoffs. To obtain food, each pair of jays had to
play several rounds with their partner before obtaining the payoffs. This
setup therefore removed immediate temptation and also allowed each
player to observe the other's responses. Under these conditions, jays cooperated with a cooperator playing tit-for-tat. They solved the repeated prisoner's dilemma, opting to cooperate rather than defect.

Clements and Stephens concluded their original paper on jays as
follows: "[T]here is no empirical evidence of non-kin cooperation in a situation, natural or contrived, where the payoffs are known to conform to a
Prisoner's Dilemma." The follow-up studies with jays led Stephens and
colleagues to a different conclusion, but one that is consistent with the
idea that animals are incapable of maintaining reciprocal relationships
under natural conditions: "Our work suggests that the timing of benefits
can be the difference between stable cooperation and cooperation that
erodes to mutual defection . . . [But] the experimental machinations required to stabilize cooperation . . . are special." In other words, nature
may never provide animals with the right conditions for reciprocally stable relationships, even if under rather extreme and artificial conditions,
some animals have the brain power to engage in reciprocity.

A third test of reciprocity involving food exchange comes from a New
World primate—the capuchin monkey. Capuchins live in large, stable social groups, mate polygamously, and inhabit the rain forests of South and
Central America. Due to their large brains, exceptional dexterity, and
highly social character, they have appeared in numerous television shows,
even portrayed—inaccurately—as the source of the *African* ebola virus in
the movie *Outbreak*. De Waal capitalized on their apparent social intelligence by conducting a series of experiments designed to address earlier
failures with this and other species.[23] In the first experiment, de Waal
trained female capuchins to work for food, either on their own or with
another unrelated individual. The task was simple: pull a rod to bring a
cup of food within reach. When there were two capuchins, and therefore
two rods, each individual had to pull at the same time in order to bring the
cups within reach. When the experimenter placed food in both cups, both

capuchins pulled. Although their joint action looked like cooperative behavior, we can more readily explain it as selfish, with each individual pulling for him- or herself. This is a case of mutualism. When the experimenter placed food in only one cup, the individual facing the food cup almost always pulled, whereas the other individual pulled less frequently. When the player lined up with an empty cup pulled, she was more likely to obtain food from the player with food than when she failed to help; I say "she" because cooperation of this kind only emerged among females. Individuals with access to the food cup rarely handed food to helpers. Instead, they allowed helpers to approach and take pieces of food through the wire mesh that separated them.

Is this reciprocal altruism? The helper provides a benefit to its partner by pulling the rod. Pulling the rod involves some cost. By definition, then, the helper is acting altruistically. The currency is energy invested in pulling. The recipient benefits in terms of food. The benefit is returned by allowing the helper to take food. It appears, then, that an initially altruistic action is repaid by another. But in terms of Trivers's account of reciprocal altruism, there is one important difference: When the individual facing the food cup reaps the rewards of the helper's pull, the helper benefits at almost the exact same time. There is virtually no delay between the initially altruistic act of helping to pull and the reciprocated act of kindness that provides helpers with access to the food. The benefit given is returned at almost the same time. This looks like mutualism. Moreover, although helpers are more likely to obtain food than nonhelpers, we cannot yet conclude that helping causes tolerated taking. We don't yet have evidence for a contingent response, one in which help is the cause of another's motivation to reciprocate.

To further explore these issues, de Waal and colleagues ran other experiments involving manipulations of food value, the sexual composition of the pair playing, and the number of opportunities to play with the same partner. Individuals were more likely to tolerate food-taking when lower-quality food items were at stake. This suggests that if reciprocation is involved, it is most often supported when the costs of food exchange are low. Among female-female pairs, but not male-male or male-female, individual A was more likely to allow individual B to take food if, on the previous run, B allowed A to take food. This addresses the issue of delay, but such exchanges accounted for less than 10 percent of the variation in behavior, suggesting that many other factors influence whether or not two females tolerate food-taking.

De Waal's work nicely shows that capuchins have the capacity to engage in joint cooperative action, can tolerate food-taking from others, and do so on the basis of help received on the rod-pulling task. Capuchins clearly cooperate. There are, however, several reasons why the capuchin work provides only limited support for reciprocal altruism: When it happens, it is infrequent and restricted to female-female pairs; because there is little cost to pulling the rod, and food exchange occurs most frequently when food quality is poor (costs of exchange are low), it is not clear that pulling the tool is altruistic; there are no situations in nature where two capuchins work together for a common goal, and where there are opportunities to reciprocate. Paralleling the conclusion for vampire bats, reciprocal altruism is at best a weak force in capuchin social relationships.

A final example comes from another New World monkey—the cotton-top tamarin.[24] Unlike capuchins, who live in large social groups, characterized by polygamous mating relationships, tamarins live in small groups characterized by monogamy; of course, like all other monogamous animals, tamarin males and females commonly look for matings outside the sanctimonious pair bond. Within groups, which consist of the breeding pair and, typically, one to two generations of offspring, older offspring help rear the younger ones. Part of the help comes in the form of food sharing. Like blue jays, therefore, tamarins are cooperative breeders. To explore the possibility of reciprocal altruism in tamarins, I designed a series of experiments with the economist Keith Chen and two honors thesis students, Emmeline Chuang and Frances Chen. In each experiment, we set up a game between unrelated tamarins, in which one animal—the actor—could pull a tool to give food to an unrelated recipient without getting any food for self; we therefore considered pulling the tool an altruistic act. Why would unrelated tamarins give each other food?

In the first test, we trained two tamarins to masquerade as actors playing diametrically opposite roles: A unilateral altruist always pulled the tool to give food to its partner and a unilateral defector never pulled the tool. You can think of these actors as Mother Teresa and Niccolò Machiavelli, respectively. The reason for training was simple: If tamarins give food to others based on previous acts of kindness, then they should give most to the altruist and least or nothing to the defector.

Tamarins followed the playbook, pulling the tool most often with the altruistic Teresa and infrequently with the defecting Niccolò. This shows two things: Tamarins give food to unrelated others, and do so contingent on acts of giving in the past. Is this reciprocal altruism? Not yet. Perhaps tamarins feel more generous when they eat more? When the altruist plays, she gives food on every trial. Getting food all the time must make her partner feel good, certainly sated. When a tamarin is sated, it is more likely to pull the tool and give food back. What looks like reciprocation based on an altruistic act of food-giving is actually the by-product of feeling good—feeling sated.

To test the feel-good explanation, we ran other experiments, using only untrained tamarins. In one game, if player A pulled the tool, it obtained one piece of food but delivered three pieces to player B. On the next trial, if player B pulled the tool, it obtained no food but delivered two pieces to player A. Given these payoffs, reciprocal pulling would pay as each player would obtain three pieces of food after a complete round. Animals in the player A role should always pull, out of selfish interest to get food, and they did. But animals in the player B role never pulled. Though B players were always rewarded by A players—as in the first experiment with Mother Teresa—and thus were always highly sated and feeling good, they didn't pull for the A players. Feeling good isn't enough to set reciprocation in motion. For food-giving to count, it can't be an accidental by-product of selfish behavior.

Tamarins give food to unrelated others, but a closer look at the patterns of giving reveal the signature of an unstable system. As each game progressed, the amount of food-giving dropped. This decline represents the signature of most games of cooperation developed by economists. If I know that this is going to be the last opportunity we have to cooperate, then it pays for me to defect if I can benefit and if there are no costs to our relationship because the relationship is ending. But if I think through this logic right before the last opportunity to interact, then I will surely think about defecting on the second-to-last opportunity, and the third to last, and so on. Cooperation unravels as the temptation to cheat surfaces. For tamarins, reciprocity unravels as the game proceeds. Further, if one of the players defects on two consecutive opportunities to pull, the entire cooperative ballet ends. Like the blue jays, tamarins can maintain some level of reciprocation under some restricted conditions. Overall, however, it is an unstable system.

There are other studies of reciprocal altruism that one could add, including games of tit-for-tat among guppies during predator inspection, cooperative territorial defense in lions, grooming among impala and a number of nonhuman primates, coalitions among male baboons and among dolphins for access to females, and food-sharing among chimpanzees.[25] These studies show one of three things: Animals don't reciprocate, apparent cases of reciprocation can be explained in a different way (e.g., mutualism), or, like vampire bats, capuchins, and tamarins, indicate that reciprocal altruism is uncommon, unstable, or generated under artificial conditions.[26] Although many animals may be motivated to reciprocate, either they are too dim, or the temptation to defect is too great, or the selective pressure is too weak.

Studies of reciprocity in animals lead to one further conclusion. For each of the species discussed—with the exception of work on vervet monkeys, where grooming and coalition support are apparently exchanged—every case of reciprocal altruism involves a single commodity, within a single context, and the time span for exchange is remarkably short. Thus, vampire bats exchange blood, but they do not exchange any other commodity. Capuchins tolerate food-taking if they have received help in getting food, but there is no evidence that they return the favor by giving predator warnings, or helping in coalitions, or even in exchanging other goods, such as grooming or care of offspring. De Waal describes chimpanzees grooming and exchanging food, but his analyses reveal no relationship between the amount of food one animal gives and the amount of grooming they receive in return. And all of these cases entail exchanges over a few minutes, or at best one or two days. Reciprocation in animals, if it exists, is based on a highly scripted text for how to interact in a particular context with a particular commodity over a short window of opportunity. As such, it lacks the generality and abstractness that typifies human reciprocation, as well as the potential to maintain the relationship with relatively long delays between reciprocated acts. Why the gap between us and them? The answer will emerge once we dissect reciprocity

into its component pieces and look for those that are unique to our own
species.

KEEPING TABS

The Age of Innocence, Edith Wharton's Pulitzer Prize–winning novel
about New York in the late 1800s, centers on the tension between the old
family wealth of the past and the postwar eruption of a nouveau riche
class. A centerpiece for this conflict was the concept of marriage, both its
proper arrangement as well as its conduct. Among the elite, the roles of
men and women were clear within the marriage. Women played second
fiddle in a highly asymmetric relationship. Men brought home the money,
or had it from family wealth. Men were free to do as they pleased with
their time, whereas women were there to support their husbands, look
glamorous, produce children, and entertain. Inequity defined these relationships. Early in the book, Archer imagines a traditional marriage where
he is the teacher, his wife a student. This daydream carries forward to a
honeymoon in which he takes his wife to the Italian lakes to introduce her
to poetry. Archer soon experiences a transformation in his idealistic views
about marriage when he hears about his friend Ellen's disastrous marriage
to Count Olenski. He finds himself arguing in Ellen's favor, arguing that
women should have greater freedom in a marriage—that there should be
greater equity and fairness: " 'Women ought to be free—as free as we are,'
he declared, making a discovery of which he was too irritated to measure
the terrific consequences."[27]

Notions of fairness extend far beyond marriage, permeating almost all
aspects of human life. They enter into legal decisions, sports, games, employment opportunities, salaries, war, and less formal day-to-day interactions involving cooperation. At the root of John Rawls's political philosophy
is the tenet that a just society is one that is fair. The psychology of fairness in
our own species is rich, including some ability to keep tabs, to place subjective values on different entities and actions, to judge when an inequity has
transpired, to distinguish accidental from intentional giving and reneging,
and to determine when an unfair act is worthy of retribution. Is all or part
of this psychology shared with other animals?

A nice example of *fair play* comes from observations by the biologist Marc Bekoff of juvenile play in canids—dogs, coyotes, and wolves.[28] Anyone who has watched dogs knows that they are playful creatures, especially when young. Dogs have stylized play gestures that guide the bout, including invitations to play and to signal that the intent is friendly as opposed to aggressive. Similar gestures appear among wolves, as might be expected given the evolutionary ancestry of our domestic breeds. Among the canids, there appear to be rules that dictate how play bouts proceed. When there are asymmetries in the age, size, and strength of the players, the one with the advantage is expected to assume a handicap—letting the little ones take the top position, not using their full strength, and running at half speed. Handicapping in a play bout is a social norm. Individuals who violate this norm are effectively told to go sit in a corner, left alone while everyone plays with more reasonable partners. And such ostracism has a cost. Among coyotes, those who fail to adhere to the norms of play are much more likely to leave their pack than those who play according to the rules. Differences between the norm followers and norm violators create variation. Variation provides the necessary fuel for natural selection. In Bekoff's study population, individuals who violate the norms of play and leave their packs early experience a twofold increase in mortality over those who adhere to the norms. It pays to play by the book.

Bekoff's observations, together with other studies of cooperative interactions, suggest that animals may have some sense of inequity. At the root of this sense is a capacity to set up expectations and then respond negatively when the expectation is violated by someone or something. Tinkelpaugh's experiments, mentioned in the last chapter, are suggestive. When a rhesus monkey watched as an experimenter concealed a banana under one of two cups, and then reached for the cup with the banana only to find a piece of lettuce, they responded with anger and apparent frustration. To show this kind of response, rhesus must set up an expectation, hold it in mind, and then detect whether or not the outcome matches or violates their expectation. They must know, at some level, when things are wrong. This is the first ingredient in the evolution of an animal that can detect inequities.

A second ingredient entails some measure of the value of the reward. If we assume that one currency of value lies in the number of food items available, then the work on number-processing in animals provides one piece of evidence that animals can detect numerical violations.[29] For example, if you show a tamarin or a rhesus monkey a simple addition operation of one grape plus a second grape placed behind a screen, and then reveal an outcome of one, two, or three grapes, they look longest at the incorrect sums of one and three. Tinkelpaugh's experiments, along with many others since, show that in addition to numerical labels of value, animals can also rank-order their preferences for some foods over others. In the experiments using mink, a cage with a pool of water is valued ahead of a cage with more space, food, or toys. The main point is that animals value some items over others, and in some cases can use their number system to place values on outcomes, as required by games such as the prisoner's dilemma.

A third ingredient, also mentioned earlier, is that animals must live in societies in which they can express their preferences, at least some of the time. In societies with dominance hierarchies, there will always be inequities in the distribution of valued resources. This is what defines a dominance hierarchy. In some animals, hierarchies show massive inequities, as when a male elephant seal takes literally all of the matings while the others sit on the sidelines, waiting for and dreaming about the day when they might take over. Other societies are more egalitarian: High-ranking animals take more than low-ranking individuals, but subordinates have a say, and are anything but excluded from food, space, and sex. In fact, dominants depend on subordinates in such societies: Without them, they would be weaker in intergroup battles and more susceptible to predation, given the many-heads-are-better-than-one principle. Subordinates therefore have some leverage against dominants.

Based on these observations, the biologists Sarah Brosnan and Frans de Waal designed a series of experiments to explore whether inequity plays any role in chimpanzee and capuchin monkey food exchange.[30] In parallel with some of the games played by economists with humans, the goal was to see whether either species would reject offers that seemed unfair, in some sense.

Brosnan and de Waal designed an inequity game using prior knowledge of capuchin and chimpanzee food preferences. Both species preferred grapes to many other foods, including cucumber and celery. Individuals first learned to exchange tokens for food, a small rock for capuchins and a piece of pipe for chimpanzees. Next, an experimenter tested each subject in four different conditions, three involving a pair of individuals and the fourth involving an animal tested alone. The first social condition required each individual to hand over a token for the lower-quality cucumber. Although individuals may not care too much for cucumber, the exchange rates were equitable, each individual getting one piece of cucumber for one token. The second social condition set up an inequity: The experimenter handed one individual a highly valued grape for a token but handed the other player a cucumber for the same token exchange. The third condition looked at effort. The experimenter handed one individual a gift—a grape without any token exchange; the other player, in contrast, traded in a token for a cucumber. Like in condition 2, there is an inequity. And if chimpanzees and capuchins take effort or investments into account in working out what is equitable, then there is an added boost to the inequity as one individual receives a highly valued reward without paying any costs. Lastly, in condition 4, the experimenter required an individual to hand over a token for a cucumber, while a highly valued but unattainable grape looked them in the eye.

Overall, both species were less likely to engage in an exchange or accept the reward when their partner got the better deal, either paying for a grape or getting it for free. Both species also rejected more opportunities for exchange over the course of several inequity trials than when they played alone but in the presence of an unattainable grape. In no case did the individual receiving the better deal share with their unlucky partner. Given de Waal's previous studies of tolerated food-taking in capuchins and chimpanzees, one might have expected the lucky grape-winners to equalize the situation by sharing some with their partner; the absence of an equity response may be due to the small size of the reward and its lack of divisibility. Among chimpanzees, rejections were more likely when both players were from a recently formed group than when they were from a long and stable social group; this suggests that history plays a key role in tolerance. If my best friend rips me off, I might be angry, but I wouldn't want the relationship to fall apart. I will therefore tolerate the inequity, at least for a while. If someone I barely know rips me off, then there is no reason to tolerate the inequity unless I think there is some

Inequity

Free gift

future to the relationship. Chimpanzees seem to show a similar sense of tolerance, although it is important to remember that an individual's rejections have no effect on their partner. For capuchins, only female pairings showed differences in their response to each condition; among chimpanzees, there were no sex differences.

Brosnan and de Waal conclude that both species recognize when there has been an inequity in the context of exchange. They further argue that although the psychology underlying this sense of inequity may not be identical to what humans experience in comparable situations, they represent a starting point for looking at the evolutionary precursors of our sense of fairness.

These are fascinating experiments, opening the door to many questions concerning the nature of exchange, the degree of inequity accepted or rejected, the capacity to tally up exchanges over longer bouts, the role of reputation, and the relationship between the costs of a token and the rewards they bring. And, like all fascinating experiments, Brosnan and de Waal's work has received intense scrutiny and criticism. For example, when individuals reject a reward, only they incur a fitness cost. Their actions have no bearing on their partner's psychological well-being or genetic fitness. Thus, and in striking contrast to the ultimatum game described in chapter 2, where a rejected offer causes the donor to go home empty-handed, a rejected cucumber simply means no cucumber for the rejector; the partner still gets to eat the grape, perhaps enjoying it even more given that the partner has turned down the only available reward. When an individual rejects a reward and then threatens the experimenter, this might represent a response to inequity or to an expectancy violation. When chimpanzees and capuchins are aggressive toward an experimenter, it is because they expected one thing and got something else. They never think of the interaction in terms of inequities.

Unlike food exchange among humans, or even other animals, chimpanzees and capuchins playing these games pay nothing for their tokens. "Exchange" may, therefore, be too strong a label for the interaction. Individuals do hand over their tokens, but this is no different than an individual pecking a key to get food. In the absence of pecking, rats and pigeons don't get food. In the absence of handing over a stone or piece of pipe, chimpanzees and capuchins don't get food. For both species, exchanges of the kind set up by Brosnan and de Waal never arise in nature. Neither species has anything like the barter system set up in these experiments. Chimpanzees do tolerate food-taking in the wild, especially following a hunt. And, as mentioned in the last section, those who allow more food to be taken tend to receive more grooming, suggesting a possible exchange across currencies. Among wild capuchins, there is only limited evidence of food-sharing, and none in the context of joint cooperation or reciprocation. Of the experimental evidence, only females engage in this kind of bartering, with no evidence of an inequity response in either male-male or male-female pairs. The mismatch between experimentally triggered behavior and spontaneous natural behavior is important. The fact that these animals appear to respond to something like inequity in captivity shows that the relevant psychology is in play, under certain conditions. That is an important discovery. But it raises deep puzzles about how the capacity evolved, in what context, and under what selective regimes.

RECIPROCAL RENEGING

Playgrounds throughout the world ring out with "Cheater, cheater!" So do sports arenas, the ivory towers of academe, and the glitzy empires of the business world. When rules are broken, emotions soar, and revenge is sought. But it only seems fair to punish those who are aware of what they are doing, showing lenience toward those who violate societal rules by accident, perhaps ignorant of the local norms.

Mathematical models—which help reveal the plausibility of particular phenomenon—show that cooperation can evolve and remain stable if individuals punish both cheaters and those who fail to punish cheaters.[31] In the absence of such punishment, cooperation deteriorates as individuals defect. Human societies have clearly evolved these psychological tools. What about animals? Do they cheat, fabricating scurrilous tissues of lies? And if they are caught, are they punished? There are two ways to cheat: falsify information or withhold it. This distinction maps onto a more familiar one, that between actions and omissions. Animals have mastered both types of deception: Tropical birds give alarm calls when there are no predators in sight, thereby causing their competitors to look up and lose the race for food; roosters give food calls in the absence of food, thereby seducing females to approach, expecting a meal and receiving amorous stares instead; rhesus monkeys suppress their copulatory and food calls when competition is high, thereby withholding information about valuable resources.[32] When animals deceive, they give in to temptation, violating social norms in an attempt to take more food, sex, or power than their competitors.

In some of these examples, deception is rare, as theory would predict.[33] For example, in studies of primates, there are hundreds of examples of presumed deception, including individuals giving predator alarm calls when there are no predators, and hiding behind rocks in order to conceal food or a mating opportunity. Though the total number of field observations is high, the number for any given species, within a population, is minute. These are rare events. In other cases, deception is quite common, presenting a problem for current theory. For example, roosters give food calls to nonfood items about 45 percent of the time; their calls recruit potential mates who either never catch on or show little concern about the costs of walking over to a rooster expecting food but finding a rock instead. Theoretically, if an individual frequently abuses the actual meaning of a signal or its expected usage, crying wolf or staying silent, then either this individual will lose all credibility or, if the abuse spreads, the signal itself will lose credibility. In this context, lies of commission and omission are different. Omission of information is a crime, but it is harder to detect and confirm such cheaters. The failure to signal could be deceptive. It could also be due to a failure to recognize the current context as an appropriate one for signaling or, because of some pact with another, to remain silent. The philosopher Sissela Bok puts it nicely by noting that "While all deception requires secrecy, all secrecy is not meant to deceive."[34] Identifying these individuals as cheaters thus becomes a matter of sampling, of observing enough cases where an individual is in the appropriate signaling context and fails to do so. In contrast, giving false information is easier to detect. If an individual gives an alarm call with absolutely no predator about, then this counts as a lie of commission. As in the fable, however, it would be unfair to label the poor boy a liar after one occurrence. To lose trust in someone requires repeated offenses.

There is overwhelming evidence that animals are sometimes dishonest, attempting to cheat in order to get ahead. If the incidence of cheating within a group of animals is low, however, then a mechanism for both recognizing and punishing cheaters is unlikely to evolve. On the other hand, if the incidence of cheating is relatively high, then natural selection should favor skeptical observers who, having detected a cheater, punish them, perhaps by refusing to engage in cooperative interactions in the future. As recent theoretical models of reciprocity suggest, cooperating with honest individuals and ostracizing dishonest ones may provide a robust form of immunity against defection. Oddly, no one has examined whether animals label others as honest or dishonest, refusing to play with disreputable cheaters and seeking out relationships with the tried and true.[35] We can nonetheless take the phenomenon of cheating, dissect it down to some of its core ingredients, and then look for whether animals have some or all of these properties.

To cheat, an animal must either be able to modify the typical association between some signal or behavior and its function or meaning, or be able to suppress information that is typically expressed. Let's take these two forms of deception in turn.

In many animal societies, there is a relatively fixed association between signal and function, such that a signal designed to convey information about the presence of a predator or a neighboring competitor is used only in these specific contexts. Deception requires flexibility. It requires using a signal in a different context, engaging in dissent and breaking with conformity.[36]

Richard Dawkins and John Krebs pointed out that communication in animals is not about information per se.[37] It is about signalers manipulating the behavior of their audience, and audience members skeptically responding to the signal transmitted, attempting to slice away the layer of lies to uncover the truth. In an evolutionary game, individuals who send honest signals of intent are vulnerable to cheaters. If individual A always signals an honest level of aggressive intent, B can counter with a higher level that is deceptive. B can take advantage of A's honesty, signaling higher and causing A to back down. The fact that dishonesty can invade the honest strategy reveals why at least some aspects of communication must be about manipulation. On the other hand, given that manipulation is part of the game, there will be selection for skepticism or, as Dawkins and Krebs originally put it, mind reading. But mind reading in this sense—as opposed to the folk psychology sense of others' beliefs and desires discussed earlier—requires a different capacity. It requires the ability to detect a mismatch between the purported function of the signal and some measure of reality. Some lovely experiments on honeybees and vervet monkeys reveal how biologists can look for skepticism in the animal kingdom.[38]

Honeybees perform a complicated dance when they return from foraging. The dance reveals to other foragers where food is located and how good it is. The dance is designed to recruit other foragers to the food site so that they may harvest the pollen and bring it back to the hive. Using a long-term resident hive situated next to a lake, the biologist Jim Gould trained a small number of foragers to move to a pollen-filled boat on land. Gradually, Gould moved the boat out into the middle of the lake. Once the foragers reliably moved out to the boat, he let them fly back to their hive and dance.

The potential recruits watched their hive-mates dance, but no one moved a wing in the direction of the pollen-filled boat. By lining up the forager's instructions with their spatial map of the area, the recruits expressed skepticism in the form of a sit-in. And the failure to recruit cannot be accounted for by a failure to fly over water, because a control experiment in which Gould placed the boat on the water, and at a comparable distance but near the shore, yielded a typical level of recruitment by other foragers. Honeybees have the tools for skepticism, using their ability to match what they currently know about the lay of the land with what a forager reveals in a dance.

To look at skepticism in vervet monkeys, Cheney and Seyfarth ran the equivalent of the boy-who-cried-wolf parable, using a standard habituation procedure. Vervets living on the plains of Kenya produce acoustically distinctive vocalizations when they detect members of a neighboring group. The *wrr* is given when their neighbors are in view, but neither too close nor in violation of their assigned territorial space. The *chutter* is given when the neighbors have invaded the caller's space and when an intergroup fight is in action. If you repeatedly play back either of these calls, eventually subjects stop responding. The logic is simple: If one animal keeps crying "aggressive neighbor at forty-five degrees to the north," and there is no such threat, you eventually turn them off. They are lying. But just because vervet Fred is a liar doesn't mean that vervet Joe is as well. What this means in experimental terms is that if subjects stop responding to Fred's calls, they don't maintain their skepticism once Joe starts calling about the neighbors. Their interest is renewed, because Joe doesn't yet have the reputation that Fred has as a liar. Like honeybees, vervets also have the capacity for skepticism, using the mismatch between a signal's function and reality to detect when someone is lying.

Although both honeybees and vervets have the capacity for skepticism, in neither case do we know how this influences their social relationship to the liar. If a forager repeatedly returns to the hive and provides false information about the location of food, do others merely ignore the dance or do they take a more proactive stance, punishing the liar through ostracism or physical punishment? In vervets, do group members ignore those who falsely send off alarms, or do they actively try to correct this behavior by means of ostracism or physical force? And for both species, if an individual attempts to deceive in one context, and is caught, do others conclude that this individual is a cheater across the board? Or is their labeling concentrated on the particular context in which the individual deceived? For example, if a vervet unreliably calls about neighboring groups, do others expect her to unreliably call about predators or disputes within groups? There are no answers to these questions, but they are necessary for understanding the psychology of deception in animals.

The psychology and evolution of omission is different. Unlike acts of commission, acts of omission are more difficult to detect, and thus should be more common. If an individual doesn't call when a predator passes by, this could be an act of omission or a failure to detect the predator; it will only be in relatively unusual contexts that an individual will know with certainty that a group member has seen the predator but chosen to remain silent. Although acts of omission should be relatively common, they require a different set of tools. To withhold information from others, individuals must suppress a vocalization or facial gesture. They must inhibit an action that is common. As I discussed in chapter 6, inhibitory control is not a strong point for most animals, as revealed by their steep discounting curves, their inability to succeed on reversal learning tasks, and their repetitive responses to tasks that tap either an innate or an overlearned bias. In addition to self-control, individuals tempted to omit information must recognize a set of contexts in which it will pay.

Observations and experiments on birds and primates suggest that animals are sometimes silent in contexts where they normally call. For example, Peter Marler and his colleagues discovered that when roosters find food, they are silent when alone or in the presence of another rooster, but call at high rates when either a familiar or an unfamiliar hen is in view.[39] Food calling in roosters is therefore sensitive to an audience. In contrast, roosters and hens give alarm calls regardless of whether the audience is male or female, young or old. As long as it is a chicken, alarm calls follow; they are silent, however, if the audience is composed of a different species, such as quail. Calling in roosters is not reflexive, inextricably linked to a context. Calling is sensitive to social context. When a rooster goes silent in the presence of food and another rooster, he appears to be withholding information in the service of gaining a competitive edge.

Rhesus monkeys can turn off their vocal cords in the context of mating and foraging. When rhesus monkeys copulate, the males often emit loud and individually distinctive vocalizations. To someone listening out of view, the sound of a copulation call carries information about the context and the caller's identity. It can also evoke competition and fighting among males. When competition for access to females rises, as is the case when there are few sexually receptive females, copulating males turn on the mute button. Of interest, though nothing comes out of their mouth, they mouth the gestures, as if gesticulating for the silent screen. This, presumably, has the effect of reducing the heat of competition.

There are alternative explanations for these cases of vocal suppression that move us toward a different component of the psychology. In each of the cases I described, it is possible that silence is not an indication of deceptive omission. Rather, without any intent to withhold information, some contexts are insufficient to trigger calling. Instead of a caller intending to withhold information because he or she knows that it will manipulate the audience's beliefs, leaving them in the dark, certain contexts either do or do not stimulate the vocal cords. When roosters find food and are alone or see another rooster, they follow a simple rule: remain silent. When they find food in the presence of a female, they call. The principle for calling is straightforward, with one parameter driving the expectation: audience composition. The same can be said about the rhesus case. Given the evidence on seeing-as-knowing in rhesus monkeys and chimpanzees, it is time to design experiments that explore whether animals commit lies of commission and omission by attending not only to what other individuals do but to what other individuals know.[40]

THE ART OF RETRIBUTION

Punishment is one way to control cheating. It is a form of external control. But to punish another requires at least two capacities. First is a sense of the range of possible or tolerable behaviors in a given context. This is necessary, because punishable actions are those that deviate in some significant way from a set of normative behaviors or emotions in the population. In *Genesis of Justice*, the legal scholar Alan Dershowitz argues that God had to find a balanced approach to punishment. His initial penalties were either too severe or not severe enough given the crime. God tells Adam that he will die if he eats from the tree of knowledge; Adam tells Eve about God's command. As we learn, God's command is an idle threat. God doesn't follow through. God punishes Eve with labor pains and with subordination to Adam. God punishes Adam by making him sweat for his bread, by ruining his crops, and by limiting his life span. God also makes these new traits heritable, such that Adam and Eve's descendants suffer in kind. But now consider the next story of sin. Cain kills his brother Abel and then tries to cover up his mess. Two strikes. Not only did Cain end another's life, but thanks to Eve's fruit-picking, he knows that some actions are morally right whereas others are wrong. While Eve had no understanding of this moral distinction, Cain did and violated it. God punishes Cain, setting him adrift, alone. Original sin number two seems worse than original sin number one. God's punishment scheme seems out of kilter.

The second capacity is an ability to distinguish between an intentional or voluntary violation and an involuntary or accidental violation. If I am angry and throw water in your face, my action is reprehensible and punishable; if I trip while carrying a glass of water and throw water in your face, the consequence is the same but the motivation is presumably not reprehensible.

Given that animals can both cheat and use their skepticism to detect cheaters, do they also have the capacity to punish, and, if so, how do they do it and does it have any effect? Moreover, is punishment something that animals do in a narrow range of contexts, or is it socially promiscuous, applied whenever a norm has been violated?

In the mid-1990s, the biologists Tim Clutton-Brock and Geoffrey Parker used evolutionary game theory to look at the nature of animal punishment. The advantage of evolutionary game theory is that it forces an appreciation of different potential strategies that compete over an evolutionary time frame, using genetic fitness as the measure of a successful and stable strategy. A further advantage is that each strategy consists of a rule for playing, thereby bringing us back to looking at the principles that potentially guide acts of punishment.

Clutton-Brock and Parker started with a fairly loose definition of punishment: An aggressive response to a fitness-reducing act that is costly to the punisher but costlier still to the punished. As defined, this looks like spite: I do something that drops my fitness but drops yours more. It differs from spite, however, in that punishment is aimed at a long-term benefit for the punisher. The goal is to reduce the instigator's fitness and to prevent future transgressions. In this sense, punishment can fuel cooperation by setting up sanctions against those who violate the norms.

Clutton-Brock and Parker developed a series of hypothetical games in which both players could either punish or refrain from doing so, and could either transgress or follow the social norm. In one game, for example, subordinates could increase their fitness at the dominant's expense by transgressing. Subordinates win if dominants pass on the opportunity to punish the transgression. From the dominant's perspective, punishment is the best strategy, as the upfront cost is outweighed by the cost to the subordinate. In most socially living animal societies, dominance hierarchies impose an asymmetry such that high-ranking animals can punish low-ranking animals for transgressions, but subordinates cannot retaliate. Punishment is therefore an evolutionary stable strategy in such situations, because subordinates have no say, and dominants win by keeping lower-ranking animals in line; it is also possible for dominants to punish as a way to recruit cooperation from subordinates. In cases where there is an asymmetry, dominants are not completely void of risk. As mentioned before, subordinates have leverage, because the health of a group in competing for resources against neighboring competitors depends on its composition, and not the dominant animals alone. Further, in several species, low-ranking animals can recruit support from others to physically attack the higher-ranking animals. Thus, cooperative alliances among subordinate animals can function to punish dominants who step outside the norms of the hierarchy.

To what extent is there empirical evidence that animals do punish? We have already encountered a few examples that fall within the framework set up by Clutton-Brock and Parker. Among territorial species, residents chase and often attack individuals who intrude, attempting to sequester space of their own. This is punishment in the sense that intruders, by attempting to set up camp, have violated a social norm that confers property rights to the resident. In species with strict dominance hierarchies, dominants will often attack a subordinate attempting to sequester an unclaimed piece of food, or to sneak a copulation. Here again, the norm is that dominant animals have priority of access to food, space, and mates. When subordinates attempt to take what is in practice not theirs for the taking, dominants often lash out, attacking the subordinates in an act of punishment. In both the territorial and dominance cases, the attacker minimally incurs the cost of chasing, and may incur an added cost if the transgressor fights back. The cost is outweighed if the transgressor backs off. In most situations, transgressors do back down. When they don't, then the punisher must assess the economics again, either pushing harder or backing down.

Among rhesus monkeys, animals sometimes produce distinctive calls when they find food, and at other times they remain silent. Paralleling the case of copulation-call suppression, these observations suggest that rhesus can withhold information about their food discoveries. There are, however, costs associated with such apparent deception. Under experimental conditions, silent discoverers are attacked more often than vocal ones, and end up with less food than those who announce their discoveries. The fact that both high- and low-ranking discoverers are attacked if they remain silent about the discovery of food suggests that rhesus monkeys punish those who have violated a social norm or convention. These observations also bring into play another aspect of the Clutton-Brock and Harvey models. In most cases, punishment will be asymmetric, with dominant animals targeting subordinate animals. In cases where the asymmetry is significant, either because of a substantial difference in size between low- and high-ranking individuals or because of a convention, low-ranking animals may recruit coalition partners to overcome the asymmetry and punish dominant animals that transgress a social norm. When a low-ranking rhesus monkey catches a silent dominant animal with food, screaming helps recruit others who join in on the chase. We, of course, don't know whether rhesus think about such norms, contemplating what it means to follow and break them. We also don't know how accurately, if at all, individuals tally the number or severity of such violations. More important, this study fails to show that aggression changes the cheater's behavior. To show that this kind of aggression functions as punishment, it would be necessary to show that it acts as a deterrent, reducing or eliminating subsequent attempts to conceal the location of food, or causing the cheater to emigrate out of the current social group.[41]

In each of the cases above, the punisher stands to gain something, directly. The territory resident is fighting for his space. The rhesus monkey is fighting for access to food, and, in many cases, chasing a silent discoverer yields some food. But there are a few cases where the benefit is less direct, leading some to suggest that there is punitive *policing* in animals: aggressive attacks that appear to be motivated by maintaining the social norm in the absence of gaining a direct material benefit. In ravens, who hide food in caches and defend them, caches are private property and generally respected. If an individual attempts to pilfer an owner's cache, the owner, and in some cases nonowners, will attack the thief. The attacks by nonowners could yield direct material benefits if they deflect future attempts to pilfer a cache.

The best example of a police force comes from observations of social insects, specifically bees, wasps, and ants. Consider honeybees.[42] Queen bees lay the majority of eggs. The other eggs are laid by workers who only produce males. Those reproducing are therefore more closely related to the queens and young males in the colony. Given the advantages of being a queen who controls reproduction, it pays for larvae to develop into queens, as opposed to workers. However, as the number of individuals who attempt to become queens goes up, colony efficiency goes down. Queens have no work skills. With a poor workforce, colony productivity plummets. Here's where the police force enters the scene. By tightly controlling larval development through their allocation of food, workers regulate the number of individuals developing into queens; those destined to become queens get more food than those entering the working class. In addition, when they find a surplus of larvae headed for a queenship, or find that the workforce is producing too many of their own eggs, they kill their hopes by ending their lives. As a general rule among the social insects, the better the police force, the lower the rate of transgression. Predictably, the working class has at least one trick up its sleeve: In some species, individuals chemically mark their eggs to mimic the odor of the queen's egg. This foil seems to pay off by lowering the odds of detection by the police force.

None of the examples of punishment discussed thus far arises, in response to individuals who cheat during a cooperative relationship. But, as I noted in the beginning of this discussion, it is precisely in such contexts that punishment appears to be so critical, especially if work on human cooperation provides any guidelines. Is there any evidence of punishment in the context of animal cooperation? In brief, no. A vampire bat who fails to reciprocate is never ostracized, slapped around, or chased from the colony. A lion who lags behind during a covert attack on an intruder is not attacked by the leader or kept away from the next kill. A chimpanzee who fails to join in on a dangerous border patrol, opting to stay behind and court a sexually receptive female, is not chased out of the community. The absence of such observations certainly does not rule out the possibility that animals might punish those who defect in a cooperative venture. Biologists may be looking in the wrong places or may not have set up the appropriate experimental conditions. For example, animals might not punish by direct physical aggression but rather by withholding resources or opportunities from those who cheat. These theoretical possibilities will be difficult to confirm, as it is extremely hard to show why an animal failed to behave in a particular way—why, for example, they remained silent.

A different explanation for the lack of physical punishment among animals is that they elicit cooperation through nonviolent means, including the use of psychological harassment.[43] Chimpanzees and squirrel monkeys incessantly beg from those who have food, using tactics that range from more subtle shadowing and staring to hand-gesturing toward the consumer's face. The goal seems to be to annoy those who have food, waiting for them to give up and hand some over or tolerate food theft. In both species, harassment yields more food than doing nothing, suggesting that it may represent a relatively cost-free way to get cooperation going. Though important, we are still left with a stark conclusion: Among animal societies, punishment represents a weak or nonexistent force in deterring those seduced by the temptation to cheat on a cooperative relationship.

ADAPTIVE NORMS

A couple of years ago, a friend of mine e-mailed a series of snapshots of a duck and her ducklings walking along a city sidewalk. The initial shot in the sequence reminded me of Robert McCloskey's classic children's book

Make Way for Ducklings. As the series progressed, the ducks approached a subway grating. The proud mother walked right over it. Her ducklings, obediently following, fell, one by one, through the grate, meeting their fate on the hard surface below. Following Mom is a rule hardwired into the brains of all ducklings and many other species where the young are precocial and able to move on their own early in development. It is typically a good move, a rule that has strong evolutionary legs.[44] But sometimes the environment throws a curveball, making the old rule either obsolete or at least in need of modification—assuming, that is, that modification is part of the developmental program.

A similar situation arises among baboons living in the Okavango Delta of Botswana. It is a more surprising case, however, as the environmental change is not as extreme as the placement of a subway grating, nor is it artificial. Every year, when the rains fall on the delta, one of the primary rivers that runs through Cheney and Seyfarth's baboon site swells. Sometimes individuals are caught on one side of the river, and, for a variety of reasons, they need to cross. Swimming is dangerous, not only because baboons don't swim as a regular hobby or pastime, but because there are crocodiles. When they cross, they do so swiftly and cautiously. But there is one thing that they sometimes forget to check as they are crossing: Whether or not they are carrying an infant on their bellies. In some cases, mothers cross with infants hanging on below. When they arrive on the other side, their infants are dead, having drowned on the journey. Baboon infants apparently lack the ability to scramble up and to the safety of their mother's backs prior to the swim, and mothers lack the wherewithal to hoist the baby up to safety before crossing. Belly-carrying is not a transportation rule for baboons. Infants will happily ride on their mother's back, and mothers will happily tolerate this mode. This pattern of behavior is clearly not adaptive, and yet it persists as a pattern in the Okavango baboons. Sometimes generally successful patterns of behavior run into novel contexts, and the outcome is negative. What is generally adaptive occasionally has maladaptive consequences when the environment changes, especially unpredictably.

The duckling and baboon examples show that typical patterns of behavior can lead to negative consequences. But as the last two chapters revealed, there are many cases where individuals follow rules of action, including certain social norms, and the consequences are positive. Dozens of animals learn what and how to eat by watching others, creating local culinary traditions. This has the advantage of saving time and reducing the costs of eating something nasty and potentially lethal. Hierarchies have similar advantages in that individuals need not obsess whether or when they should compete for resources. By learning the hierarchy and following the rules that dictate resource competition and access, individuals can save time and avoid unnecessary injuries. Norms therefore play an adaptive function in animal societies, setting up reliable expectations concerning helping and harming others. Norms need not be consciously accessed to be adaptive. In fact, they are optimally effective when they operate covertly, as they may well do most of the time in animals, including humans.

There are several reasons why I am focusing on norms concerning cooperation, and especially reciprocity. Reciprocity is at the core of the Golden Rule. The Golden Rule appears in one form or another in all cultures, through either explicit religious doctrine or implicit social norms. Universals often provide the signature of a common biological mechanism, part of the species' genetic heritage. Universals also often show up as legacies of our past, pieces of psychological machinery that we inherited from our ancestors. Unlike other forms of cooperation, which can be explained by Hamilton's rule or mutualism, reciprocity requires a different account and different psychological ingredients, including rules for when harming another individual is allowable. Although there is much more to moral judgment than solving the problem of cooperation, it is one problem that we share with other animals. Gnawing at the desire to be nice is the temptation to be selfish. The temptation to defect is rooted in the organism's competitive desires to optimize the acquisition of food, sex, or power. Animals are nice to each other when their genes benefit directly through kinship. Based on the regularity with which this happens, animals can expect help from their close relatives. When non-kin have the opportunity to engage in a long-term, reciprocally altruistic relationship, the temptation to defect overwhelms the systems of control, except under some highly constrained conditions. Other factors, unrelated to control, may also perturb the stability of a reciprocal relationship. Given the lack of punishment in the context of cooperation, those who defect may well escape without any costs. The lack of costs may fuel further attempts to cheat. The lack of control, the power of temptation, and the limits of their psychology combine to preempt enduring reciprocal relationships.

We have come full circle: Humans appear to be uniquely endowed with a capacity that enables large-scale cooperation among unrelated individuals, and to support stable relationships that rely on reciprocity. Here, in this final section, I provide an explanation for why the taxonomic gap exists; a portion of this explanation will take us back to part I, where I described a puzzling dissociation between people's judgments and their justifications for moral dilemmas. In order to run with this explanation, however, I first return to the logic of the linguistic analogy and what it demands in terms of evidence. To understand what is unique about our moral faculty, we must determine both which aspects are unique to humans and which are unique to the moral domain. This involves running two separate subtraction operations: One subtracts what is shared with other animals to leave what is uniquely human, and the other subtracts what is shared with other domains of knowledge to leave what is uniquely moral. I synthesize what we know about this problem, and note several areas where we are on terra incognita, allowing only speculation to surface.

At the core of the Rawlsian creature is an appraisal mechanism that extracts the relevant properties from an event. These properties are represented by physical actions together with their causes and consequences. Infants are endowed with some components of this system, showing early sensitivity to the hierarchical structure of events, the importance of distinguishing inanimate from animate objects, the role of goals and intentions, and the relationship between actions and emotional consequences. When we look to animals, we find striking parallels in their processing of action primitives and basic emotions. The burgeoning research on mind reading in animals also presents an increasingly similar picture. If we include distantly related animals, such as birds and dogs, as well as more closely related species, such as monkeys and apes, we find evidence of joint attention, reading intentions and goals, and using seeing to draw inferences about knowing. These are the beginning steps in evolving a theory of mind, and ongoing studies are hot on the trail of discovering other capacities. How far these studies will go is anyone's guess. But if the current trend is any indication, it will force an about-face for philosophers such as Christine Korsgaard, who think that because only we have the ability to turn our attention on what we believe, desire, and intend, only we have a sense of the normative or prescriptive: "Our capacity to turn our attention on to our own mental activities is also a capacity to distance ourselves from them, and to call them into question . . . Shall I believe? Is this perception really a *reason* to believe? . . . Shall I act? Is this desire really a *reason* to act?"[45] Without reflection, animals can't distance themselves from their actions, and thus cannot contemplate alternative reasons for action. So the story goes.

A central aspect of our capacity to reciprocate and to engage with moral dilemmas where the long-term merits of particular actions outweigh the short-term and selfishly tasty alternatives is the ability to delay gratification. Granted, young children are more vulnerable to immediate temptations, as are patients with damage to the frontal lobes. And all of us will occasionally break down in the face of waiting for a larger reward. But when normal human adults are contrasted with normal adult pigeons, rats, tamarins, marmosets, and macaques, we are outliers, operating on a different timescale. Even the most patient animals wait for only a few seconds before the immediate and smaller reward pulls them in. Humans, in contrast, can wait for days or even weeks before caving to the smaller and more immediate temptation. One reason for this difference is a substantially larger and more architecturally differentiated frontal lobe that plays a central role in inhibitory control. We can not only inhibit a desire for immediate gratification, but we can use our capacity for self-control to block previously learned rules, thereby opening up opportunities to set up new social norms. For most if not all animals, socially learned rules tend to acquire a rather fixed status. When animals evolve rules for handling statistical regularities in the environment, they tend to hang on to them even if they backfire due to changes in the environment. Once acquired, they tend to stay, providing assembly-line actions. Another difference is that we, perhaps uniquely, are aided by a bag of physical and psychological tricks, prophylactics that keep temptation at bay. Young children can use language, and especially metaphor, to transform a delicious marshmallow into a tasteless cloud. Adults can throw out their credit cards to avoid the lure of shopping, or commit to a week of diet boot camp to fend off the aromas of a local bakery. These are some of the consequences of patience in nonmoral contexts. Patience is equally potent as a force in the moral domain, enabling us to fend off the temptation to cheat and hang on to our reciprocal relationships.

Patience is only one of the necessary guardians against cooperative instability. All cooperative relationships are vulnerable to cheaters. Humans are no different from animals on this front. What appears to be different, in part, is the psychology that emerges when we catch a cheater. Though animals lash out when someone has violated a rule in the context of resource defense, there is no evidence that animals attack those who cheat on a cooperative venture. There are two possible explanations for this gap. Animals either fail to detect cheaters in the context of cooperation, or fail to apply the logic of aggression in the context of resource defense to the context of cooperation. Either way, they lack one of the central mechanisms enabling humans to engage in reciprocity as well as large-scale cooperation among unrelated individuals: punishment. We don't know when this capacity evolved in our species. We also don't understand which aspect of the environment changed, creating pressure for punishment to evolve. Hunter-gatherers punish through ostracism. But did pre–hunter-gatherers have this ability and, if so, why? Whatever happened, and whenever it happened, the landscape for cooperation changed. I will return to this point in a moment.

If we run the subtraction operation, taking away those aspects of our moral psychology that we share with other animals, we are left with a suite of traits that appear uniquely human: certain aspects of a theory of mind, the moral emotions, inhibitory control, and punishment of cheaters. There may be others, and some of those remaining from the subtraction operation may be more developed in animals than currently believed. If I have learned anything from watching and studying animals, as well as reading about their behavior from my colleagues, it is that reports of human uniqueness are often shot down before they have much of a shelf life. Consider the proposed set of nonoverlapping abilities as an interim report. But also keep in mind that something will be left over, something uniquely human.

I warned at the beginning of part III that those looking for an answer to the question of whether animals are moral would be left unsatisfied. I don't have an answer, because the question is ill-posed, depending as it does on what one thinks is most interesting or important about morality. What I have done instead is to use the linguistic analogy to pose specific questions about our moral faculty, including what its components are, how they develop, and, most relevant here, how they evolved. Parts of our moral faculty are shared with other animals, and parts appear to be uniquely human. Can we therefore conclude, as did Darwin, that "Any animal whatever, endowed with well-marked social instincts, the parental and filial affections being here included, would inevitably acquire a moral sense or conscience, as soon as its intellectual powers had become as well developed, or nearly as well developed, as in man." We, of course, can draw this conclusion, but it leaves unanswered the most important part: What are these intellectual powers, such that they tip the scale or put us on to a different measuring device? One answer lies in a distinction that often enters legal discussion: the difference between a *moral agent* and a *moral patient*. Although much has been written about this distinction, especially given its centrality in debates about legal guardianship for children, adults with mental handicaps, and animals, we can distill the details down to one issue: Does the individual understand and respect others' rights and assume responsibility for his or her actions? If the answer is yes, then the individual is a moral agent. If the answer is no, but the individual can suffer, then he or she is a moral patient; moral agents are responsible in some sense for the welfare of moral patients. When philosophers, biologists, and psychologists joined forces to extend the Bill of Rights to the great apes—orangutans, gorillas, bonobos, and chimpanzees—they were forced to argue that these animals should have certain basic rights while recognizing that someone else would have to defend these rights. They had to place the great apes into the category of moral patients. That seems perfectly reasonable, although many have argued that the implications of this classification are complicated, including questions about where current zoo animals should go, why we should stop with the great apes, why only the particular rights are considered, and what we should do if any of these animals commits a crime, perhaps killing a human?

The real challenge with the moral agent-patient distinction is that it hinges on a test that doesn't yet exist: How do you determine whether nonlinguistic creatures understand and respect others' rights, and assume responsibility for what they do? The respect part is potentially the easiest, as we can see whether individuals follow certain kinds of principles: tolerating another's property, harming and helping others in specific, societally defined contexts. But it is not clear how one tests for comprehension. Nor is it clear how one determines whether they understand the concept of responsibility. That animals take care of others who require such care is one thing. It is something else altogether to feel the weight of obligation, cognizant of the consequences of breaking a commitment to help. The reason I dwell upon this distinction here is because for many scholars writing about our moral sense, and especially our sense of justice, the cooperative actions of animals are *merely* coordinated social behaviors. What they lack is the explicit recognition of why there are rules for cooperation and why there must be a group-level acknowledgment and adherence to such rules. In discussing his guiding principle of justice as fairness, Rawls makes this point explicit: "Social cooperation is distinct from merely socially coordinated activity—for example, activity coordinated by orders issued by an absolute central authority. Rather, social cooperation is guided by publicly recognized rules and procedures which those cooperating accept as appropriate to regulate their conduct."[46]

What can be said is that we inherited a suite of abilities from our primate ancestors. We use many of these capacities in both moral and nonmoral contexts. It appears, however, that those capacities that evolved uniquely within our species may have played a pivotal role in our capacity to sustain large-scale cooperation with unrelated individuals. One of the most comprehensive accounts of this recent evolutionary development comes from the biologists Robert Boyd and Peter Richerson. They start from one of Darwin's early intuitions about the evolution of morality. Recall that Darwin argued in favor of one group acquiring a larger and more stable set of moral norms than their neighbors, providing an edge in intergroup competition. In this sense, Darwin explained moral evolution by appealing to group selection. Boyd and Richerson follow this argument, but supplement it with more intricate mathematical models, experimental data, and cross-cultural observations.

The first point of note is that humans, in contrast to any other animal, show more marked differences between groups. Neighboring groups can have different languages, dress codes, marital principles, rules for punishment, and beliefs about the supernatural as well as the hereafter. Among animals, neighboring groups rarely like each other, but the differences between them are trivial. Even in cases where the contrasting groups are further apart geographically, the differences remain trivial. In a collaborative project among primatologists studying chimpanzees distributed across East and West Africa, several dozen social traditions emerged. Most of the traditions involved tool technologies, with some populations using stones to crack open nuts, others using fishing sticks to extract termites. Less frequently, observers noted differences in grooming techniques and a few other social gestures. As the cognitive scientist David Premack and I have noted, these differences don't carry much emotional weight in chimpanzee life. If a small group of chimpanzees decided not to hand-clasp groom, or to fish for termites using their arm rather than a stick, members of the community wouldn't excommunicate them. In this sense, the traditions we observe among different chimpanzee communities are more like the differences between countries where people drive on the right or the left side of the road than whether a person from Rwanda is a Tutsi or a Hutu, an Irishman is a Catholic or a Protestant, or, in the days of the American Civil War, whether a person was from the North or the South. Within a culture, we, of course, want everyone to drive on the appropriate side of the road, and would fine those in violation. But getting a population to shift its driving practice—as Sweden did in the 1970s—hardly constitutes an ideological shift worthy of note. The Swedes were still the Swedes after they changed from one side of the road to the other, and did so without any change in the accident rate. Try telling the Swedes to replace their Lutheran ministers with Orthodox rabbis! Not only would warfare most likely break out in this historically peaceful country, but the transition would be slow, effortful, and emotionally painful.

The lack of variation between animal groups, together with the relatively high migration rates, virtually eliminates the possibility of group selection. In contrast, the significant variation between human groups creates an opportunity for group selection.

Added to intergroup differences are two mechanisms that create greater homogeneity within groups: imitation and a conformity bias. Though chimpanzees and perhaps some other animals may be close contenders in terms of their capacity to imitate, imitation in humans differs in at least three ways: its immediate and reflexive appearance in development, its independence from a particular sensory modality, and its close connection with the mental states of others, including their intentions and goals. Imitation, together with our capacity to teach and transmit precise information, enables high-fidelity copying. It enables psychological cloning within groups, creating a potential for differences between groups. Added to these mechanisms is a conformity bias, a tendency to do what others do, leaving dissent to the outliers. As I mentioned in part II, dozens of experiments in social psychology reveal our chameleon-like minds, activated without our consent. We speak with someone who scratches his head a lot, and we have the urge to scratch. We see an elderly gentleman moving slowly, and we creep. Like it or not, our minds are constantly playing Simon Says, and playing quite well. From these pieces of neural hardware and software we create fads and, as Boyd and Richerson note, groups with distinctive markers. Given our strong interest in helping the in-group and harming the out-group, we now have the basis for large-scale cooperation among genetically unrelated individuals. This, together with our powerful systems of punishment, give us a solution to the paradox of human cooperation.

Whether, and to what extent, we carry forward our hunter-gatherer mind-sets, life is different today than it was then. Norms that were once adaptive may no longer be so. Moral dilemmas that we face today are, in some cases, wildly different from those we confronted on the savanna. Many of us confront opportunities to help genetically unrelated foreigners that live thousands of miles away on a different continent. Individuals are invited to contribute to aid organizations, and nations are invited to eliminate AIDS or abuses against human rights. The principles that evolved to guide helping didn't evolve in this context. If hunter-gatherer societies provide clues, the principles underlying helping were designed to resolve dilemmas in the near field, such as a sick relative or injured leader lying a few feet away. Helping at a distance, with its probabilistic uncertainty, wasn't in the cards. The same issues hold for harming. The hunter-gatherer artillery for violence entails, at best, bows and arrows that can shoot several dozen feet. When a hunter-gatherer kills another member of his group or a neighboring group, he is directly responsible for the death. Cases like the trolley problem, in which an action indirectly harms another and, often, many others, were never under consideration.

Similar issues arise in the context of utilitarian concerns about utility or value. The utilitarian perspective takes as its central tenet that we evaluate moral dilemmas in terms of consequences, where "consequences" translates to maximizing some notion of "good." As Tetlock[47] has noted, however, we are "struggling to protect sacred values from secular encroachments by increasingly powerful societal trends." When people are pushed to engage in these kinds of trade-offs—paying for the purchase of a child, selling organs in an auction—they are uncomfortable, often responding with moral outrage to the mere suggestion that this is a valid question.

The conclusion is straightforward: The systems that generate intuitive moral judgments are often in conflict with the systems that generate principled reasons for our actions, because the landscape of today only dimly resembles our original state. The Rawlsian creature will, therefore, fire off its intuitions about moral rights and wrongs, the Kantian will fire back principled arguments against these intuitions, and sometimes caught in the middle will be the Humean, generating angst, attempting to tilt the weight of the evidence toward one of the moral poles.

EPILOGUE:
THE RIGHT IMPULSE

I am convinced that a vivid consciousness of the primary importance of moral principles for the betterment and ennoblement of life does not need the idea of a law-giver, especially a law-giver who works on the basis of reward and punishment.

—ALBERT EINSTEIN[1]

UNLIKE ANY OTHER SPECIES that has existed on earth, humans are at once both highly variable and virtual clones. People in different parts of the world speak mutually unintelligible languages, practice different sexual rituals, listen to different music, are emotionally excited by different events, believe in different gods, engage in different sports, and have different social norms for helping and harming others. This landscape highlights our ingenuity for innovation and cultural variation, as well as our disrespect for conformity. But for each of these dimensions of human existence, there is reason to believe that there are universal properties of the human mind that constrain the range of cultural variation.

Our expressed languages differ, but we generate each one on the basis of a universal set of principles. Our artistic expressions vary wildly, but the biology that underpins our aesthetics generates universal preferences for symmetry in the visual arts and consonance in music. The idea I have developed in this book is that we should think of morality in the same way. Underlying the extensive cross-cultural variation we observe in our expressed social norms is a universal moral grammar that enables each child to grow a narrow range of possible moral systems. When we judge an action as morally right or wrong, we do so instinctively, tapping a system of unconsciously operative and inaccessible moral knowledge. Variation between cultures in their expressed moral norms is like variation between cultures in their spoken languages: Both systems enable members of one group to exchange ideas and values with each other, but not with members of another group. Whether the process of creating intergroup differences in intelligibility is adaptive or the by-product of isolation and historical contingencies is presently unclear, for both language and morality.

The idea that morality is grounded in our biology runs counter to three classically held beliefs. First, biological perspectives are inherently evil, as they create predetermined outcomes, thereby eliminating free will. As philosophers and psychologists, such as Daniel Dennett, Steven Pinker, and Daniel Wegner, have argued, however, nothing about an evolutionary or biological perspective leads inextricably to the notion of a determined, fixed, or immutable set of judgments or beliefs. Biology doesn't work this way. Our biology, and the biology of all species on earth, sets up a range of possible behaviors. The range we observe is only a limited sampling of the potential variation. This is because our biology interacts with the environment, and environments change. But from the fact that environments change we are not licensed to assume that cultures will change in parallel, entirely unconstrained. If there is a universal moral grammar, the principles are fixed, but the potential range of moral systems is not. The potential range is only constrained by the logical possibilities of the brain and some degree of historical inertia.

Second, if a biological perspective on morality is true, then the moral principles must be encoded in the DNA. Different amino acid sequences link to different deontological rules, some for harming and some for helping. This idea is indeed incoherent, but it has no resemblance to what I have argued. To say that we are endowed with a universal moral grammar is to say that we have evolved general but abstract principles for deciding which actions are forbidden, permissible, or obligatory. These principles lack specific content. There are no principles dictating which particular sexual, altruistic, or violent acts are permissible. Nothing in our genome codes for whether infanticide, incest, euthanasia, or cooperation are permissible, and, if permissible, with which individuals. And the simplest way to see that this must be so is to recognize that each child, depending upon his or her cultural origins, will acquire a distinctive moral system. The universal moral grammar is a theory about the principles that enable children to build a large but finite range of distinctive moral systems.

Third, even if biology contributes something to our moral psychology, only religious faith and legal guidelines can prevent moral decay. These two formal systems, with their explicitly articulated rules, must step up to the plate, knocking back those self-interested impulses. Darwin held a position close to this, stating that "man who has no assured and ever present belief in the existence of a personal God or of future existence with retribution and reward, can have for his rule of life, as far as I can see, only to follow those impulses and instincts which are the strongest or which seem to him the best ones."[2] In more recent times, Adolf Hitler followed Darwin's logic, but with explicitly nefarious goals, arguing that only religion can shape our moral character, and thus "Secular schools can never be tolerated because such schools have no religious instruction, and a general moral instruction without a religious foundation is built on air; consequently, all character training and religion must be derived from faith . . . we need believing people."[3]

Though equating morality with religion is commonplace, it is wrong in at least two ways: It falsely assumes that people without religious faith lack an understanding of moral rights and wrongs, and that people of religious faith are more virtuous than atheists and agnostics. Based on studies of moral judgments in a wide range of cultures, atheists and agnostics are perfectly capable of distinguishing between morally permissible and forbidden actions. More important, across a suite of moral dilemmas and testing situations, Jews, Catholics, Protestants, Sikhs, Muslims, atheists, and agnostics deliver the same judgments—and with the same level of incoherence or insufficiency when it comes to their justifications. Admittedly, the sample is limited. But within this range, which includes people who say that they are highly religious and others who say that they are not at all, there is no difference. These observations suggest that the system that unconsciously generates moral judgments is immune to religious doctrine.

The idea that religion is necessary for generating moral judgments fails on another level. Most, if not all religions, rely on relatively simple deontological rules—don't kill, lie, steal, break promises. These rules will not, however, explain the pattern of moral judgments that I described in the previous chapters. We feel the weight of a moral dilemma when simple deontological or utilitarian principles fail us. Religion may compel people to say that euthanasia and abortion are morally wrong, but when confronted with similar but less familiar and emotionally charged cases, their intuitions tilt them in a different direction.

Accompanying the religious perspective is the idea that our universally shared moral attitudes derive from shared experiences and tutoring, and not our common biology. Instead of an impoverished environment, each child develops within a relatively homogeneous and enriching environment that hands off universal rules for deciding when helping and harming are permissible. Children copy the morally permissible actions, but not the forbidden ones. Parents and educators correct the moral wrongs and applaud the moral rights.

This response both falsely assumes that the moral-instinct perspective denies all learning, and also fails to explain the growth of moral knowledge. Like language, the specifically expressed and culturally variable moral systems are learned in the sense that the detailed contents of particular social norms are acquired by exposure to the local culture; the abstract principles and parameters are innate. The role of experience is to instruct the innate system, pruning the range of possible moral systems down to one distinctive moral signature. This type of instructive learning is characteristic of countless biological processes, ranging from the immune system to language. When the genome has constructed a mechanism for generating a virtually limitless range of meaningful variation, the role of experience is to set a series of options so that the outcome is constrained. The immune system could have responses to a massive number of molecules, but due to early experience, ends up locking on to only a few. Similarly, the language faculty could build a massive number of expressed languages, but due to experience with the native language, switches on a few parameters in order to generate a single language. Morality, I have argued, works the same way.

Invoking experience as the sole determiner of the child's moral knowledge also can't explain the range of morally relevant actions expressed by children without copying and without parental instruction.

Children do all sorts of things that parents and their peers never do. In the same way that parents never say "I goed to the market," parents never grab all the toys in a display of extreme possessiveness, nor do they hit or bite their playmates when they are frustrated. The child's moral repertoire is not a parental clone. Moreover, if you observe all the things that children *don't* do, these do not correspond to all the things that parents and peers corrected in the past. Parental instruction of moral norms can only explain a small fraction of the child's knowledge. The main reason for this is because many of the principles underlying our moral judgments are inaccessible to conscious reflection. Adults use the fact that intended harms are worse than foreseen harms (pushing the fat man in front of the trolley is worse than flipping the switch), but are not consciously aware that they are using this distinction. Parents can't teach what they don't know.

The moral intuitions that drive many of our judgments often conflict with the guidelines dictated by law, religion, or both. Nowhere has this been more important than in the domain of bioethics, and, especially, some of the recent battles concerning euthanasia and abortion. To see how this dynamic works, and why a deeper understanding of our moral faculty is necessary if we want to navigate between descriptive and prescriptive principles, let me return to euthanasia and the case of Terri Schiavo that started this book. As a case of euthanasia, involving termination of life support, it seemed rather unexceptional. On June 15, 2005, a Google search of "Terri Schiavo" returned 1,220,000 hits; on the same day, Googling "euthanasia" returned 1,480,000 hits while "world hunger" returned 8,780,000. These stats suggest anything but unexceptional.

When this case hit the news, between 60 and 70 percent of people polled thought that the doctors should terminate life support by removing the feeding tube. These same respondents also noted that they would do the same in this situation, both for themselves and for a spouse. Those opposing this majority position were largely aligned with the religious right, arguing that life is precious and in God's hands. Cardinal José Saraiva Martins of the Vatican's Office for Sainthood said that the decision to remove Terri's feeding tube was "an attack against God."

The cause of the hype is still unclear. It may have stemmed from the current climate in the United States, where the division between government

and religion has been increasingly blurred. It may have stemmed from
the tortuous facts of the case, including the cycles of life-support termination and reinstantiation, the feud between family members, the in-
volvement of high-level government officials with no personal connection
to Terri or her family, and the apparent ability of the executive branch to
override law according to its personal or religious beliefs. Hype aside,
most people sided with Terri's own wishes for ending life support, a view
that generally coincides with most philosophical, legal, and medical arguments and policy but goes against much of religious doctrine. Through
the lenses of reason and intuition, many of us believe that if a person is
suffering from a disease with little to no hope of a cure, the most humane
response is to terminate life, either by removing support or by assisted
dying. This is a case in which many see harm as permissible, and some
may even wish to see it as obligatory, especially when the pain and suffering are excruciating and there are no cures. Where some people disagree is whether there is a meaningful difference between allowing the
patient to die (terminate life support) and helping the patient to die (overdose of a drug).

A purely economic analysis of this case forces us to push harder on the
nature of our intuitions concerning life support, while tying back to the
moral faculty's act/omission distinction and our moral obligations to others. Consider: a federal government can either spend $2 million on con-
tinued life support for a patient in a vegetative state or spend this money
on famine relief, saving the lives of fifty thousand people; if the money is
sent to the patient, the fifty thousand will die, whereas if the funds go to
famine relief, the patient will die. Is it permissible, obligatory, or forbidden for this federal government to send the money to the famine-relief
fund? Intuitions are likely to differ for this question. In general, legal policy rarely enforces a moral obligation to help, but is constantly looking for
ways to forbid actions that will cause harm. Spending money on the patient results—directly, in this case—in the death of fifty thousand; the
omission of aid is effectively murder, if we only attend to consequences.
Pushing the utilitarian position further, if only consequences matter, it
should be permissible to send the money to the famine-relief fund, and
the government's moral obligation to do so, given the options; sending the
money to the patient should be forbidden.

Do people ever have the moral obligation to die, to cause themselves harm, in effect? Though it may be hard to see how natural selection could ever result in a suicide instinct, further reflection suggests that this sacrifice might pay off in the service of preserving valuable resources for kin. Self-sacrifice may have been selected in certain circumstances. Such selection may, in turn, have generated a psychology of moral obligation.

Once we open the door to such moral maneuvers for euthanasia, admittedly abhorrent to many, parallel issues arise for abortion and infanticide, and for notions of harming others more generally. What appear to be different moral cases on the surface may reduce to the same set of principles with small, yet significant changes in parametric settings. This is where the idea of a universal moral grammar may have its most significant impact, highlighting how an understanding of descriptive, and possibly universal, moral principles bears on our approach to the prescriptive principles of what ought to be.

The signature of progress in any science is an increasingly rich set of explanatory principles to account for the phenomenon at hand, as well as the delivery of new questions that could never have been contemplated in the past. The science of morality is only at the earliest stages of growth, but with exciting new discoveries uncovered every month, and rich prospects on the horizon. If the recent history of linguistics is any guide, and if history provides insights for what is in store, then by raising new questions about our moral faculty, as I have done here, we can anticipate a renaissance in our understanding of the moral domain. Inquiry into our moral nature will no longer be the proprietary province of the humanities and social sciences, but a shared journey with the natural sciences.

I started this exploration into our moral psychology by building on an analogy inspired by John Rawls—the idea that we are endowed with a moral instinct, a faculty of the human mind that unconsciously guides our judgments concerning right and wrong, establishing a range of learnable moral systems, each with a set of shared and unique signatures. Like Rawls, I favor a pluralistic position, one that recognizes different moral systems, and sees adherence to a single system as oppressive. The notion of a universal moral grammar with parametric variation provides one way to think about pluralism. It requires us to understand how, in development, particular parameters are fixed by experience. It also requires us to

appreciate that, once fixed, we may be as perplexed by another community's moral system as we are by their language. Appreciating the fact that
we share a universal moral grammar, and that at birth we could have acquired any of the world's moral systems, should provide us with a sense
of comfort, a sense that perhaps we can understand each other.

NOTES

PROLOGUE: RIGHTEOUS VOICES

1. Chomsky on language (1957; 1965; 1986; 1988; 1995; 2000).
2. Although the United States generally holds to the AMA's distinction, some states have shifted certain elements of the policy. In a June 1, 2004, piece in the *New York Times*, J. Schwartz and J. Estrin report that in Oregon, there has been a steady increase in the number of prescriptions written by doctors to enable patients with terminal illnesses to take their own lives through a drug overdose. From 1998 to 2003, the number of patients dying due to overdose tripled.

1. WHAT'S WRONG?

1. Quoted on http://www.foodreference.com/html/qadamandeve.html.
2. (Rachels, 2003), p. 12.
3. The slippery path from IS to OUGHT (May, Friedman, & Clark, 1996; Moore, 1903; Sober, 1994; Wilson, Diertrich, & Clark, 2003). Though Moore was clearly responsible for this distinction, he used it to attack certain aspects of virtue theory, and, in particular, the idea that the notion of good can be reduced to a set of defining features. For Moore, one should not confuse the property of goodness with the set of objects or events that are good, or constitute things that we desire because they are good.
4. Rachels, 2000.
5. For discussions of the naturalistic fallacy, see (Greene, 2003; Wilson et al., 2003); for

discussion of the relationship between moral theory and the biological sciences, see (Ridley, 1996; Singer, 2000; Sober & Wilson, 1998; Wilson, 1998).

6. See Philip Kitcher (1985) for a trenchant critique of some of the early ambiguities associated with the sociobiology revolution and its forecasts for a theory of morality.

7. The dicey, complicated, intriguing, challenging world of moral dilemmas (Greenspan, 1995; Kamm, 1998, 2001a; Mason, 1996; Singer, 1993; Unger, 1996). There is a parallel set of discussions emanating from psychology and, in particular, from the elegant work of Elliot Turiel (1983), who coined the distinction between transgressions or violations that break social conventions as distinct from moral rules. In contrast with social conventions, violations of moral rules are more serious, independent of authority figures, and generalizable to other people living in different cultures; for example, pulling someone's hair for fun is wrong, everywhere, even if your parents or God say it is okay. I will return to this distinction later on, but as several authors have pointed out, most noticeably Shaun Nichols (2004), the conceptual dividing line between these two forms of social norms is not always crisp.

8. (Unger, 1996).

9. Baron's guidebook to intuition blindness (1994; 1998).

10. Throughout this book, I will either explicitly state or imply that moral philosophers—as a species of intellectual—have largely ignored or trivialized the potential importance of scientific findings for understanding the nature of moral beliefs and judgments. This could be stated with almost complete confidence about ten years ago. Today, a new generation of moral philosophers is well versed in the mind and brain sciences. Like their predecessors, they begin papers by laying out the logic of their arguments and intuitions, often flavored with vivid examples. But then, and in contrast to their predecessors, they present the results of experiments. This new generation of empirically minded moral philosophers includes Joshua Greene, Joshua Knobe, and Shaun Nichols.

11. *Paradise Lost*, Book 2, line 910.

12. (Hobbes, 1651/1968), Part I, chapter 13.

13. Immanuel Kant's thinking about morality (1785/1959; 2001).

14. Kantian scholars debate whether we should think of the categorical imperative as a method that prods a decision procedure or as a particular test of the principles. For those who see it as a test, the issue is not so much whether one can imagine a world in which promises are broken, but whether this is a rationally intelligible world. I thank Susan Dwyer for helping me clarify this point.

15. (Kohlberg, 1981; Piaget, 1932/1965, 1954).

16. (Kohlberg, 1981); p. 181.

17. Let me note here that over the course of developing his theoretical framework, Kohlberg backed off some of the more stringent aspects of his final stage of moral maturity, and its reliance on a Kantian architecture. Though his last formulation of the stages of moral development did not include Kant's primary principles, Kohlberg held to a strong rational and consciously reasoned perspective on moral development. In many ways, therefore, Kohlberg out-Kanted Kant in terms of a reliance on deliberate reasoning.

18. Problems, problems, and more problems for the stage theory of moral development (Gibbs, 2003; Macnamara, 1990).

19. These cases come from the work of the social psychologist Jonathan Haidt (2001; 2003), whose observations on moral dumbfounding and emotion figure prominently throughout this section.

20. For two of the most recent Humean applications, one from psychology and one from philosophy, see Haidt (2001) and Nichols (2004).

21. (Hume, 1739/1978), p. 474, 500.

22. (Rachels, 2003), p. 12.

23. (Hoffman, 2000), p. 3. Hume used the term "sympathy," whereas Hoffman uses "empathy." Some authors make a distinction between these two terms, with sympathy centered on a concern for others without a necessary feeling of what another feels, whereas empathy centers explicitly on the "what it is like" part—imagining oneself in someone else's shoes. For purposes of discussion, I will use the terms interchangeably, sticking largely with empathy.

24. The chameleon effect and how our perception of the social actions reflexively triggers empathy, altruism, and even some unwanted behavior (Ferguson & Bargh, 2004).

25. Ibid., p. 48.

26. http://moral.wjh.harvard.edu.

27. See the many discussions of trolley-esque problems, originating with Foot and Thomson (Fischer & Ravizza, 1992; Mikhail, 2000).

28. (Hare, 1981), p. 139.

29. Frances Kamm (1992; 1998) develops this point beautifully in her two books, *Creation and Abortion* and *Morality, Mortality*.

30. *Vanity Fair*, November 2005, p. 377.

31. Cross-cultural similarities in moral intuitions (Mikhail, 2000; O'Neill & Petrinovich, 1998; Petrinovich & O'Neill, 1996).

32. (Chomsky, 1968; Chomsky, 1988; Hume, 1739/1978; Hume, 1741/1875; Hume, 1748/1975; Rawls, 1950; Rawls, 1951; Rawls, 1971; Sidgwick, 1907; Harman, 1999; Mikhail, 2000; Dwyer, 1999; Dwyer, 2004; Jackendoff, 2005).

33. Unconscious knowledge systems of the mind (Dehaene, 1997; Lerdahl & Jackendoff, 1996; Mikhail, 2000; Rawls, 1971; Spelke, 1994).

34. Some linguists who adhere to the principles-and-parameters perspective think that there may be default settings for some parameters, set up during fetal development but open to change upon delivery.

35. From *is* to *'s* (Anderson & Lightfoot, 2000).

36. (Rawls, 1971); pp. 46–47.

37. It is perhaps worth noting here that even Kant appreciated the significance of the relationship between moral judgments and action perception: "When moral worth is in question, it is not a matter of actions which one sees but of their inner principles which one does not see" (p. 23).

38. As discussed later in the book, my comment here concerning the lack of information on act perception is specific to the moral domain. Social psychologists and vision scientists have provided a good deal of information on how we perceive actions and events (Heider & Simmel, 1944; Scholl & Nakayama, 2002; Scholl & Tremoulet, 2000; Wegner, 2002; Zacks et al., 2001; Zacks & Tversky, 2001).

39. The philosopher Bernard Gert (Gert, 1998, 2004) has developed one of the most complete accounts of these rules, as well as some of the exceptions that follow from them.

40. (Knobe, 2003a, 2003b; Leslie, Knobe, & Cohen, in press; Mele, 2001).

41. What is guilt? (Baumeister, Stillwel, & Heatherton, 1994; Haidt, 2003).

2. JUSTICE FOR ALL

1. http://www.quotegallery.com/asp/cquotes.asp?parent=Virtue+%2F+Good+%2F+Evil & child=Justice.

2. Huxley's trouncing of Wilberforce: http://www.uta.edu/english/danahay/debate.html.

3. Dawkins, 1976, p. 205.

4. (Jefferson, 1787/1955), ME 6:257, Paper 12:15.

5. This is a modification of Peter Singer's principle (Singer, 1993), p. 230.

6. Much of the discussion that follows is based on the work of John Mikhail, who has attempted to resuscitate Rawls's linguistic analogy and carry it forward into modern cognitive science (Mikhail, 2000, 2002; Mikhail, Sorrentino, & Spelke, 2002).

7. As discussed earlier, there is a voluminous literature on domain-specific principles for acquiring knowledge, starting perhaps with Chomsky's work in linguistics and continuing into the present for domains including knowledge of artifacts, social relationships, and mathematics (Caramazza, 1998; Chomsky, 1957, 1986, 2000; Cosmides & Tooby, 1994; Dehaene, 1997; Gelman, 2003; Pinker, 1997).

8. (Chomsky, 1965), pp. 8–9.

9. (Rawls, 1950), pp. 45–46.

10. Rawls's second principle of distribution is problematic because it creates larger inequities among the population even though it improves the well-being of those worse off. Imagine one state where the potential is for the richest third of the population to earn $100 a month, the middle class $80, and the poorest (least well off) $60. In a second state, the potential is for the richest to obtain $100, the middle class $61, and the poorest $61. Based on Rawls's second principle, people should favor the regime for the second state as the poorest do better. The logic works, but the inequity between the richest third and the rest is far greater than in the first state.

11. In a wonderfully lucid book, Baker (Baker, 2001) lays out the principles-and-parameters view of language, a position initially developed by Chomsky (1981). My focus here is not meant as an endorsement for the P&P view. I fully acknowledge the many important alternatives, most recently reviewed by Jackendoff (Jackendoff, 2002). Rather, I believe that the P&P view offers a simple way of describing some of the relevant issues in the study of language and, important, provides a simple way of setting up the argument for our moral faculty.

12. Linguistic conceptions of fairness (Lakoff, 1996); for integration of these ideas into a Darwinian political framework, see Rubin (P. Rubin, 2002).

13. (Bowles & Gintis, 1999; D. Rubin, 2002).

14. Games economists play (Kagel & Roth, 1995).

15. (Nowak & Sigmund, 2000), p. 819.

16. Fair play and reputation (Kagel & Roth, 1995; Nowak, Page, & Sigmund, 2000).

17. Almost all of the literature in experimental economics and evolutionary biology focusing on games such as the ultimatum and dictator presume that the relevant psychology concerns issues of fairness. Why fairness, as distinct from generosity? Consider the following example. A mother gives her eldest son, Billy, a box of ten chocolates and suggests that he might share some with his younger brother Joey. Billy walks over and says to Joey, "You can have one piece of candy." It would seem odd for Joey to reply, "No. That's not fair. You have an entire box. I should get at least three pieces." Joey might think that Billy isn't generous, but certainly not unfair. Fairness seems to be associated, more often than not, with issues concerning justice. For example, in an elegant legal treatise titled "Fairness Versus Welfare," the lawyers Louis Kaplow and Steven Shavell state: "Notions of fairness . . . include ideas of justice, rights, and related concepts [which] provide justification and language for legal policy decisions." For example, if someone has been hit by a car and seriously injured, a fair trial would see to it that the victim is compensated for the driver's recklessness. The psychology of generosity may be a more appropriate target in these experiments than the psychology of fairness.

18. Tragedy of the commons, or the temptation to overuse public goods (Hardin, 1968).

19. (Duncan, 1993).

20. The notion of indirect reciprocity was first introduced in a brilliant, and in some sense clairvoyant, book by Richard Alexander (1987) titled *The Biology of Morality*.

21. Punishment and reputation represent the core ingredients of stable cooperation among humans (Fehr & Gachter, 2002; Milinski, Semmann, & Krambeck, 2002; Wedekind & Milinski, 2000).

22. Strong reciprocity is pure altruism and uniquely human (Fehr & Fischbacher, 2003; Fehr & Henrich, 2003; Gintis, 2000; Gintis, Bowles, Boyd, & Fehr, 2003).

23. Different strokes for different folks (Henrich et al., 2001; Henrich et al., in press).

24. (Binmore, 1998; Boehm, 1999).

25. The Ache of Paraguay and a model system of sharing (Hill, 2002).

26. Bringing Rawlsian philosophy to the experimental table (Frohlich & Oppenheimer, 1993).

27. Personal communication from Norman Frohlich, October 21, 2003.

28. Kahneman is, of course, not alone in making such stabs. See (Gilovich, Griffin, & Kahneman, 2002).

29. Although most in political philosophy interpreted Rawls's initial thesis as providing a universalist account of distributive justice, he subsequently clarified his position, suggesting that it may only have applications to Westernized democracies.

30. (Kaplow & Shavell, 2002).

31. Evolving norms (Axelrod, 1984; Kameda, Takezawa, & Hastie, 2003; Kaplow & Shavell, 2002; McElreath, Boyd, & Richerson, 2003).

32. (Posner, 2000), p. 3.

33. Recent brain-imaging work by Ernst Fehr reveals that during a bargaining game, the areas of the brain involved in reward are active when people punish. Punishing feels good, rewarding.

34. Theoretical models of punishment (Bowles & Gintis, 2003; Boyd, Gintis, Bowles, & Richerson, 2003).

35. Scapegoating controls (Russell, 1965).

36. Hadza think that it is better to receive than to give (Marlowe, 2004).
37. Scarlet-letter punishments on the rise (Garvey, 1998; Litowitz, 1997; Whitman, 1998).
38. (Simon, 1990), p. 7; for further developments of Simon's early insights, see (Todd & Gigerenzer, 2003).
39. (Whitman, 2003).
40. (Goldman, 1979), p. 54.
41. What our legal systems think they are accomplishing with punitive measures and how juries actually decide (Sunstein, Hastie, Payne, Schkade, & Viscusi, 2002).
42. (Daly & Wilson, 1988), p. 251.
43. (Kaplow & Shavell, 2002), p. 328.
44. Exodus, 22:1.
45. (Dolinko, 1992), p. 1656.

3. GRAMMARS OF VIOLENCE

1. http://www.bartleby.com/100/139.5.html.
2. Foot's trolley problem and its alternative incarnations (Fischer & Ravizza, 1992; Foot, 1967; Thomson, 1970).
3. (Fischer & Ravizza, 1992; Kamm, 1998, 2001).
4. (O'Neill & Petrinovich, 1998; Petrinovich & O'Neill, 1996; Petrinovich, O'Neill, & Jorgensen, 1993).
5. (O'Neill & Petrinovich, 1998), p. 364.
6. (Mikhail, 2000; Mikhail et al., 2002).
7. Universals of violence (Daly & Wilson, 1988, 1998; Hirschi & Gottfredson, 1983; Wilson, Daly, & Pound, 2002).
8. Thanks to Richard Wrangham for bringing these points to my attention.
9. (Wilson et al., 2002), p. 289.
10. The origins and cultural evolution of partiality (Stingl & Colier, 2005).
11. Macho cultures (Nisbett & Cohen, 1996).
12. City names reflect psychology of violence (Kelly, 1999).
13. The broader implications of discounting (Ainslie, 2000; Daly & Wilson, 1988; Elster, 1979, 2000; Rogers, 1994, 1997; Wilson & Herrnstein, 1985; Wilson et al., 2002).
14. Shocking experiments (Milgram, 1974).
15. Cultures of honor and the IS-OUGHT distinction (Vandello & Cohen, 2004).
16. (Vandello & Cohen, 2004).
17. (Gordon, 2001).
18. Killing for honor (Emery, 2003; Kulwicki, 2002; Mayell, 2002).
19. Amnesty International on honor killings in Pakistan: http://web.amnesty.org/library/Index/engASA330062002?OpenDocument.
20. cited in Vandello and Cohen, 2003, p. 998.
21. The psychology and law behind crimes of passion (Abu-Odeh, 1997; Engel, 2002; Milgate, 1998).
22. (Kundera, 1985), p. 34.
23. (Nourse, 1997), pp. 1340–1342.
24. (Horder, 1992), p. 20.
25. (Harris, 1989).

26. (Horder, 1992), p. 70.
27. (Hobbes, 1642/1962), p. 25, cited in Horder, ibid.
28. (Horder, 1992), p. 98.
29. Mikhail (Mikhail, 2000) gives a much more formal analysis of this and other problems, while Kamm (Kamm, 2000) provides many of the important extensions, pointing out limitations to the principle of double effect.

4. THE MORAL ORGAN

1. (Hutcheson, 1728/1971), preface.
2. (Leibniz, 1704/1966), book I, chapter ii.
3. (Darwin, 1958), p. 94.
4. Prinz's [in press] attack on a nativist moral psychology.
5. For general discussions of expectancy in the philosophy of action and developmental psychology, see (Kagan, 2002b; Olson et al., 1996; White, 1995).
6. A number of developmental psychologists have discussed the importance of expectation in understanding the minds of young infants, while alluding to parallels with animals (Kagan, 2002b).
7. Learning a simple rule: pick B, not A (Baillargeon et al., 1990; Diamond et al., 1994; Diamond & Gilbert, 1989; Harris, 1986; Marcovitch & Zelazo, 1999; Munakata, 1997; Piaget, 1954; Smith et al., 1999; Wellman et al., 1986; Zelazo et al., 1998).
8. Looking and reaching for hidden objects (Baillargeon, 1995; Baillargeon & DeVos, 1991; Baillargeon et al., 1985; Bogartz et al., 1997).
9. Self-propelled motion as a signature of living things (Gelman, 1990; Leslie, 1994; Premack, 1990; Schlottmann & Surian, 1999).
10. Inferring goals from action (Johnson, 2003; Santos & Hauser, 1999; Woodward et al., 2001). Johnson shows that even an amorphous blob that babbles back to a human with nonspeech sounds can have goals.
11. On theoretical expectations, see (Gergely & Csibra, 2003; Leslie, 1994; Premack, 1990; Premack & Premack, 1995, 1997); for theory and, especially, brilliant experimental tests, see (Csibra & Gergely, 1998; Csibra et al., 1999; Gergely et al., 1995).
12. (Bassili, 1976; Heider & Simmel, 1944; Shimizu & Johnson, 2004).
13. (Johnson, 2000, 2003).
14. Action and emotion (Kuhlmeier et al., 2003; Premack & Premack, 1997).
15. (Abell et al., 2000; Kuhlmeier et al., 2003).
16. The philosophy of actions and events (Casati & Varzi, 1996).
17. (Goldman, 1971).
18. The units of event perception, from behavior to brain (Newtson, 1973; Newtson et al., 1977; Newtson et al., 1987; Zacks et al., 2001; Zacks & Tversky, 2001).
19. How babies cut up events into actions (Baldwin & Baird, 2001).
20. The extraordinary world of identical twins reared apart (Bouchard et al., 1990; Deary, 2001; McClearn, 1997; Pinker, 2002).
21. Self in culture (Boehm, 1999; Haidt, 2003; Markus & Kitayama, 1991).
22. (Flanagan, 1996).
23. Emerging sense of self (Kalnins & Bruner, 1973; Lewis et al., 1985; Rochat & Striano, 1999).

24. It is important to note here that even before a self-reflective or self-conscious sense emerges, individuals obtain much information about what is happening to them—to their own body—based on a rich set of internal mechanisms, including systems that monitor body position, temperature, heart rate, touch, sight, hearing, and so forth. As Damasio (Damasio, 2000) has clearly expressed, this is a proto-self, one that emerges in evolution and development before the self-reflective and conscious self.

25. Damaged self-recognition with intact self-awareness (Ellis & Lewis, 2001).

26. Hidden knowledge in prosopagnosia (Tranel & Damasio, 1985, 1993; Tranel et al., 1988; Tranel et al., 1995; Young & DeHaan, 1988).

27. Capgras or the impostor delusion (Ellis & Lewis, 2001).

28. Slaves are socially dead (Patterson, 1982).

29. Emotional variants (Fessler, 1999; Fessler & Haley, 2003; Haidt, 2003).

30. (Bentham, 1789/1948); p. 1; Socrates, quoted in Plato's *Phaedo*, section 68c–69d.

31. Prinz's *Gut Reactions* (Prinz, 2004) provides a lucid analysis of the current theoretical landscape on the nature of emotions. It illustrates the kinds of issues that are at stake and how studies by psychologists and neuroscientists have started to disentangle the causes and consequences of emotion for our behavior.

32. Much of my discussion of emotion draws on the following excellent texts and theoretical papers (Damasio, 1994, 2000; Davidson et al., 2003; Eisenberg et al., 2003; Ekman, 1992; Fessler & Haley, 2003; Fiske, 2002; Frank, 1988; Griffiths, 1997; Haidt, 2001; Haidt & Joseph, 2004; LeDoux, 1996; Nesse, 2001; Prinz, 2004).

33. (Eisenberg et al., 2003), p. 789; for a detailed and spirited discussion of empathy, including its evolutionary and developmental origins, see (Preston & de Waal, 2002). More classic treatments are found in (Gibbs, 2003; Hoffman, 2000).

34. Me, my reflection, my yawns, and my helping (Gallup, 1991; Platek et al., 2003).

35. Disgust in the brain (Moll et al., 2005; Sprengelmeyer et al., 1996; Sprengelmeyer et al., 1997).

36. Darwin's disgust (Darwin, 1872); p. 253.

37. Beyond Darwin's disgust (Fessler, in press; Haidt et al., 1994; Rozin, 1997; Rozin & Fallon, 1987; Rozin et al., 2000; Rozin et al., 1986; Rozin et al., 1989).

38. The chicken-and-egg problem of disgust and moral beliefs (Fessler, in press).

39. Incest taboos as a human universal (Brown, 1991; van den Berghe, 1983).

40. (Westermark, 1891).

41. From familiarity to disgust (Lieberman et al., in press; Shepher, 1971; Wolf, 1995).

42. (Frazer, 1910).

43. The will to act (Dennett, 2003; Wegner, 2002).

44. The breakdown of the social mind into its component parts was perhaps first discussed by the cognitive scientist Alan Leslie (Leslie, 1994, 2000; Leslie & Keeble, 1987), especially in terms of a modular faculty. Baron-Cohen (1995), Karmiloff-Smith (1992), and many others subsequently added on and modified these earlier views.

45. Although blind children appear to understand perspective-taking, some authors have suggested that they are either delayed on tasks that require an understanding of false beliefs, or fail to acquire such understanding altogether (Hobson, 2002; McAlpine & Moore, 1995; Minter et al., 1998). Because *some* blind children do acquire an understanding of false beliefs, however, it is clear that visual input is not necessary for acquisition.

46. Linking pretend play to the child's developing theory of mind (Harris, 2000; Leslie, 1987).

47. Contrary to Piaget and Kohlberg, who believed that the child's understanding of another's intent develops late, around the ninth birthday, a suite of studies, most recently by Leslie (Leslie et al., in press) and Siegal (Siegal & Peterson, 1998), indicate a much earlier appreciation, including the relationship between intention and harmful or helpful consequences, as well as the difference between lies, innocent accidents, and negligent mistakes.

48. The connections between the vast but independent literatures on the development of theory of mind and moral judgment have only recently met face-to-face (Baird & Astington, 2004; Chandler et al., 2000; Kahn, 2004; Moore & Macgillivray, 2004; Wainryb, 2004).

49. Sally, Ann, and the possession of false belief. As Alan Leslie pointed out to me, the original version of this task, developed by Wimmer and Perner (Wimmer & Perner, 1983), involved a character named Maxi and his mother. With this version of the task, all three-year-olds failed and four-year-olds were about fifty-fifty; only six-years-olds got it. Due to the complexity of the task, Baron-Cohen, Leslie, and Frith (1985) simplified things and gave birth to Sally and Ann; this task led to three-year-olds failing and most four-year-olds succeeding.

50. Conceptual revolutions in science and infants (Carey, 1985; Kuhn, 1970).

51. A study using the expectancy-violation looking method suggests that fifteen-month-old babies may have an early form of the capacity to attribute mental states to others (Onishi & Baillargeon, 2005).

52. I address the topic of animal intentions in chapters 6 and 7, including some work on dogs and chimpanzees.

53. Williams syndrome and an impoverished folk biology (Johnson & Carey, 1998).

54. Race in mind (Hirschfeld, 1996); http://www.umich.edu/~newsinfo/MT/96/Jun96/mta1j96.html.

55. Bad biases (Allport, 1954; Dovidio et al., 2005; Fiske, 1998).

56. *Miller v. California*, 413 U.S. 15, 93 S. Ct. 2607 (1973); cited in Nussbaum (2004) p. 2.

57. Unconscious attitudes toward *l'autre* (Banaji, 2001; Cunningham et al., 2005; Phelps et al., 2000).

58. (Hazlitt, 1805/1969), p. 3.

59. Waiting for one marshmallow now or five later (Ayduk et al., 2000; Metcalfe & Mischel, 1999; Mischel, 1966, 1974; Mischel et al., 1989; Mischel et al., 1974; Peake et al., 2002; Sethi et al., 2000). This exceptional body of work is joined by another, collected over an equally long period of time by my colleague and friend Jerome Kagan on temperament (Kagan, 2002a; Kagan et al., 1988). Kagan's work also provides an exquisite example of how detailed longitudinal studies of children reveal the power of biological mechanisms to guide the child's responses to its environment.

60. (Kochanska et al., 1996).

61. Waiting for rewards and controlling violence (Ayduk et al., 1999; Ayduk et al., 2000; Downey & Feldman, 1996; Downey et al., 2000).

62. Prudence and altruism (Moore & Macgillivray, 2004; Thompson et al., 1997).

63. Virtue ethics (Hursthouse, 1999; Sherman, 1989).

64. Sometimes it pays not to delay (Kacelnik, 1997, 2003; Rogers, 1994, 1997; Wilson et al., 2002).

65. Oscar Wilde, *The Critic as Artist*, part 1, "Intentions," 1891.

66. Your brain caught in a trolley problem (Greene et al., 2004; Greene et al., 2001).

67. I note here that Greene wouldn't specifically endorse the view of Kant that I have described here. Under his view, which I take to be quite unorthodox with respect to Kant's writings and those who have followed his moves, Kant's philosophy is mediated by emotions and then rationalized by deliberate reasoning. For Greene, therefore, the kind of deontological principles that Kant advocated are more closely aligned with our emotions than with our reason. Although we clearly do have emotions associated with deontological principles, such as killing is wrong, it is also possible that our emotions follow from abstract and emotionless rules. For purposes of exposition, and in keeping with more traditional views of Kant's writings, I will maintain my characterization of the Kantian creature.

68. Inside our clockwork orange (Berthoz et al., 2002; Moll et al., 2001; Moll, Oliveira-Souza et al., 2002; Moll et al., 2005; Moll, Olivier-Souza et al., 2002; Oliveira-Souza & Moll, 2000).

69. Simulation and the mirror neuron system (Gallese & Goldman, 1998; Gallese et al., 2004; Goldman, 1989, 1992; Gordon, 1986; Rizzolatti et al., 1999; Rizzolatti et al., 1996).

70. Frontal lobe damage, control, emotion, and moral behavior (Adolphs, 2003; Damasio, 1994, 2000; Fuster, 1997; Goldberg, 2001; Shallice, 1988).

71. Breakdown of the social brain (Adolphs, 2003; Adolphs et al., 1998; Anderson et al., 1999; Damasio, 1994).

72. Gambling, prefrontal damage, and the failure to recruit emotions in the service of making wise decisions for the future (Bechara et al., 1994; Bechara et al., 1997; Damasio, 2000; Tranel et al., 2000).

73. Brain-damaged patients and the discounting problem of temptation and control (Ainslie, 2000; Elster, 1979, 2000).

74. A forceful paper by the philosopher Adina Roskies (2003) argues that Damasio's patient data knock out a classic position in ethics (belief-internalism) that sees our moral knowledge as inextricably linked to our motivational systems. In other words, moral beliefs are just motivational states, such that a person who believes a particular action is morally right is by necessity motivated to carry out this action. For Roskies, the frontal-lobe patients show that this view is bankrupt, as the system that supports motivation is knocked out, and yet these same individuals have normal moral beliefs. A problem with Roskies's view, however, is that the patient deficit is not in the motivation to act but in the emotive systems that regulate action. There is no evidence that these patients are unmotivated to do the right thing, but rather that whether or not they are is unchecked by a normal suite of emotions that either promote or diffuse particular moral actions. This is a performance view of morality, and leaves untouched issues of moral competence.

75. Damage to the orbitofrontal cortex in infancy blocks the acquisition of a moral sense but no other higher cognitive function, including language, planning, and reasoning (Anderson et al., 1999; Driscoll et al., 2004).

76. Our ancestry for violence (Wrangham & Peterson, 1996).

77. Out-of-control aggression (Hare, 1993; Hollander & Stein, 1995).

78. A clinician's guide to psychopathy (Hare, 1993).

79. The emotional profile of a psychopath (Blair, 1995, 1997; Blair & Cipolotti, 2000; Blair et al., 1999; Blair et al., 1995; Stevens et al., 2001). Although Blair's work shows that an emotional deficit of some sort is responsible for psychopathy, psychopaths also lack the critical connection between emotion and a normative theory of why some actions are right and others are wrong, a point made by the philosopher Shaun Nichols (2002). Blair has since backed off the strong version of his claim and argued instead that in addition to their inability to recognize distress cues, psychopaths also have difficulty reversing a particular pattern of responses, suggesting that a key connection between emotion and decision-making has been severed.

80. The brain of a psychopath (Intrator et al., 1997; Kiehl et al., 1999; Kiehl et al., 2001; Muller et al., 2003; Raine et al., 2004).

81. The heat of a moral transgression and the cool of a conventional violation (Nichols, 2002, 2004; Nucci, 2001; Smetana, 2005; Turiel, 2005).

5. PERMISSIBLE INSTINCTS

1. (Molière, 1673), Act 1, scene 1.

2. (Barry, 1995), p. 44.

3. Life outside of gestational Eden (Hrdy, 1999), p. 440.

4. Zahavi's handicapped signals (Zahavi, 1975, 1987).

5. (Baird & Rosenbaum, 2001; Dworkin, 1993).

6. Thinking about abortion in the more general context of the principles of harm (Kamm, 1992; Thomson, 1971).

7. I say "presciently" here because Rawls wrote this paragraph (pp. 503–504) well before Dawkins and Wilson each wrote their seminal and synthetic publications in the field of sociobiology. At least some of his insights derived from conversations with Robert Trivers, then only a graduate student at Harvard but in the throes of his own contribution to morality, especially the publication of his theory of reciprocal altruism published in the same year as Rawls's *Theory of Justice*.

8. From *Homo economicus* to *Homo reciprocans* (Bowles & Gintis, 1998).

9. The evolution and ontogeny of the human number sense. This vast literature is reviewed in much greater detail elsewhere (Butterworth, 1999; Dehaene, 1997; Gallistel & Gelman, 2000; Hauser, 2000; Hauser & Spelke, 2004).

10. How the mind computes numbers without language (Dehaene, 1997; Gallistel & Gelman, 2000; Gordon, 2004; Hauser, 2000; Hauser & Carey, 1998; Pica et al., 2004; Pylyshyn & Storm, 1998).

11. How children allocate a fair share (Huntsman, 1984).

12. Indirect reciprocity (Alexander, 1987), p. 85.

13. Games kids play (Harbaugh, Krause, & Liday, 2001; Harbaugh et al., 2003; Harbaugh, Krause, & Vesterlund, 2001; Keil, 1986; Murnighan & Saxon, 1998).

14. When kids play games (Harbaugh, Krause, & Liday, 2001).

15. Although the literature on deception in child development makes little connection with the work in moral psychology, it seems that the action/omission bias, which appears early in development, may play an essential role (Baron, 1998; Baron et al., 1993).

16. Little white lies (Bok, 1978; Sweetser, 1987).
17. (Russell et al., 1994), p. 301.
18. Coordinating the eyes to coordinate intentions and goals (Tomasello et al., 2005).
19. Developing the deceptive mind (Carlson et al., 1998b; Chandler et al., 1989; DePaulo et al., 1982; Freire et al., 2004; Hala et al., 1991; Lee et al., 2002; Polak & Harris, 1999; Ruffman et al., 1993; Sodian et al., 1991; Talwar & Lee, 2002; Talwar et al., 2002). There is presently considerable debate among developmental psychologists about when the child's deceptive competences come on line and what capacities are required. I am skirting much of this debate here to make two simple points: There is an early competence for deception that comes on line before children can justify or explain their behavior, and this competence appears to emerge in all children, at about the same time, independent of culture or the teachings of parents, siblings, and peers.
20. (Austen, 1814).
21. Different interpretations of the Wason selection task (Almor & Sloman, 1996; Cheng & Holyoak, 1985, 1989; Fiddick et al., 2000; Lieberman & Klahr, 1996; Sperber & Girotto, in press).
22. A patient who understands precaution rules but not social contracts (Stone et al., 2002).
23. Permission rules include social contract and precaution rules, but not all permission rules are social contracts or precautions. In early discussions of Cosmides's work on the Wason task, Cheng & Holyoak argued that the results said little about an evolved module for detecting cheaters, but rather spoke to the more general and less novel claim, that context can influence performance, especially when the logic is framed in the form of social rules of permission (Cheng & Holyoak, 1985, 1989; Cosmides & Tooby, 2000).
24. Logical inference in children (Harris & Nunez, 1996; Nunez & Harris, 1998).
25. Quoted in J. Epstein (Epstein, 2003), p. 7; for the science of envy, see (Fessler & Haley, 2003).
26. J. Epstein (Epstein, 2003), p. xix; (Gaylin, 2003), p. 64.
27. Oscar Wilde's *Lady Windermere's Fan*, Act 3.
28. Emotions and the commitment problem (Fessler & Haley, 2003; Frank, 1988; Nesse, 2001; Schelling, 1960).
29. Trust someone who looks like you (DeBruine, 2002).
30. Guilty glue (Trivers, 1971; Fessler & Haley, 2003; Ketelaar & Au, 2003).
31. The brains behind cooperation (McCabe et al., 2001; Riling et al., 2002; Sanfey et al., 2003; Zak et al., 2003).
32. Punishment gives the brain an electrical high (de Quervain et al., 2004).
33. Returning wallets (West, 2003).
34. Herman Melville, *The Writings of Herman Melville*, vol. 7, eds. Harrison Hayford, Hershel Parker, and G. Thomas Tanselle (1971); http://www.bartleby.com/66/21/38921.html.
35. Norms of responsibility and reciprocity (Berkowitz & Daniels, 1964; Durkin, 1961; Peterson et al., 1977).
36. Splitting the varieties of permission rules (Fiddick, 2004).
37. Different rules for different problems (Smetana, 1995, 2005; Turiel, 1998, 2005).
38. How data may inform the debate between moral objectivism and relativism, or, as some see it, between response dependence and independence (Darwall, 1998; Mackie, 1977; Nichols & Folds-Bennett, 2003; Nucci, 2001).

39. This passage by Antipoff was quoted by Piaget (1932/1954; p. 228) in his discussion of the child's moral development. To be perfectly fair to Piaget, he never claimed that the child enters the world a blank slate, with relationships doing all the etchings. As he stated the case, he regarded instinctive tendencies "a necessary but not a sufficient condition for the formation of morality" (p. 344), and that "the child's behavior towards persons shows signs from the first of those sympathetic tendencies and affective reactions in which one can easily see the raw material of all subsequent moral behavior. But an intelligent act can only be called logical and a good-hearted impulse moral from the moment that certain norms impress a given structure and rules of equilibrium upon this material" (p. 405). Nonetheless, he devoted most of his work to fleshing out the details of the child's experiences and how they build toward moral maturity.
40. (Hume, 1739/1978), p. 500.
41. This section owes much to the works of the philosopher Jesse Prinz (in press), who has written the most detailed recent critique of the nativist position. Much of my response here, therefore, bounces off his ideas, but many of the issues have been raised before.

6. ROOTS OF RIGHT

1. Mark Twain, "What Is Man?" http://www.twainquotes.com/Moral_Sense.html.
2. Rousseau's phylogeny of morals, http://www.bartleby.com/66/49/47349.html.
3. T. H. Huxley, *Evolution and Ethics,* 1893, http://aleph0.clarku.edu/huxley/CE9/E-E.html.
4. Ernst Mayr, Frans de Waal, and others have discussed this point in the context of criticizing recent attempts by Williams, Dawkins, and Wright to account for the evolution of morality by restricting the focus to our dark side.
5. (Darwin, 1871), p. 163.
6. For wide-ranging discussions of the evolution of culture and, especially, issues that concern social norms, see Boyd and Richerson (Boyd & Richerson, 2005; Richerson & Boyd, 2005).
7. How the brain represents expectations (Hassani & Cromwell, 2001; Schultz & Dickinson, 2000; Tinkelpaugh, 1928, 1932; Tremblay & Schultz, 1999; Watanabe, 1996; Schultz et al., 1997).
8. Expectations about where an object can go (Hauser, 1998; Santos et al., in prep.).
9. Using attention as a guide to goals (Santos & Hauser, 1999).
10. Chimpanzees' detection of goals (Uller, 2004).
11. (De Waal, 1989a, 2000a; de Waal & Berger, 2000; Hauser et al., 2003; Milinski, 1981; Stephens et al., 2002).
12. How chimpanzees read human goals (Call et al., 2004b; Premack, 1976; Premack & Premack, 1983, 2002).
13. For a video clip of a female looking at herself in the handheld mirror, see http://www.wjh.harvard.edu/~mnkylab/media/mirror.html.
14. Mirror, mirror on the wall, the person who put this red mark on my ear has got gall (Gallup, 1970, 1991; Hauser, 2000; Povinelli et al., 1993; Tomasello & Call, 1997).
15. Rhesus know what they know (Hampton, 2001; Hampton et al., 2005).
16. (Mead, 1912).

17. (Bekoff, 2001a; Darwin, 1872; de Waal, 1996; Haidt, 2001; Hauser, 2000; Kagan, 1998; LeDoux, 1996).

18. What's fear? (Kagan, 1998; LeDoux, 1996).

19. Minds prepared to hate snakes (Mineka et al., 1984; Mineka et al., 1980; Ohman et al., 2001).

20. (Kagan, 1998), p. 20.

21. Animal peaceniks (Aureli & de Waal, 2000; de Waal, 1982, 1989b, 1996, 2000b; Silk, 2002a).

22. The psychology of reconciliation (D. Cheney et al., 1995; D. L. Cheney et al., 1995; Silk, 2002a).

23. The ethology of animal needs (Dawkins, 1983, 1990; Mason et al., 2001).

24. Do animals have a theory of mind? What the past reveals (D. Cheney et al., 1995; Hauser, 2000; Premack & Premack, 2002; Tomasello & Call, 1997). Theory of mind in animals, circa 2005 (Call et al., 2004a, 2004b; Call & Tomasello, 1999; Flombaum & Santos, 2005; Hare et al., 2000; Hare et al., 2001; Heyes, 1998; Povinelli, 2000; Povinelli & Eddy, 1996; Premack & Premack, 2002; Tomasello et al., 2003).

25. Mind blind primates (Cheney & Seyfarth, 1990a; Povinelli & Eddy, 1996).

26. As David Premack has rightly pointed out, the failure here is not only despite massive training, but perhaps *because* of massive training. Prior to testing, Povinelli trained the chimpanzees to beg from *one* trainer who attended to them, and rewarded them for each act of begging. As countless studies of animal-learning reveal, when an experimenter provides an animal with 100 percent reinforcement, he effectively wipes out discrimination. By reinforcing each of the chimpanzee's begging actions with one attentive trainer, Povinelli may have wiped out his subject's ability or interest in attending to the key discriminative cue: where one of the two trainers was looking!

27. Primates deception, brain size, and eye gaze (Byrne & Whiten, 1990; Byrne & Corp, 2004; Gomez, 2005; Whiten & Byrne, 1988); bird brains, but capable of deception using eye gaze (Emery & Clayton, 2001; Ristau, 1991).

28. Hare's insight into chimpanzee mind reading (Call, 2001; Hare et al., 2000; Hare et al., 2001).

29. Natural telepathy run wild (Call et al., 2004b; Emery & Clayton, 2001; Flombaum & Santos, 2005; Hare & Tomasello, 2004; Kaminski et al., 2004; Tomasello et al., 2003).

30. The asymmetry between cooperative and competitive contexts is hard to evaluate, as the methods are not directly comparable.

31. Laser-beam intelligence or context effects on thought (Hurley, 2003).

32. Discounting and the economics of animal choice (Ainslie, 2000; Kacelnik, 2003; Rachlin, 2000). Although much of the early work focused on rats and pigeons, it has since expanded to other species (Bateson & Kacelnik, 1997; Gibbon et al., 1988; Mazur, 1987; Rosati, 2005; Stevens et al., 2005; Tobin et al., 1996). In this literature, most make a distinction between three types of discounting models: exponential, hyperbolic, and rate maximization. For simplicity, I have collapsed hyperbolic and rate maximization as they generate similar predictions. More detailed discussions of this literature can be found in both Rachlin and Kacelnik.

33. As in so much work comparing humans and other animals, my arrogance in claiming victory may well not hold. Studies that attempt to mimic the animal experiments by using commodities like food or water, as opposed to money, appear to show steeper dis-

counting effects (Reynolds & Schiffbauer, 2004). When it comes to looking at currencies that tap our core, evolved motivational systems, our ancestral rat may be lurking.

34. The economics of impatience (Fehr, 2002).

35. Social impulsivity and serotonin (Fairbanks et al., 2001; Higley et al., 1996; Kaplan et al., 1995; Mehlman et al., 1994; Mehlman et al., 1995).

36. Risk-taking, testosterone, and serotonin (Mehlman et al., 1994; Mehlman et al., 1995).

37. T is trouble (Sapolsky, 1998).

38. Domesticating aggressive impulses (Belyaev, 1979; Hare et al., 2005; Kruska, 1988; Leach, 2003; Nikulina, 1991; Trut, 1999; Wrangham et al., in press).

39. Catching a yawn in chimpanzees (Anderson et al., 2004).

40. Rats are social eaters (Galef, 1996).

41. A broad view of empathy (Preston & de Waal, 2002).

7. FIRST PRINCIPLES

1. (Chomsky, 1979).

2. The Golden Rule through time and cultures (Ebbesen, 2002; Kane, 1998; Wattles, 1996).

3. A growing group of evolutionary biologists, led to a large extent by the work of David Sloan Wilson (Sober & Wilson, 1998; Wilson, 2002b), have articulated a different view of group selection, what they refer to as "multilevel selection theory." The basic idea here is that group selection refers to the differential success of one group over another due to differences in the proportion of altruists or cooperators. This view acknowledges the importance of selfish behavior but also argues that when the distribution of genes within a group promotes acts that are mutually beneficial to all or most members of the group, then these genes will preferentially spread, thereby leading to between-group differences in fitness. This view is different from the one targeted by the likes of Dawkins, Hamilton, Maynard Smith, Trivers, and Williams in the 1970s.

4. Hamilton's rule (1964a; 1964b).

5. Although this is not a radically new approach, those who have supported the idea that we need to look at the multiple causes of altruism (Cosmides & Tooby, 1992; Fehr & Fischbacher, 2003; Pinker, 1997; Sober & Wilson, 1998; Tooby & Cosmides, 1998; Wilson, 2002a; Wilson, 1998) have not, in my opinion, swung the net widely enough.

6. These are descriptions of principles—rules of thumb, in most cases—that influence what animals do, bypassing what, if anything, animals think about their actions and the principles that govern them. They are principles that evolved to solve particular problems associated with group life. When there is variation in a principle, the variations are learnable by members of the species. Principles that are favored by selection set up expectations about how individuals typically respond in particular situations. Actions that are unexpected, given the principles in play, should be treated as violations. Whether this characterization is appropriate, and leads to new discoveries, we shall see. The important point for this chapter is to see these principles as descriptive characterizations of what animals do.

7. For a fascinating and lucid discussion of parenting, especially the role of cooperative mating systems in the evolution of human maternal behavior, see Sarah Hrdy's wonderful book *Mother Nature*.

8. Much of this section draws on the original theoretical papers by Robert Trivers (1972; 1974), the detailed summary of work on parental care by Tim Clutton-Brock (1991), and the wonderfully lucid synthesis of the complexities of family dynamics by Doug Mock (2004).

9. Cited in Mock (2004), pp. 24–25.

10. For the sake of simplicity, the sense of inclusive fitness described here glosses over one important piece of the concept. In particular, once one adds up the number of offspring that A produces and helps to produce in relatives, it is necessary to subtract the help A received in producing her own offspring.

11. What grannies do for our genes (Hawkes, 2003; Hrdy, 1999). There is continued debate over whether we are the only species to live well past our reproductive years; the short-finned pilot whale example comes from Richard Connor. What is clear is that we do live on for decades after our last births. And this is the puzzle that must be explained; I thank Sarah Hrdy for clarifying this controversy.

12. Owners, intruders, winners, and losers (Sih & Mateo, 2001; Stamps & Krishnan, 1999).

13. (Kummer & Cords, 1992).

14. The world of animal dominance (Lewis, 2002; Packer & Pusey, 1985; Preuschoft, 1999; Smuts, 1987).

15. Political smarts among our chimpanzee cousins (de Waal, 1982, 1989b, 1996).

16. When savvy outranks rank (Stammbach, 1988).

17. Knowing your rank in the world (Bergman et al., 2003; D. L. Cheney et al., 1995; Seyfarth & Cheney, 2003; Silk, 2002b).

18. Baboons detect rank violations (D. L. Cheney et al., 1995).

19. When and why animals are nice to each other (Axelrod, 1984; Dugatkin, 1997; Sachs et al., 2004; Stevens et al., 2006). I leave out of this discussion the many fascinating cases that arise between species, including cleaner fish and their hosts, as well as the greater honeyguide bird of Africa, who has evolved a mutually beneficial relationship with humans, first leading them to beehives, and then sharing the spoils once humans destroy the hive.

20. Cooperation on the savannah and in the sea: lions and dolphins do it (Connor et al., 1999, 2000; Connor et al., 1992a, 1992b; Packer & Ruttan, 1988; Scheel & Packer, 1991).

21. Modeling the dynamics of alliances: why, when, and how (Johnstone & Dugatkin, 2000; Whitehead & Connor, 2005).

22. Jays play games (Clements & Stephens, 1995; Stephens et al., 2002).

23. (Chalmeau et al., 1997; de Waal, 2000a; de Waal & Berger, 2000; Mendres & de Waal, 2000).

24. Tamarin food exchange (Hauser et al., 2003; McGrew & Feistner, 1992).

25. Reciprocation, kind of (Barrett & Henzi, 2001; Bercovitch, 1988; Connor, 1996; Connor et al., 1992a; de Waal, 1989a; Hart & Hart, 1989; Heinsohn & Packer, 1995; Milinski, 1981; Noe, 1990; Noe et al., 2001; Packer, 1977; Seyfarth & Cheney, 1984).

26. I am ignoring some of the studies that have looked at reciprocal altruism between species, such as the relationship between cleaner fish and their hosts. Bshary (2001) provides some elegant results on this relationship, which will need shoring up with a population of known individuals that can be followed over time.

27. *The Age of Innocence*, chapter 5, p. 40.

28. Dogs on the playground (Bekoff, 2001b, 2004; Spinka et al., 2001).

29. There is a vast literature on number quantification in animals, including evidence that both birds and mammals, and possibly fish and amphibians, have access to an analog magnitude system that is limited by the ratio between two quantities as distinct from their absolute values (Gallistel & Gelman, 2000; Hauser, 2000).

30. Chimpanzees and capuchins seek a square deal (Brosnan & de Waal, 2003; Brosnan et al., 2005).

31. Punishment solves the cooperation problem (Boyd & Richerson, 1992; Clutton-Brock & Parker, 1995).

32. Animal cheaters (Caldwell, 1986; Ducoing & Thierry, 2003; Gyger & Marler, 1988; Hauser, 1992, 1993; Munn, 1986).

33. Infrequent cheaters expected (Dawkins & Krebs, 1978; Hauser, 1996; Krebs & Dawkins, 1984).

34. Secrets Pantheon 83, quoted by Frank Trippett, "The Public Life of Secrecy," *Time*, January 17, 1985.

35. Honesty badges and punishment (Clutton-Brock & Parker, 1995; Gintis, 2000; Gintis et al., 2001; Noe et al., 2001).

36. Reviewed in Hauser, 1996; Searcy & Nowicki, 2005.

37. Manipulation and mind reading as a way of communicating (Dawkins & Krebs, 1978; Krebs & Dawkins, 1984).

38. Skeptic bees and monkeys (Cheney & Seyfarth, 1990b; Gould, 1990).

39. Why did the chicken go mute? (Marler et al., 1991).

40. Cheney and Seyfarth pioneered this line of experiments. Their results suggest that female Japanese macaques do not take into account what their offspring do or do not see when making decisions about whether they should or should not call alarm in the face of a predator. These results suggest that at least some monkeys may not take into account what others know when they call. However, given the evidence for seeing-knowing in rhesus monkeys and chimpanzees in a competitive task, I wish to hold open the door for new experiments to show greater sensitivity-to-knowledge states in the context of communication (e.g., Hare et al., in press).

41. Punishing cheaters (Foster & Ratnieks, 2000; Frank, 1995; Hauser, 1992, 1997; Ratnieks & Visscher, 1989).

42. The origins of a police force (Ratnieks & Visscher, 1989; Ratnieks & Wenseleers, 2005).

43. Harassment as cheap route to cooperation (Stevens, 2004).

44. For the aficionados, the rule is actually to follow the first thing that moves in front of you. As Lorenz and those who followed his imprinting work showed, chicks will follow what moves in front of them, even if it is a red ball. As it turns out, of course, the most common thing to move in front of them when they open their eyes is their mother. The rule works almost all the time, except, of course, when scientists wish to find out the details of the learning mechanism.

45. Reflection, awareness, and a sense of the normative (Korsgaard, 1996), p. 93.
46. (Rawls, 2001), p. 6.
47. (Tetlock et al., 2000), p. 853.

EPILOGUE: THE RIGHT IMPULSE

1. Letter to M. Berkowitz, October 25, 1950; Einstein Archive 59–215.
2. *The Autobiography of Charles Darwin.* New York: Totem Books, 1958/2004, p. 94.
3. Adolf Hitler, April 26, 1933, from a speech made during negotiations leading to the Nazi-Vatican Concordant of 1933: http://atheism.about.com/library/quotes/bl_q_AHitler.htm.

REFERENCES

———

Abell, F., Happe, F., & Frith, U. (2000). Do triangles play tricks? Attribution of mental states to animated shapes in normal and abnormal development. *Cognitive Development, 15*, 1–16.

Abu-Odeh, L. (1997). Comparatively speaking: the "honor" of the "east" and the "passion" of the "west." *Utah Law Review, 281*, 287–307.

Adolphs, R. (2003). Cognitive neuroscience of human social behavior. *Nature Neuroscience Reviews, 4*, 165–178.

Adolphs, R., Tranel, D., & Damasio, A. (1998). The human amygdala in social judgment. *Nature, 393*, 470–474.

Ainslie, G. (2000). *Break-Down of Will.* New York: Cambridge University Press.

Alexander, R. D. (1987). *The Biology of Moral Systems.* New York: Aldine de Gruyter.

Allport, G. W. (1954). *The Nature of Prejudice.* Cambridge: Addison-Wesley.

Almor, A., & Sloman, S. (1996). Is deontic reasoning special? *Psychological Review, 103*, 374–380.

Altmann, J. (1980). *Baboon Mothers and Infants.* Cambridge, MA: Harvard University Press.

Anderson, J. R., Myowa-Yamakoshi, M., & Matsuzawa, T. (2004). Contagious yawning in chimpanzees. *Proceedings of the Royal Society, London, Biology Letters, 4*, S1–S3.

Anderson, S. R., & Lightfoot, D. (2000). The human language faculty as an organ. *Annual Review of Physiology, 62*, 697–722.

Anderson, S. W., Bechara, A., Damasio, H., Tranel, D., & Damasio, A. R. (1999). Impairment of social and moral behavior related to early damage in human prefrontal cortex. *Nature Neuroscience, 2*, 1032–1037.

Aureli, F., & de Waal, F. B. M. (2000). *Natural Conflict Resolution*. Berkeley: University of California Press.

Austen, J. (1814). *Mansfield Park*. New York: Oxford University Press.

Axelrod, R. (1984). *The Evolution of Cooperation*. New York: Basic Books.

Ayduk, O., Downey, G., Testa, A., Yen, Y., & Shoda, Y. (1999). Does rejection elicit hostility in high rejection sensitive women? *Social Cognition, 17*, 245–271.

Ayduk, O., Mendoza-Denton, R., Mischel, W., Downey, G., Peake, P. K., & Rodriguez, M. L. (2000). Regulating the interpersonal self: strategic self-regulation for coping with rejection sensitivity. *Journal of Personality and Social Psychology, 79*, 776–792.

Baird, J. A., & Astington, J. W. (2004). The role of mental state understanding in the development of moral cognition and moral action. *New Directions for Child and Adolescent Development, 103*, 37–49.

Baird, R. M., & Rosenbaum, S. E. (2001). *The Ethics of Abortion: Pro-Life vs Pro-Choice*. New York: Prometheus Books.

Baker, M. C. (2001). *The Atoms of Language*. New York: Basic Books.

Baldwin, D. A., & Baird, J. A. (2001). Discerning intentions in dynamic human action. *Trends in Cognitive Science, 5*, 171–178.

Banaji, M. R. (2001). Implicit attitudes can be measured. In H. L. Roediger & J. S. Nairne & I. Neath & A. Surprenant (Eds.), *The Nature of Remembering: Essays in Honor of Robert G. Crowder*. Washington: Amerian Psychological Association.

Baron, J. (1994). Nonconsequentialist decisions. *Behavioral and Brain Sciences, 17(1)*, 1–10.

Baron, J. (1998). *Judgment Misguided: Intuition and Error in Public Decision Making*. Oxford: Oxford University Press.

Baron, J., Granato, L., Spranca, M., & Teubal, E. (1993). Decision making biases in children and early adolescents: exploratory studies. *Merrill-Palmer Quarterly, 39*, 23–47.

Baron-Cohen, S. (1995). *Mindblindness*. Cambridge, MA: MIT Press.

Baron-Cohen, S., Leslie, A. M., & Frith, U. (1985). Does the autistic child have a "theory of mind"? *Cognition, 21*, 37–46.

Barrett, L., & Henzi, S. P. (2001). The utility of grooming in baboon troops. In R. Noe & J. A. R. A. M. van Hoof & P. Hammerstein (Eds.), *Economics in Nature*. (pp. 119–145). Cambridge: Cambridge University Press.

Barry, D. (1995). *Complete Guide to Guys*. New York: Ballantine Book, Random House.

Bassili, J. N. (1976). Temporal and spatial contingencies in the perception of social events. *Journal of Personality and Social Psychology, 33*, 680–685.

Bateson, M., & Kacelnik, A. (1997). Starlings' preferences for predictable and unpredictable delays to food. *Animal Behaviour, 53*, 1129–1142.

Baumeister, R. F. (1999). *Evil. Inside human violence and cruelty*. New York: W.H. Freeman.

Baumeister, R. F., Stillwel, A., & Heatherton, T. F. (1994). Guilt: An interpersonal approach. *Psychological Bulletin, 115*, 243–267.

Bechara, A., Damasio, A., Damasio, H., & Anderson, S. W. (1994). Insensitivity to future consequences following damage to human prefrontal cortex. *Cognition, 50*, 7–15.

Bechara, A., Damasio, H., Tranel, D., & Damasio, A. (1997). Deciding advantageously before knowing the advantageous strategy. *Science, 275*, 1293–1295.

Beck, A. T. (1999). *Prisoners of Hate: The cognitive basis of anger, hostility and violence*. New York: HarperCollins.

Bekoff, M. (2001a). *The Dolphin's Smile.* New York: Discover.

Bekoff, M. (2001b). Social play behaviour: cooperation, fairness, trust, and the evolution of morality. *Journal of Consciousness Studies, 8,* 81–90.

Bekoff, M. (2004). Wild justice, cooperation, and fair play: minding manners, being nice, and feeling good. In R. Sussman & A. Chapman (Eds.), *The Origins and Nature of Sociality.* (pp. 53–79). Chicago: Aldine Press.

Belyaev, D. K. (1979). Destabilizing selection as a factor in domestication. *Journal of Heredity, 70,* 301–308.

Bentham, J. (1789/1948). *Principles of Morals and Legislation.* New York: Haffner.

Bercovitch, F. B. (1988). Coalitions, cooperation and reproductive tactics among adult male baboons. *Animal Behaviour, 36,* 1198–1209.

Bergman, T. J., Beehner, J. C., Cheney, D., & Seyfarth, R. (2003). Hierarchical classification by rank and kinship in baboons. *Science, 302,* 1234–1236.

Berkowitz, L., & Daniels, L. R. (1964). Affecting the salience of the social responsibility norm: effects of past help on the response to dependency relationships. *Journal of Abnormal and Social Psychology, 68,* 275–281.

Berthoz, S., Armony, J. L., Blair, R. J. R., & Dolan, R. J. (2002). An fMRI study of intentional and unintentional (embarrassing) violations of social norms. *Brain, 125,* 1696–1708.

Binmore, K. G. (1998). *Game theory and the social contract. II. Just Playing.* Cambridge, MA: MIT Press.

Blair, R. J. R. (1995). A cognitive developmental approach to morality: investigating the psychopath. *Cognition, 57,* 1–29.

Blair, R. J. R. (1997). Moral reasoning and the child with psychopathic tendencies. *Personality and Individual Differences, 22,* 731–739.

Blair, R. J. R., & Cipolotti, L. (2000). Impaired social response reversal. A case of 'acquired sociopathy.' *Brain, 123,* 1122–1141.

Blair, R. J. R., Morris, J. S., Frith, C. D., Perrett, D. I., & Dolan, R. J. (1999). Dissociable neural responses to facial expressions of sadness and anger. *Brain, 122,* 883–893.

Blair, R. J. R., Sellars, C., Strickland, I., Clark, F., Williams, A. O., Smith, M., & Jones, L. (1995). Emotion attributions in the psychopath. *Personality and Individual Differences, 19,* 431–437.

Boehm, C. (1999). *Hierarchy in the Forest: The evolution of egalitarian behavior.* Cambridge, MA: Harvard University Press.

Bok, S. (1978). *Lying: Moral Choice in Public and Private Life.* London: Quarter Books.

Bolig, R., Price, C. S., O'Neill, P. L., & Suomi, S. J. (1992). Subjective assessment of reactivity level and personality traits of rhesus monkeys. *International Journal of Primatology, 13,* 287–306.

Bouchard, T. J., Lykken, D. T., McGue, M., Segal, N. L., & Tellegen, A. (1990). Sources of human psychological differences: the Minnesota Study of Twins Reared Apart. *Science, 250,* 223–228.

Bowles, S., & Gintis, H. (1998). Is equality passé? Homo reciprocans and the future of egalitarian politics. *Boston Review, Fall,* 1–27.

Bowles, S., & Gintis, H. (1999). Is equality passé? Homo reciprocans and the future of egalitarian politics. *Boston Review, 23,* 4–35.

Bowles, S., & Gintis, H. (2003). *The evolution of cooperation in heterogeneous populations.* Unpublished manuscript, Santa Fe.

Boyd, R., Gintis, H., Bowles, S., & Richerson, P. J. (2003). The evolution of altruistic punishment. *Proceedings of the National Academy of Sciences, USA, 100,* 3531–3535.

Boyd, R., & Richerson, P. J. (1992). Punishment allows the evolution of cooperation (or anything else) in sizeable groups. *Ethology and Sociobiology, 113,* 171–195.

Boyd, R., & Richerson, P. J. (2005). *The Origin and Evolution of Cultures.* New York: Oxford University Press.

Brewer, M. B. (1999). The psychology of prejudice: ingroup love or outgroup hate? *Journal of Social Issues, 55,* 429–444.

Brosnan, S. F., & de Waal, F. B. M. (2003). Monkeys reject unequal pay. *Nature, 425,* 297–299.

Brosnan, S. F., Schiff, H. C., & de Waal, F. B. M. (2005). Tolerance for inequity may increase with social closeness in chimpanzees. *Proceedings of the Royal Society, London, B, 1560,* 253–258.

Brothers, L. (1997). *Friday's Footprints.* New York: Oxford University Press.

Brothers, L., & Ring, B. (1992). A neuroethological framework for the representation of minds. *Journal of Cognitive Neuroscience, 4,* 107–118.

Brown, D. E. (1991). *Human Universals.* New York: McGraw Hill.

Bshary, R. (2001). The cleaner fish market. In R. Noe & J. A. R. A. M. van Hoof & P. Hammerstein (Eds.), *Economics in Nature.* (pp. 146–172). Cambridge: Cambridge University Press.

Butterworth, B. (1999). *What Counts: How every brain is hardwired for math.* New York: Free Press.

Byrne, R., & Whiten, A. (1990). Tactical deception in primates: The 1990 database. *Primate Report, 27,* 1–101.

Byrne, R. W., & Corp, N. (2004). Neocortex size predicts deception rate in primates. *Proceedings of the Royal Society, London, B, 271,* 1693–1699.

Caldwell, R. L. (1986). The deceptive use of reputation by stomatopods. In R. W. Mitchell & N. S. Thompson (Eds.), *Deception: perspectives on human and nonhuman deceit.* (pp. 129–146). New York: SUNY Press.

Call, J. (2001). Chimpanzee social cognition. *Trends in Cognitive Science, 5,* 388–393.

Call, J., Hare, B., Carpenter, M., & Tomasello, M. (2004a). Do chimpanzees discriminate between an individual who is unwilling to share and one who is unable to share? *Developmental Science, 4,* 488–498.

Call, J., Hare, B., Carpenter, M., & Tomasello, M. (2004b). 'Unwilling' versus 'unable': chimpanzees' understanding of human intentional action. *Developmental Science, 7,* 488–498.

Call, J., & Tomasello, M. (1999). A nonverbal theory of mind test. The performance of children and apes. *Child Development, 70,* 381–395.

Caramazza, A. (1998). The interpretation of semantic category-specific deficits: what do they reveal about the organization of conceptual knowledge in the brain? *Neurocase, 4,* 265–272.

Caramazza, A., & Shelton, J. (1998). Domain-specific knowledge systems in the brain: the animate-inanimate distinction. *Journal of Cognitive Neuroscience, 10,* 1–34.

Carey, S. (1985). *Conceptual Change in Childhood.* Cambridge, MA: MIT Press.

Carey, S. (in press). *The Origins of Concepts.* Cambridge, MA: MIT press.

Carlson, S. M., & Moses, L. J. (2001). Individual differences in inhibitory control and children's theory of mind. *Child Development, 72,* 1032–1053.

Carlson, S. M., Moses, L. J., & Hix, H. R. (1998). The role of inhibitory processes in young children's difficulties with deception and false belief. *Child Development, 69(3),* 672–691.

Casati, R., & Varzi, A. C. (1996). *Events.* Brookfield: Dartmouth Publishing Company.

Chalmeau, R., Visalberghi, E., & Galloway, A. (1997). Capuchin monkeys, *Cebus apella,* fail to understand a cooperative task. *Animal Behaviour, 54,* 1215–1225.

Chandler, M., Fritz, A. S., & Hala, S. (1989). Small scale deceit: deception as a marker of two-, three-, and four-year olds' early theories of mind. *Child Development, 60,* 1263.

Chandler, M. J., Sokol, B. W., & Wainryb, C. (2000). Beliefs about truth and beliefs about rightness. *Child Development, 71[1],* 91–97.

Cheney, D., Seyfarth, R., & Silk, J. (1995). Reconciliatory grunts by dominant female baboons influence victims' behaviour. *Animal Behaviour, 54,* 409–418.

Cheney, D. L., & Seyfarth, R. M. (1990a). Attending to behaviour versus attending to knowledge: Examining monkeys' attribution of mental states. *Animal Behaviour, 40,* 742–753.

Cheney, D. L., & Seyfarth, R. M. (1990b). *How Monkeys See the World: Inside the mind of Another Species.* Chicago: University of Chicago Press.

Cheney, D. L., Seyfarth, R. M., & Silk, J. (1995). The responses of female baboons (*Papio cynocephalus ursinus*) to anomalous social interactions: Evidence for causal reasoning? *Journal of Comparative Psychology, 109,* 134–141.

Cheng, P. W., & Holyoak, K. W. (1985). Pragmatic reasoning schemas. *Cognitive Psychology, 17,* 391–416.

Cheng, P. W., & Holyoak, K. W. (1989). On the natural selection of reasoning theories. *Cognition, 33,* 285–313.

Chomsky, N. (1957). *Syntactic Structures.* The Hague: Mouton.

Chomsky, N. (1965). *Aspects of the Theory of Syntax.* Cambridge, MA: MIT Press.

Chomsky, N. (1979). *Language and Responsibility.* New York: Pantheon.

Chomsky, N. (1986). *Knowledge of language: Its nature, origin, and use.* New York: Praeger.

Chomsky, N. (1988). *Language and Problems of Knowledge.* Cambridge, MA: MIT Press.

Chomsky, N. (1995). *The Minimalist Program.* Cambridge, MA: MIT Press.

Chomsky, N. (2000). *On Nature and Language.* New York: Cambridge Univesity Press.

Clements, K. C., & Stephens, D. W. (1995). Testing models of non-kin cooperation: mutualism and the Prisoner's Dilemma. *Animal Behaviour, 50,* 527–535.

Clements, W. A., & Perner, J. (1994). Implicit understanding of belief. *Cognitive Development, 9,* 377–395.

Clutton-Brock, T. H. (1991). *The Evolution of Parental Care.* Princeton, NJ: Princeton University Press.

Clutton-Brock, T. H., & Parker, G. A. (1995). Punishment in animal societies. *Nature, 373,* 209–216.

Connor, R. C. (1996). Partner preferences in by-product mutualism and the case of predator inspection in fish. *Animal Behaviour, 51*, 451–454.

Connor, R. C., Heithaus, M. R., & Barre, L. M. (1999). Superalliance in bottlenose dolphins. *Nature, 371*, 571–572.

Connor, R. C., Heithaus, M. R., & Barre, L. M. (2000). Complex social structure, alliance stability and mating access in a bottlenose dolphin 'super-alliance.' *Proceedings of the Royal Society, London, B, 268*, 263–267.

Connor, R. C., Smolker, R. A., & Richards, A. F. (1992a). Dolphin alliances and coalitions. In A. H. Harcourt & F. B. M. d. Waal (Eds.), *Coalitions and Alliances in Humans and Other Animals* (pp. 415–443). Oxford: Oxford University Press.

Connor, R. C., Smolker, R. A., & Richards, A. F. (1992b). Two levels of alliance formation among male bottlenose dolphins. *Proceedings of the National Academy of Sciences, USA, 89*, 987–990.

Coppieters, B., & Fotion, N. (2002). *Moral Constraints on War. Principles and Causes.* Lanham: Lexington Books.

Cosmides, L., & Tooby, J. (1992). Cognitive adaptations for social exchange. In J. Barkow & L. Cosmides & J. Tooby (Eds.), *The Adapted Mind* (pp. 163–228). New York: Oxford University Press.

Cosmides, L., & Tooby, J. (2000). The cognitive neuroscience of social reasoning. In M. Gazzaniga (Ed.), *The New Cognitive Neurosciences* (pp. 1259–1270). Cambridge, MA: MIT Press.

Csibra, G., & Gergely, G. (1998). The teleological origins of mentalistic action explanation: A developmental hypothesis. *Developmental Science, 1*, 255–259.

Csibra, G., Gergely, G., Biro, S., Koos, D., & Brockbank, M. (1999). Goal attribution without agency cues: The perception of "pure reason" in infancy. *Cognition, 72*, 237–267.

Cummins, D. (1996a). Evidence for the innateness of deontic reasoning. *Mind and Language, 11*, 160–190.

Cummins, D. (1996b). Evidence of deontic reasoning in 3- and 4-year-old children. *Memory and Cognition, 24*, 823–829.

Cunningham, W. A., Nezlek, J. B., & Banaji, M. R. (2005). Implicit and explicit ethnocentrism: revisiting the ideologies of prejudice. *Personality and Social Psychology Bulletin, 31*, 105–132.

Daalder, I. H., & Lindsay, J. M. (2004). *America Unbound: The Bush Revolution in Foreign Policy.* Washington: Brookings Institution Press.

Daly, M., & Wilson, M. (1988). *Homicide.* Hawthorn, NY: Aldine de Gruyter.

Daly, M., & Wilson, M. (1998). *The Truth About Cinderella.* New Haven, CT: Yale University Press.

Damasio, A. (1994). *Descartes' Error.* Boston: Norton.

Damasio, A. (2000). *The Feeling of What Happens.* New York: Basic Books.

Darwall, S. (1998). *Philosophical Ethics.* Boulder: Westview Press.

Darwin, C. (1871). *The Descent of Man and Selection in Relation to Sex.* Princeton, NJ: Princeton University Press.

Darwin, C. (1872). *The Expression of the Emotions in Man and Animals.* London: John Murray.

Darwin, C. (1958). *The morality of evolution. Autobiography.* London: Norton.

Davidson, R. J., Scherer, K. R., & Goldsmith, H. H. (2003). *Handbook of Affective Sciences.* New York: Oxford University Press.

Dawkins, M. S. (1983). Battery hens name their price: consumer demand theory and the measurement of ethological 'needs.' *Animal Behaviour, 31*, 1195–1205.

Dawkins, M. S. (1990). From an animal's point of view: motivation, fitness and animal welfare. *Behavioral and Brain Sciences, 13*, 1–61.

Dawkins, R., & Krebs, J. R. (1978). Animal signals: information or manipulation. In J. R. Krebs & N. B. Davies (Eds.), *Behavioural Ecology* (pp. 282–309). Oxford: Blackwell Scientific Publications.

de Quervain, D. J.-F., Fischbacher, U., Treyer, V., Schellhammer, M., Schnyder, U., Buck, A., & Fehr, E. (2004). The neural basis of altruistic punishment. *Science, 305*, 1254–1258.

de Waal, F. B. M. (1982). *Chimpanzee Politics.* Baltimore: Johns Hopkins University Press.

de Waal, F. B. M. (1989a). Food sharing and reciprocal obligations among chimpanzees. *Journal of Human Evolution, 18*, 433–459.

de Waal, F. B. M. (1989b). *Peacemaking among primates.* Cambridge: Cambridge University Press.

de Waal, F. B. M. (1996). *Good Natured.* Cambridge, MA: Harvard University Press.

de Waal, F. B. M. (2000a). Attitudinal reciprocity in food sharing among brown capuchin monkeys. *Animal Behaviour, 60*, 253–261.

de Waal, F. B. M. (2000b). Primates: a natural heritage of conflict resolution. *Science, 289*, 586–590.

de Waal, F. B. M., & Berger, M. L. (2000). Payment for labour in monkeys. *Nature, 404*, 563.

Deary, I. J. (2001). *Intelligence: A Very Short Introduction.* Oxford: Oxford University Press.

DeBruine, L. M. (2002). Facial resemblance enhances trust. *Proceedings of the Royal Society, London, B, 269*, 1307–1312.

Dehaene, S. (1997). *The Number Sense.* Oxford: Oxford University Press.

Dennett, D. C. (2003). *Freedom Evolves.* New York: Viking.

DePaulo, B. M., Jordan, A., Irvine, A., & Laser, P. S. (1982). Age changes in the detection of deception. *Child Development, 53*, 701–709.

Dewey, J. (1922/1988). Human nature and conduct. In J. Dewey (Ed.), *Collected Works: The Middle Works, Volume 14.* Carbondale, IL: Southern Illinois University Press.

Dienes, Z., & Perner, J. (1999). A theory of implicit and explicit knowledge. *Behavioral and Brain Research, 22*, 735–755.

Dolinko, D. (1992). Three mistakes of retributivism. *UCLA Law Review, 39*, 1623–1657.

Dovidio, J. F., Glick, P., & Rudman, L. (2005). *On the Nature of Prejudice: 50 Years after Allport.* New York: Blackwell Publishing.

Downey, G., & Feldman, S. (1996). Implications of rejection sensitivity for intimate relationships. *Journal of Personality and Social Psychology, 70*, 1327–1343.

Downey, G., Feldman, S., & Ayduk, O. (2000). Rejection sensitivity and male violence in romantic relationships. *Personal Relationships, 7*, 45–61.

Driscoll, D. M., Anderson, S. W., & Damasion, H. (2004). *Executive function test performances following prefrontal cortex damage in childhood.* San Diego: Society for Neurosciences Abstract.

Ducoing, A. M., & Thierry, B. (2003). Withholding information in semifree-ranging tonkean macaques. *Journal of Comparative Psychology, 117*, 67–75.

Dugatkin, L. A. (1997). *Cooperation Among Animals: An Evolutionary Perspective*. New York: Oxford University Press.

Duncan, D. (1993). *Miles from Nowhere: Tales from America's Contemporary Frontier*. New York: Viking/Penguin Press.

Durkin, D. (1961). The specificity of children's moral judgments. *Journal of Genetic Psychology, 98*, 3–13.

Dworkin, R. (1993). *Life's Dominion: An argument about abortion, euthanasia, and individual freedom*. New York: Alfred A. Knopf.

Ebbesen, M. (2002). *The Golden Rule and Bioethics*. Unpublished Master's, University of Aarhus, Aarhus, The Netherlands.

Eisenberg, N., Losoya, S., & Spinrad, T. (2003). Affect and prosocial responding. In R. J. Davidson & K. R. Scherer & H. H. Goldsmith (Eds.), *Handbook of Affective Sciences* (pp. 787–803). New York: Oxford University Press.

Ekman, P. (1985). *Telling lies: Clues to deceit in the marketplace, marriage, and politics*. New York: W.W. Norton.

Ekman, P. (1992). An argument for basic emotions. *Cognition and Emotion, 6*, 169–200.

Ekman, P., Friesen, W. V., & O'Sullivan, M. (1988). Smiles while lying. *Journal of Personality and Social Psychology, 54*, 414–420.

Ekman, P., & O'Sullivan, M. (1991). Who can catch a liar? *American Psychologist, 46*, 913–920.

Ekman, P., O'Sullivan, M., Friesen, W. V., & Scherer, K. R. (1991). Face, voice, and body in deceit. *Journal of Nonverbal Behavior, 15*, 125–135.

Ellis, H. D., & Lewis, M. B. (2001). Capgras delusion: a window on face recognition. *Trends in Cognitive Science, 5*, 149–156.

Elster, J. (1979). *Ulysses and the Sirens*. Cambridge: Cambridge University Press.

Elster, J. (2000). *Ulysses Unbound*. New York: Cambridge University Press.

Emery, J. (2003). *Reputation is everything*. Available: http://www.worldandi.com/newhome/public/2003/may/clpub.asp.

Emery, N. J., & Clayton, N. S. (2001). Effects of experience and social context on prospective caching strategies by scrub jays. *Nature, 414*, 443–446.

Engel, H. (2002). *Crimes of Passion: An Unblinking Look at Murderous Love*. Westport, CT: Firefly Books.

Epstein, J. (2003). *Envy*. Oxford: Oxford University Press.

Fairbanks, L. A. (1996). Individual differences in maternal style of Old World monkeys. *Advances in the Study of Behaviour, 25*, 579–611.

Fairbanks, L. A., Melega, W. P., Jorgensen, M. J., Kaplan, J. R., & McGuire, M. T. (2001). Social impulsivity inversely associated with CSF 5-HIAA and Fluoxetine exposure in vervet monkeys. *Neuropsychopharmacology, 24*, 370–378.

Fehr, E. (2002). The economics of impatience. *Nature, 415*, 269–272.

Fehr, E., & Fischbacher, U. (2003). The nature of human altruism—Proximate and evolutionary origins. *Nature, 425*, 785–791.

Fehr, E., & Gachter, S. (2002). Altruistic punishment in humans. *Nature, 415*, 137–140.

Fehr, E., & Henrich, J. (2003). Is strong reciprocity a maladaptation? On the evolutionary foundations of human altruism. In P. Hammerstein (Ed.), *The Genetic and Cultural Evolution of Cooperation*. (pp. 55–82). Cambridge, MA: MIT Press.

Ferguson, M. J., & Bargh, J. A. (2004). How social perception can automatically influence behavior. *Trends in Cognitive Science, 8*, 33–38.

Fessler, D. M. T. (1999). Toward an understanding of the universality of second order emotions. In A. Hinton (Ed.), *Beyond Nature or Nurture: Biocultural approaches to the emotions.* (pp. 75–116). New York: Cambridge University Press.

Fessler, D. M. T., Arguello, A. P., Mekdara, J. M., & Macias, R. Disgust sensitivity and meat consumption: A test of an emotivist account of moral vegetarianism. *Appetite, 41(1),* 31–41.

Fessler, D. M. T., & Haley, K. J. (2003). The strategy of affect: emotions in human co-operation. In P. Hammerstein (Ed.), *Genetic and Cultural Evolution of Cooperation.* (pp. 7–36). Cambridge, MA: MIT Press.

Fiddick, L. (2004). Domains of deontic reasoning: resolving the discrepancy between the cognitive and moral reasoning literatures. *The Quarterly Journal of Experimental Psychology, 57A,* 447–474.

Fiddick, L., Cosmides, L., & Tooby, J. (2000). No interpretation without representation: the role of domain-specific representations in the Wason selection task. *Cognition, 77,* 1–79.

Fischer, J. M., & Ravizza, M. (1992). *Ethics: Problems and Principles.* New York: Holt, Rinehart & Winston.

Fiser, J., & Aslin, R. N. (2001). Unsupervised statistical learning of higher-order spatial structures from visual scenes. *Psychological Science, 12,* 499–504.

Fiske, A. P. (2002). Socio-moral emotions motivate action to sustain relationships. *Self and Identity, 1,* 169–175.

Fiske, A. P. (2004). Four modes of constituting relationships. In N. Haslam (Ed.), *Relational Models Theory: A contemporary overview.* Mahwah, NJ: Lawrence Erlbaum.

Fiske, A. P., & Tetlock, P. E. (2000). Taboo trade-offs: constitutive prerequisites for political and social life. In S. A. Renshon & J. Duckitt (Eds.), *Political Psychology: Cultural and Crosscultural Foundations* (pp. 47–65). London: Macmillan Press.

Fiske, S. T. (1998). Stereotyping, prejudice, and discrimination. In D. T. Gilbert & S. T. Fiske & G. Lindzey (Eds.), *The Handbook of Social Psychology.* (pp. 357–411). New York: McGraw Hill.

Fitch, W. T., Hauser, M. D., & Chomsky, N. (2005). The evolution of the language faculty: clarifications and implications. *Cognition, 97,* 179–210.

Flanagan, O. (1996). *Self Expressions. Mind, morals, and the meaning of life.* New York: Oxford University Press.

Flombaum, J., & Santos, L. (2005). Rhesus monkeys attribute perceptions to others. *Current Biology, 15,* 1–20.

Foot, P. (1967). The problem of abortion and the doctrine of double effect. *Oxford Review, 5,* 5–15.

Foster, K. R., & Ratnieks, F. L. W. (2000). Facultative worker policing in a wasp. *Nature, 407,* 692–693.

Frank, R. H. (1988). *Passion Within Reason: The Strategic Role of the Emotions.* New York: Norton.

Frank, S. A. (1995). Mutual policing and repression of competition in the evolution of co-operative groups. *Nature, 377,* 520–522.

Frazer, J. G. (1910). *A Treatise on Certain Early Forms of Superstition and Society*. London: Macmillan.

Freire, A., Eskritt, M., & Lee, K. (2004). Are eyes windows to a deceiver's soul? Children's use of another's eye gaze cues in a deceptive situation. *Developmental Psychology, 40*, 1093–1104.

Frohlich, N., & Oppenheimer, J. A. (1993). *Choosing Justice: An Experimental Approach to Ethical Theory*. Berkeley: University of California Press.

Fuster, J. (1997). *The Prefrontal Cortex: Anatomy, Physiology, and Neuropsychology of the Frontal Lobe*. Philadelphia: Lippincott-Raven.

Galef, B. G., Jr. (1996). Social enhancement of food preferences in Norway rats: a brief review. In C. M. Heyes & J. B.G. Galef (Eds.), *Social Learning in Animals: The Roots of Culture* (pp. 49–64). San Diego: Academic Press.

Gallese, V., & Goldman, A. (1998). Mirror neurons and the simulation theory of mind-reading. *Trends in Cognitive Science, 12*, 493–501.

Gallese, V., Keysers, C., & Rizzolatti, G. (2004). A unifying view of the basis of social cognition. *Trends in Cognitive Science, 8*, 398–403.

Gallistel, C. R., & Gelman, R. (2000). Non-verbal numerical cognition: from reals to integers. *TICS, 4*, 59–65.

Gallup, G. G., Jr. (1970). Chimpanzees: self-recognition. *Science, 167*, 86–87.

Gallup, G. G., Jr. (1991). Toward a comparative psychology of self-awareness: Species limitations and cognitive consequences. In G. R. Goethals & J. Strauss (Eds.), *The Self: An Interdisciplinary Approach* (pp. 121–135). New York: Springer-Verlag.

Garvey, S. P. (1998). Can shaming punishments educate? *University of Chicago Law Review, 65*, 733.

Gaylin, W. (2003). *Hatred: The psychological descent into violence*. New York: Public Affairs.

Gelman, R. (1990). First principles organize attention to and learning about relevant data: number and the animate-inanimate distinction as examples. *Cognitive Science, 14*, 79–106.

Gelman, S. A. (2003). *The Essential Child: Origins of Essentialism in Everyday Thought*. London: Oxford University Press.

Gentner, T., Margoliash, D., & Nussbaum, H. (in press). Recursive syntactic pattern learning in a songbird. *Nature,* still in press.

Gergely, G., & Csibra, G. (2003). Teleological reasoning in infancy: the naive theory of rational action. *Trends in Cognitive Science, 7*, 287–292.

Gergely, G., Nadasdy, Z., Csibra, G., & Biro, S. (1995). Taking the intentional stance at 12 months of age. *Cognition, 56*, 165–193.

Gert, B. (1998). *Morality: Its Nature and Justification*. New York: Oxford University Press.

Gert, B. (2004). *Common Morality*. New York: Oxford University Press.

Gibbon, J., Church, R. M., Fairhust, S., & Kacelnik, A. (1988). Scalar expectancy theory and choice between delayed rewards. *Psychological Review, 95*, 102–114.

Gibbs, J. C. (2003). *Moral Development and Reality*. Thousand Oaks, CA: Sage Publications.

Gilovich, T., Griffin, D., & Kahneman, D. (2002). *Heuristics and Biases: The psychology of intuitive judgment*. Cambridge: Cambridge University Press.

Gintis, H. (2000). Strong reciprocity and human sociality. *Journal of Theoretical Biology, 206*, 169–179.

Gintis, H., Bowles, S., Boyd, R., & Fehr, E. (2003). Explaining altruistic behavior in humans. *Evolution and Human Behavior, 24*, 153–172.

Gintis, H., Smith, E., & Bowles, S. (2001). Costly signaling and cooperation. *Journal of Theoretical Biology, 213*, 103–119.

Girotto, V., Light, P., & Colbourn, C. J. (1988). Pragmatic schemas and conditional reasoning in children. *Quarterly Journal of Experimental Psychology, 40A*, 469–482.

Goldberg, E. (2001). *The Executive Brain.* New York: Oxford University Press.

Golding, W. (1999). *Lord of the Flies.* New York: Penguin Books.

Goldman, A. (1971). The individuation of action. *Journal of Philosophy, 68*, 761–774.

Goldman, A. (1979). The paradox of punishment. *Philosophy and Public Affairs, 9*, 42–58.

Goldman, A. (1989). Interpretation psychologized. *Mind and Language, 4*, 161–185.

Goldman, A. (1992). In defense of the simulation theory. *Mind and Language, 7*, 104–119.

Gomez, J.-C. (2005). Species comparative studies and cognitive development. *Trends in Cognitive Science, 9*, 118–125.

Gordon, N. (2001). Honor killings: An editorial. *Iris, 42*, http://iris.virginia.edu/archives/42/nonfiction.html.

Gordon, P. (2004). Numerical cognition without words: evidence from Amazonia. *Science, 306*, 496–499.

Gordon, R. (1986). Folk psychology as simulation. *Mind and Language, 1*, 158–171.

Gould, J. L. (1990). Honeybee cognition. *Cognition, 37*, 83–103.

Greene, J. D. (2003). From neural 'is' to moral 'ought': what are the moral implications of neuroscientific moral psychology. *Nature Neuroscience Reviews, 4*, 847–850.

Greene, J. D., Nystrom, L. E., Engell, A. D., Darley, J. M., & Cohen, J. D. (2004). The neural bases of cognitive conflict and control in moral judgment. *Neuron, 44*, 389–400.

Greene, J. D., Sommerville, R. B., Nystrom, L. E., Darley, J. M., & Cohen, J. D. (2001). An fMRI investigation of emotional engagement in moral judgment. *Science, 293*, 2105–2108.

Greenspan, P. S. (1995). *Practical Guilt.* New York: Oxford University Press.

Griffiths, P. E. (1997). *What Emotions Really Are.* Chicago: University of Chicago Press.

Gyger, M., & Marler, P. (1988). Food calling in the domestic fowl (*Gallus gallus*): the role of external referents and deception. *Animal Behaviour, 36*, 358–365.

Haidt, J. (2001). The emotional dog and its rational tail: A social intuitionist approach to moral judgment. *Psychological Review, 108*, 814–834.

Haidt, J. (2003). The moral emotions. In R. J. Davidson & K. R. Scherer & H. H. Goldsmith (Eds.), *Handbook of Affective Sciences* (pp. 852–870). Oxford: Oxford University Press.

Haidt, J., & Joseph, C. (2004). Intuitive ethics: how innately prepared intuitions generate culturally variable virtues. *Daedalus*, 55–66.

Haidt, J., McCauley, C. R., & Rozin, P. (1994). Individual differences in sensitivity to disgust: a scale sampling seven domains of disgust elicitors. *Personality and Individual Differences, 16*, 701–713.

Hala, S., Chandler, M., & Fritz, A. S. (1991). Fledgling theories of mind: deception as a marker of three year olds' understanding of false belief. *Child Development, 62*, 83–97.

Hamilton, W. D. (1964a). The evolution of altruistic behavior. *American Naturalist, 97*, 354–356.

Hamilton, W. D. (1964b). The genetical evolution of social behavior. *Journal of Theoretical Biology, 7,* 1–52.

Harbaugh, W. T., Krause, K., & Liday, S. (2003). Children's bargaining behavior: differences by age, gender, and height. Working pages.

Harbaugh, W. T., Krause, K., Liday, S., & Vesterlund, L. (2003). Trust in children. In E. Ostrom & J. Walker (Eds.), *Trust and Reciprocity* (pp. 302–322). New York: Russell Sage Foundation.

Harbaugh, W. T., Krause, K., & Vesterlund, L. (2001). Risk attitudes of children and adults: choices over small and large probability gains and losses. *Experimental Economics, 5(1),* 53–84.

Hardin, G. (1968). The tragedy of the commons. *Science, 162,* 1243–1248.

Hare, B., Call, J., Agnetta, B., & Tomasello, M. (2000). Chimpanzees know what conspecifics do and do not see. *Animal Behaviour, 59,* 771–785.

Hare, B., Call, J., & Tomasello, M. (2001). Do chimpanzees know what conspecifics know? *Animal Behaviour, 61,* 139–151.

Hare, B., Plyusnina, I., Ignacio, N., Schepina, O., Stepika, A., Wrangham, R. W., & Trut, L. N. (2005). Social cognitive evolution in captive foxes is a correlated by-product of experimental domestication. *Current Biology, 15(3),* 226–230.

Hare, B., & Tomasello, M. (2004). Chimpanzees are more skilful in competitive than in cooperative cognitive tasks. *Animal Behaviour, 68,* 571–581.

Hare, R. D. (1993). *Without Conscience.* New York: Guilford Press.

Hare, R. M. (1981). *Moral Thinking: Its levels, method, and point.* Oxford: Clarendon Press.

Harman, G. (1977). *The Nature of Morality: An Introduction to Ethics.* New York: Oxford University Press.

Harris, P. L. (2000). *The Work of the Imagination.* Oxford: Blackwell Publishers.

Harris, P. L., & Nunez, M. (1996). Understanding of permission rules by preschool children. *Child Development, 67,* 1572–1591.

Harris, R. (1989). *Murder and Madness: Medicine, Law and Society in the Fin de Siecle.* Oxford: Clarendon Press.

Hart, B. L., & Hart, L. A. (1989). Reciprocal allogrooming in impala, *Aepyceros melampus. Animal Behaviour, 44,* 1073–1084.

Hassani, O. K., & Cromwell, H. C. (2001). Influence of expectation of different rewards on behavior-related neuronal activity in the striatum. *Journal of Neurophysiology, 85,* 2477–2489.

Hasselmo, M. E., Rolls, E. T., & Baylis, G. C. (1989). The role of expression and identity in the face-selective responses of neurons in the temporal visual cortex of the monkey. *Behavioural Brain Research, 32,* 203–218.

Hauser, M. D. (1992). Costs of deception: cheaters are punished in rhesus monkeys. *Proceedings of the National Academy of Sciences, 89,* 12137–12139.

Hauser, M. D. (1993). Rhesus monkey (*Macaca mulatta*) copulation calls: Honest signals for female choice? *Proceedings of the Royal Society, London, 254,* 93–96.

Hauser, M. D. (1996). *The Evolution of Communication.* Cambridge, MA: MIT Press.

Hauser, M. D. (1997). Minding the behavior of deception. In A. Whiten & R. W. Byrne (Eds.), *Machiavellian Intelligence II* (pp. 112–143). Cambridge: Cambridge University Press.

Hauser, M. D. (1998). Expectations about object motion and destination: Experiments with a nonhuman primate. *Developmental Science, 1,* 31–38.

Hauser, M. D. (2000). *Wild Minds: What Animals Really Think*. New York: Henry Holt.

Hauser, M. D., & Carey, S. (1998). Building a cognitive creature from a set of primitives: Evolutionary and developmental insights. In D. Cummins & C. Allen (Eds.), *The Evolution of Mind* (pp. 51–106). Oxford: Oxford University Press.

Hauser, M. D., Chen, M. K., Chen, F., & Chuang, E. (2003). Give unto others: genetically unrelated cotton-top tamarin monkeys preferentially give food to those who altruistically give food back. *Proceedings of the Royal Society, London, B, 270*, 2363–2370.

Hauser, M. D., Chomsky, N., & Fitch, W. T. (2002). The faculty of language: What is it, who has it, and how did it evolve? *Science, 298*, 1569–1579.

Hauser, M. D., & Spelke, E. (2004). Evolutionary and developmental foundations of human knowledge. In M. Gazzaniga (Ed.), *The Cognitive Neurosciences* (pp. 853–865). Cambridge, MA: MIT Press.

Hawkes, K. (2003). Grandmothers and the evolution of human longevity. *American Journal of Human Biology, 15*, 380–400.

Hazlitt, W. (1805/1969). *An Essay on the Principles of Human Action and Some Remarks on the Systems of Hartley and Helvetius*. Gainesville, FL: Scholars' Facsimiles and Reprints.

Heider, F., & Simmel, M. (1944). An experimental study of apparent behavior. *American Journal of Psychology, 57*, 243–259.

Heinsohn, R., & Packer, C. (1995). Complex cooperative strategies in group-territorial African lions. *Science, 269*, 1260–1262.

Henrich, J., Boyd, R., Bowles, S., Camerer, C., Fehr, E., Gintis, H., & McElreath, R. (2001). In search of Homo economicus: behavioral experiments in fifteen small-scale societies. *American Economics Review, 91*, 73–78.

Henrich, J., Boyd, R., Bowles, S., Camerer, C., Fehr, E., Gintis, H., McElreath, R., Barr, A., Ensminger, J., Hill, K., Gil-White, F., Gurven, M., Marlowe, F., Patton, J. Q., Smith, N., & Tracer, D. (in press). Economic Man in cross-cultural perspective: behavioral experiments in 15 small scale societies. *Behavioral and Brain Research, 28*, 795–855.

Heyes, C. M. (1998). Theory of mind in nonhuman primates. *Behavioral and Brain Sciences, 21*, 101–114.

Higley, J. D., Mehlman, P. T., Higley, S. B., Fernald, B., Vickers, J., Lindell, S. G., Taub, D. M., Suomi, S. J., & Linnoila, M. (1996). Excessive mortality in young free-ranging male nonhuman primates with low cerebrospinal fluid 5-hydroxy-indoleacetic acid concentrations. *Archives of General Psychiatry, 53*, 537–543.

Hill, K. (2002). Altruistic cooperation during foraging by the Ache, and the evolved human predisposition to cooperate. *Human Nature, 13*, 105–128.

Hinde, R. A. (2003). *Why Good Is Good: The sources of morality*. London: Routledge.

Hirschfeld, L. A. (1996). *Race in the Making: Cognition, culture and the child's construction of human kinds*. Cambridge, MA: MIT Press.

Hirschi, T., & Gottfredson, M. R. (1983). Age and the explanation of crime. *American Journal of Sociology, 89*, 552–584.

Hobbes, T. (1651/1968). *Leviathan*. New York: Penguin.

Hobbes, T. (Ed.). (1642/1962). *De Corpore*. New York: Oxford University Press.

Hobson, R. P. (2002). *The Cradle of Thought*. London: Macmillan.

Hoffman, M. L. (2000). *Empathy and Moral Development*. Cambridge: Cambridge University Press.

Hollander, E., & Stein, D. (1995). *Impulsivity and Aggression.* New York: Wiley Press.

Horder, J. (1992). *Provocation and Responsibility.* Oxford: Clarendon Press.

Hrdy, S. B. (1999). *Mother Nature: A history of mothers, infants, and natural selection.* New York: Pantheon Books.

Hume, D. (1739/1978). *A Treatise of Human Nature.* Oxford: Oxford University Press.

Hume, D. (Ed.). (1748). *Enquiry Concerning the Principles of Morals.* Oxford: Oxford University Press.

Huntsman, R. W. (1984). Children's concepts of fair sharing. *Journal of Moral Education, 13,* 31–39.

Hurley, S. (2003). Animal action in the space or reasons. *Mind and Language, 18,* 231–256.

Hursthouse, R. (1999). *On Virtue Ethics.* New York: Oxford University Press.

Hutcheson, F. (1728/1971). *Illustrations on the Moral Sense.* Cambridge, MA: Harvard University Press.

Huxley, T. H. (1888). The struggle for existence: a programme. *Nineteenth Century, 23,* 161–180.

Inagaki, K., & Hatano, G. (2002). *Young Children's Thinking about the Biological World.* New York: Psychology Press.

Inagaki, K., & Hatano, G. (2004). Vitalistic causality in young children's naive biology. *Trends in Cognitive Science, 8[8],* 356–362.

Intrator, J., Hare, R. D., Stritzke, P., Brichtswein, K., Dorfman, D., Harpur, T., Bernstein, D., Handelsman, L., Schaefer, C., Keilp, J., Rosen, J., & Machac, J. (1997). A brain imaging (single photon emission computerized tomography) study of semantic and affective processing in psychopaths. *Biological Psychiatry, 42,* 96–103.

Ivins, M. (2004). Imperialists' ball. *The Progressive, June,* 50.

Jackendoff, R. (2002). *Foundations of Language.* New York: Oxford University Press.

Jefferson, T. (1787/1955). Letter to Peter Carr, August 10. In J. P. Boyd (Ed.), *The Papers of Thomas Jefferson* (p. 15).

Johnson, S. C. (2000). The recognition of mentalistic agents in infancy. *Trends in Cognitive Science, 4,* 22–28.

Johnson, S. C. (2003). Detecting agents. *Philosophical Transactions of the Royal Society of London, B, 358,* 549–559.

Johnson, S. C., & Carey, S. (1998). Knowledge enrichment and conceptual change in folk biology: evidence from people with Williams syndrome. *Cognitive Psychology, 37,* 156–200.

Johnstone, R. A., & Dugatkin, L. A. (2000). Coalition formation in animals and the nature of winner and loser effects. *Proceedings of the Royal Society, London, B, 267,* 17–21.

Kacelnik, A. (1997). Normative and descriptive models of decision making, time discounting and risk sensitivity. In G. Bock & G. Cardew (Eds.), *Characterizing Human Psychological Adaptations* (pp. 51–70). London: Wiley.

Kacelnik, A. (2003). The evolution of patience. In G. Loewenstein (Ed.), *Time and Decision: Economics and Psychological Perspectives on Intertemporal Choice* (pp. 115–138). New York: Russell Sage Foundation.

Kagan, J. (1998). *Three Seductive Ideas.* Cambridge, MA: Harvard University Press.

Kagan, J. (2000). Human morality is distinctive. *Journal of Consciousness Studies, 7,* 46–48.

Kagan, J. (2002a). Morality, altruism, and love. In S. G. Post & L. G. Underwood &

J. P. Schloss & W. B. Hurlbut. (Eds.), *Altruism and Altruistic Love* (pp. 40–50). New York: Oxford University Press.

Kagan, J. (2002b). *Surprise, Uncertainty, and Mental Structures.* Cambridge, MA: Harvard University Press.

Kagan, J., Reznick, J. S., & Snidman, N. (1988). Biological bases of childhood shyness. *Science, 250,* 167–171.

Kagan, J., Snidman, N., & Arcus, D. (1998). Childhood derivatives of high and low reactivity in infancy. *Child Development, 69,* 1483–1493.

Kagan, S. (1988). The additive fallacy. *Ethics, 90,* 5–31.

Kagel, J. H., & Roth, A. E. (1995). *Handbook of Experimental Economics.* Princeton, NJ: Princeton University Press.

Kahn, P. (2004). Mind and morality. *New Directions for Child and Adolescent Development, 103,* 73–83.

Kahneman, D., Schkade, D., & Sunstein, C. R. (1998). Shared outrage and erratic awards: the psychology of punitive damages. *Journal of Risk and Uncertainty, 16,* 49–86.

Kalnins, I. V., & Bruner, J. S. (1973). The coordination of visual observation and instrumental behavior in early infancy. *Perception, 2,* 307–314.

Kaminski, J., Call, J., & Tomasello, M. (2004). Body orientation and face orientation: two factors controlling apes' begging behavior from humans. *Animal Cognition, 7,* 216–223.

Kamm, F. M. (1992). *Creation and Abortion: A study in moral and legal philosophy.* New York: Oxford University Press.

Kamm, F. M. (1998). *Morality, Mortality: Death and whom to save from it.* New York: Oxford University Press.

Kamm, F. M. (2000). Nonconsequentialism. In H. LaFollette (Ed.), *Ethical Theory* (pp. 205–226). Malden, MA: Blackwell Publishing.

Kamm, F. M. (2001a). *Morality, Mortality: Rights, duties, and status.* New York: Oxford University Press.

Kamm, F. M. (2001b). Ronald Dworkin on abortion and assisted suicide. *The Journal of Ethics, 5,* 221–240.

Kamm, F. M. (2004). Failures of just war theory: terror, harm, and justice. *Ethics, 114,* 650–692.

Kane, R. (1998). *Through the Moral Maze: Searching for Absolute Values in a Pluralistic World.* New York: Paragon House.

Kant, I. (1785/1959). *Foundations of the Metaphysics of Morals* (L. W. Beck, Trans.). New York: Macmillan.

Kant, I. (2001). *Lectures on Ethics.* New York: Cambridge University Press.

Kanwisher, N. (2000). Domain specificity in face perception. *Nature Neuroscience, 3,* 759–763.

Kanwisher, N., Downing, P., Epstein, R., & Kourtzi, Z. (2001). Functional neuroimaging of human visual recognition. In Kingstone & Cabeza (Eds.), *The Handbook on Functional Neuroimaging* (pp. 109–152). Cambridge, MA: MIT Press.

Kaplan, J. R., Fontenot, M. B., Berard, J., & Manuck, S. B. (1995). Delayed dispersal and elevated monoaminergic activity in free-ranging rhesus monkeys. *American Journal of Primatology, 35,* 229–234.

Kaplow, L., & Shavell, S. (2002). *Fairness versus Welfare.* Cambridge, MA: Harvard University Press.

Karmiloff-Smith, A. (1992). *Beyond Modularity.* Cambridge, MA: MIT Press.

Keil, L. J. (1986). Rules, reciprocity, and rewards: a developmental study of resource allocation in social interaction. *Journal of Experimental Social Psychology, 22,* 419–435.

Kelly, M. H. (1999). Regional naming patterns and the culture of honor. *Names, 47,* 3–20.

Ketelaar, T., & Au, W.-T. (2003). The effects of feelings of guilt on the behaviour of uncooperative individuals in repeated social bargaining games: an affect-as-information interpretation of the role of emotion in social interaction. *Cognition and Emotion, 17,* 429–453.

Kiehl, K. A., Hare, R. D., Liddle, P. F., & McDonald, J. J. (1999). Reduced P300 responses in criminal psychopaths during a visual oddball task. *Biological Psychiatry, 45,* 1498–1507.

Kiehl, K. A., Smith, A. M., Hare, R. D., Mendrek, A., Forster, B. B., Brink, J., & Liddle, P. F. (2001). Limbic abnormalities in affective processing by criminal psychopaths as revealed by functional magnetic resonance imaging. *Biological Psychiatry, 50,* 677–684.

Kitcher, P. (1985). *Vaulting Ambition: Sociobiology and the quest for human nature.* Cambridge, MA: MIT Press.

Knobe, J. (2003a). Intentional action and side effects in ordinary language. *Analysis, 63,* 190–193.

Knobe, J. (2003b). Intentional action in folk psychology: an experimental investigation. *Philosophical Psychology, 16,* 309–324.

Kochanska, G., Murray, K., Jacques, T. Y., Koenig, A. L., & Vandegeest, K. A. (1996). Inhibitory control in young children and its role in emerging internalization. *Child Development, 67,* 490–507.

Kohlberg, L. (1981). *Essays on Moral Development, Volume 1: The Philosophy of Moral Development.* New York: Harper and Row.

Korsgaard, C. M. (1996). *Sources of Normativity.* Cambridge: Cambridge University Press.

Krebs, J. R., & Dawkins, R. (1984). Animal signals: mind-reading and manipulation. In J. R. Krebs & N. B. Davies (Eds.), *Behavioural Ecology* (pp. 380–402). Sunderland, MA: Sinauer Associates Inc.

Kruska, D. (1988). Mammalian domestication and its effect on brain structure and behavior. In H. J. Jerison (Ed.), *Intelligence and Evolutionary Biology* (pp. 211–250). New York: Springer Verlag.

Kuhlmeier, V., Wynn, K., & Bloom, P. (2003). Attribution of dispositional states in 12-month-olds. *Psychological Science, 14,* 402–408.

Kuhn, T. (1970). *The Structure of Scientific Revolutions.* Chicago: University of Chicago Press.

Kulwicki, A. D. (2002). The practice of honor crimes: a glimpse of domestic violence in the Arab world. *Issues in Mental Health Nursing, 23,* 77–87.

Kummer, H., & Cords, M. (1992). Cues of ownership in long-tailed macques, *Macaca fascicularis. Animal Behaviour, 42,* 529–549.

Kundera, M. (1985). *The Unbearable Lightness of Being.* New York: HarperCollins.

Lakoff, G. (1996). *Moral Politics.* Chicago: University of Chicago Press.

Lasnik, H., Uriagereka, J., & Boeckx, C. (2005). *A Course in Minimalist Syntax: Foundations and Prospects.* Malden, MA: Blackwell Publishers.

Leach, H. M. (2003). Human domestication reconsidered. *Current Anthropology, 44*, 349–368.

LeDoux, J. (1996). *The Emotional Brain*. New York: Simon and Schuster.

Lee, K., Cameron, C. A., Doucette, J., & Talwar, V. (2002). Phantoms and fabrications: young children's detection of implausible lies. *Child Development, 73*, 1688–1702.

Leibniz, G. W. (1704/1966). *New Essays on Human Understanding*. Cambridge: Cambridge University Press.

Lerdahl, F., & Jackendoff, R. (1996). *A Generative Theory of Tonal Music*. Cambridge, MA: MIT Press.

Leslie, A. M. (1987). Pretense and representation: the origins of "theory of mind." *Psychological Review, 94*, 412–426.

Leslie, A. M. (1994). ToMM, ToBY, and Agency: core architecture and domain specificity. In L. A. Hirschfeld & S. A. Gelman (Eds.), *Mapping the Mind: Domain Specificity in Cognition and Culture* (pp. 119–148). Cambridge: Cambridge University Press.

Leslie, A. M. (2000). Theory of mind as a mechanism of selective attention. In M. Gazzaniga (Ed.), *The New Cognitive Neurosciences* (pp. 1235–1247). Cambridge, MA: MIT Press.

Leslie, A. M., & Keeble, S. (1987). Do six-month-old infants perceive causality? *Cognition, 25*, 265–288.

Leslie, A. M., Knobe, J., & Cohen, A. (in press). Acting intentionally and the side-effect effect: 'Theory of mind' and moral judgment. *Psychological Science*, still in press.

Lewis, M., Sulivan, M. W., & Brooks-Gunn, J. (1985). Emotional behavior during the learning of a contingency in early infancy. *British Journal of Developmental Psychology, 3*, 307–316.

Lewis, R. J. (2002). Beyond dominance: the importance of leverage. *Quarterly Review of Biology, 77*, 149–164.

Lieberman, D., Tooby, J., & Cosmides, L. (in press). Does morality have a biological basis? An empirical test of the factors governing moral sentiments relating to incest. *Proceedings of the Royal Society, London, B, 270*, 819–826.

Lieberman, N., & Klahr, Y. (1996). Hypothesis testing in Wason's selection task: social exchange cheating detection or task understanding? *Cognition, 58*, 127–156.

Litowitz, D. (1997). The trouble with 'scarlet letter' punishments. *Judicature, 81(2)*, 52–57.

Mackie, J. (1977). *Ethics: inventing right and wrong*. London: Penguin.

Macnamara, J. (1990). The development of moral reasoning and the foundations of geometry. *Journal for the Theory of Social Behavior, 21*, 125–150.

Markus, H. R., & Kitayama, S. (1991). Culture and self: implications for cognition, emotion, and motivation. *Psychological Review, 98*, 224–253.

Marler, P., Karakashian, S., & Gyger, M. (1991). Do animals have the option of withholding signals when communication is inappropriate? The audience effect. In C. Ristau (Ed.), *Cognitive ethology: the minds of other animals* (pp. 135–186). Hillsdale, NJ: Lawrence Erlbaum Associates.

Marlowe, F. (2004). *Better to receive than to give: How the Hadza play the game.* Unpublished manuscript, Cambridge, MA.

Martau, P. A., Caine, N. G., & Candland, D. K. (1985). Reliability of the emotions profile index, primate form, with Papio hamadryas, Macaca fuscata, and two Saimiri species. *Primates, 26*, 501–505.

Mason, G. J., Cooper, J., & Clarebrough, C. (2001). Frustrations of fur-farmed mink. *Nature, 410*, 35–36.

Mason, H. E. (1996). *Moral Dilemmas and Moral Theory.* New York: Oxford University Press.

May, L., Friedman, M., & Clark, A. (1996). *Minds and Morals.* Cambridge, MA: MIT Press.

Mayell, H. (2002). *Thousands of women killed for family "honor."* Available: http://news. nationalgeographic.com/news/2002/02/0212_020212_honorkilling.html.

Mazur, J. E. (1987). An adjusting procedure for studying delayed reinforcement. In M. L. Commons & J. E. Mazur & J. A. Nevin & H. Rachlin (Eds.), *Quantitative Analyses of Behavior* (pp. 55–73). Hillsdale, NJ: Lawrence Erlbaum Associates.

McAlpine, L. M., & Moore, C. L. (1995). The development of social understanding in children with visual impairments. *Journal of Visual Impairment and Blindness, 89*, 349–358.

McCabe, K. A., Houser, D., Ryan, L., Smith, V. L., & Trouard, T. (2001). A functional imaging study of cooperation in two-person reciprocal exchange. *Proceedings of the National Academy of Sciences, USA, 98*, 11832–11835.

McClearn, G. E. (1997). Substantial genetic influence on cognitive abilities in twins 80 or more years old. *Science, 276*, 1560–1563.

McGrew, W. C., & Feistner, A. T. C. (1992). Two nonhuman primate models for the evolution of human food sharing: chimpanzees and callitrichids. In J. H. Barkow & L. Cosmides & J. Tooby (Eds.), *The Adapted Mind* (pp. 229–249). New York: Oxford University Press.

Mead, G. H. (1912). The mechanism of social consciousness. *Journal of Philosophy, Psychology, and Scientific Methods, 9*, 401–416.

Mehlman, P. T., Higley, J. D., Faucher, I., Lilly, A. A., Taub, D. M., Vickers, J., Suomi, S. J., & Linnoila, M. (1994). Low CSF 5-HIAA concentrations and severe aggression and impaired impulse control in nonhuman primates. *American Journal of Psychiatry, 151*, 1485–1491.

Mehlman, P. T., Higley, J. D., Faucher, I., Lilly, A. A., Taub, D. M., Vickers, J., Suomi, S. J., & Linnoila, M. (1995). Correlation of CSF 5-HIAA concentration with sociality and the timing of emigration in free-ranging primates. *American Journal of Psychiatry, 152*, 907–913.

Mele, A. (2001). Acting intentionally: probing folk notions. In B. F. Malle & L. J. Moses & D. Baldwin (Eds.), *Intentions and Intentionality: Foundations of Social Cognition* (pp. 27–43). Cambridge, MA: MIT Press.

Mendres, K. A., & de Waal, F. B. M. (2000). Capuchins do cooperate: the advantage of an intuitive task. *Animal Behaviour, 60*, 523–529.

Metcalfe, J., & Mischel, W. (1999). A hot/cool-system analysis of delay of gratification: dynamics of willpower. *Psychological Review, 106*, 3–19.

Mikhail, J. M. (2000). *Rawls' linguistic analogy: A study of the 'generative grammar' model of moral theory described by John Rawls in 'A theory of justice.'* Unpublished PhD, Cornell University, Ithaca, NY.

Mikhail, J. M. (2002). Law, science, and morality: a review of Richard Posner's "The Problematics of Moral and Legal Theory." *Stanford Law Review, 54*, 1057–1127.

Mikhail, J. M. (in press). *Rawls' Linguistic Analogy.* New York: Cambridge University Press.

Mikhail, J. M., Sorrentino, C., & Spelke, E. (2002). *Aspects of the theory of moral cognition: Investigating intuitive knowledge of the prohibition of intentional battery, the rescue principle, the first principle of practical reason, and the principle of double effect.* Unpublished manuscript, Stanford, CA.

Milgate, D. E. (1998). The flame flickers, but burns on: modern judicial application of the ancient heat of passion defense. *Rutgers Law Review, 51*, 193–227.

Milgram, S. (1974). *Obedience to Authority: An experimental view.* New York: Harper and Row Publishers.

Milinski, M. (1981). Tit for tat and the evolution of cooperation in sticklebacks. *Nature, 325*, 433–437.

Milinski, M., Semmann, D., & Krambeck, H.-J. (2002). Reputation helps solve the 'tragedy of the commons.' *Nature, 415*, 424–426.

Miller, D., & Walzer, M. (1995). *Pluralism, Justice and Equality.* Oxford: Oxford University Press.

Mineka, S., Davidson, M., Cook, M., & Keir, R. (1984). Observational conditioning of snake fear in rhesus monkeys. *Journal of Abnormal Psychology, 93*, 355–372.

Mineka, S., Keir, R., & Price, V. (1980). Fear of snakes in wild and laboratory reared rhesus monkeys (*Macaca mulatta*). *Animal Learning and Behavior, 8*, 653–663.

Minter, M., Hobson, R. P., & Bishop, M. (1998). Congenital visual impairment and 'theory of mind.' *British Journal of Developmental Psychology, 16*, 183–196.

Mischel, W. (1966). Theory and research on the antecedents of self-imposed delay of reward. In B. A. Maher (Ed.), *Progress in Experimental Personality Research* (pp. 85–132). New York: Academic Press.

Mischel, W. (1974). Processes in delay of gratification. In L. Berkowitz (Ed.), *Advances in Experimental Social Psychology, Vol. 7* (pp. 249–292). New York: Academic Press.

Mischel, W., Shoda, Y., & Rodriguez, M. L. (1989). Delay of gratification in children. *Science, 244*, 933–938.

Mischel, W., Zeiss, R., & Zeiss, A. (1974). Internal-external control and persistence: validation and implications of the Stanford preschool internal-external scale. *Journal of Personality and Social Psychology, 29*, 265–278.

Mock, D. W. (2004). *More Than Kin and Less Than Kind.* Cambridge, MA: Harvard University Press.

Molière, J. B. P. (1673). *Le Misanthrope.* Oberon Books: London.

Moll, J., Eslinger, P. J., & Oliviera-Souza, R. (2001). Frontopolar and anterior temporal cortex activation in a moral judgment task: preliminary functional MRI results in normal subjects. *Archives of Neuropsychiatry, 59*, 657–664.

Moll, J., Oliveira-Souza, R., & Eslinger, P. J. (2002). The neural correlates of moral sensitivity: a functional MRI investigation of basic and moral emotions. *Journal of Neuroscience, 27*, 2730–2736.

Moll, J., Oliveira-Souza, R., Moll, F. T., Ignacio, I. E., Caparelli-Daquer, E. M., & Eslinger, P. J. (2005). The moral affiliations of disgust. Working paper. Not yet published.

Moll, J., Olivier-Souza, R., & Bramati, I. E. (2002). Functional networks in moral and nonmoral social judgments. *NeuroImage, 16*, 696–703.

Moore, C., & Macgillivray, S. (2004). Altruism, prudence, and theory of mind in preschoolers. *New Directions for Child and Adolescent Development, 103*, 51–62.

Moore, G. E. (1903). *Principia Ethica.* Cambridge: Cambridge University Press.

Muller, J. L., Sommer, M., Wagner, V., Lange, K., Taschler, H., Roder, C. H., Schuierer, G., Klein, H. E., & Hajak, G. (2003). Abnormalities in emotion processing within cortical and subcortical regions in criminal psychopaths: evidence from a functional magnetic resonance imaging study using pictures with emotional content. *Biological Psychiatry, 54,* 152–163.

Munn, C. (1986). Birds that 'cry wolf.' *Nature, 319,* 143–145.

Murnighan, J. K., & Saxon, M. S. (1998). Ultimatum bargaining by children and adults. *Journal of Economic Psychology, 19,* 415–445.

Nesse, R. M. (Ed.). (2001). *Evolution and the Capacity for Commitment.* New York: Russell Sage Foundation.

Newport, E. L., & Aslin, R. N. (2004). Learning at a distance. I. Statistical learning of non-adjacent dependencies. *Cognitive Psychology, 48,* 127–162.

Newtson, D. (1973). Attribution and the unit of perception of ongoing behavior. *Journal of Personality and Social Psychology, 28,* 28–38.

Newtson, D., Engquist, G., & Bois, J. (1977). The objective basis of behavior units. *Journal of Personality and Social Psychology, 35,* 847–862.

Newtson, D., Hairfield, J., Bloomingdale, J., & Cutino, S. (1987). The strucutre of action and interaction. *Social Cognition, 5,* 191–237.

Nichols, S. (2002). Norms with feeling: toward a psychological account of moral judgment. *Cognition, 84,* 221–236.

Nichols, S. (2004). *Sentimental Rules.* New York: Oxford University Press.

Nichols, S., & Folds-Bennett, T. (2003). Are children moral objectivists? Children's judgments about moral and response-dependent properties. *Cognition, B23-B32.*

Nikulina, E. M. (1991). Neural control of predatory aggression in wild and domesticated animals. *Neuroscience and Biobehavioral Reviews, 15,* 545–547.

Nisbett, R. E., & Cohen, D. (1996). *Culture of Honor: The Psychology of Violence in the South.* Boulder, CO: Westview Press.

Noe, R. (1990). A veto game played by baboons: a challenge to the use of the Prisoner's Dilemma as a paradigm of reciprocity and cooperation. *Animal Behaviour, 39,* 78–90.

Noe, R., van Hoof, J. A. R. A. M., & Hammerstein, P. (2001). *Economics in Nature.* Cambridge: Cambridge University Press.

Nourse, V. (1997). Passion's progress: modern law reform and the provocation defense. *Yale Law Journal, 106,* 1331–1344.

Nowak, M. A., Page, K. M., & Sigmund, K. (2000). Fairness versus reason in the ultimatum game. *Science, 289,* 1773–1775.

Nowak, M. A., & Sigmund, K. (2000). Enhanced: Shrewd investments. *Science, 288,* 819.

Nucci, L. (2001). *Education in the Moral Domain.* Cambridge: Cambridge University Press.

Nunez, M., & Harris, P. L. (1998). Psychological and deontic concepts: separate domains or intimate connection? *Mind and Language, 13,* 153–170.

Nussbaum, M. (2004). *Hiding from Humanity: Disgust, Shame, and the Law.* Princeton, NJ: Princeton University Press.

O'Neill, P., & Petrinovich, L. (1998). A preliminary cross cultural study of moral intuitions. *Evolution and Human Behavior, 19,* 349–367.

Ohman, A., Flykt, A., & Esteves, F. (2001). Emotion drives attention: Detecting the snake in the grass. *Journal of Experimental Psychology: General, 130,* 466–478.

Oliveira-Souza, R., & Moll, J. (2000). The moral brain: a functional MRI study of moral judgment. *Neurology, 54,* 1331–1336.

Olson, J. M., Roese, N. J., & Zanna, M. P. (1996). Expectancies. In E. T. Higgins &

A. W. Kruglanski (Eds.), *Social Psychology: Handbook of Basic Principles* (pp. 211–238). New York: Guilford Press.

Olsson, A., Ebert, J. P., Banaji, M. R., & Phelps, E. A. (2005). The role of social groups in the persistence of learned fear. *Science, 309*, 711–713.

Onishi, K. H., & Baillargeon, R. (2005). Do 15-month-old infants understand false beliefs? *Science, 308*, 255–257.

Packer, C. (1977). Reciprocal altruism in olive baboons. *Nature, 265*, 441–443.

Packer, C., & Pusey, A. (1985). Asymmetric contests in social mammals: respect, manipulation and age-specific aspects. In P. J. Greenwood & P. H. Harvey (Eds.), *Evolution: Essays in Honour of John Maynard Smith* (pp. 173–186). Cambridge: Cambridge University Press.

Packer, C., & Ruttan, L. (1988). The evolution of cooperative hunting. *American Naturalist, 132*, 159–198.

Paladino, M.-P., Leyens, J.-P., Rodriguez, R., Rodriguez, A., Gaunt, R., & Demoulin, S. (2002). Differential association of uniquely and non-uniquely human emotions with the ingroup and outgroup. *Group Processes and Intergroup Relations, 5*, 105–117.

Parr, L. A. (2001). Cognitive and psychological markers of emotional awareness in chimpanzees. *Animal Cognition, 4*, 223–229.

Patterson, O. (1982). *Slavery and Social Death.* Cambridge, MA: Harvard University Press.

Peake, P. K., Mischel, W., & Hebl, M. (2002). Strategic attention deployment for delay of gratification in working and waiting situations. *Developmental Psychology, 38(2)*, 313–326.

Peterson, L., Hartmann, D. P., & Gelfand, D. M. (1977). Developmental changes in the effects of dependency and reciprocity cues on children's moral judgments and donation rates. *Child Development, 48*, 1331–1339.

Petrinovich, L., & O'Neill, P. (1996). Influence of wording and framing effects on moral intuitions. *Ethology and Sociobiology, 17*, 145–171.

Petrinovich, L., O'Neill, P., & Jorgensen, M. J. (1993). An empirical study of moral intuitions: towards an evolutionary ethics. *Ethology and Sociobiology, 64*, 467–478.

Phelps, E. A., O'Connor, K. J., Cunningham, W. A., Funayama, S., Gatenby, J. C., Gore, J. C., & Banaji, M. R. (2000). Performance on indirect measures of race evaluation predicts amygdala activation. *Journal of Cognitive Neuroscience, 12*, 729–738.

Piaget, J. (1932/1965). *The Moral Judgment of the Child.* New York: Free Press.

Piaget, J. (1954). *The Construction of Reality in the Child.* New York: Basic Books.

Pica, P., Lemer, C., Izard, V., & Dehaene, S. (2004). Exact and approximate arithmetic in an Amazonian Indigene Group. *Science, 306*, 499–503.

Pinker, S. (1994). *The Language Instinct.* New York: William Morrow and Company.

Pinker, S. (1997). *How the Mind Works.* New York: Norton.

Pinker, S. (2002). *The Blank Slate.* New York: Penguin.

Pinker, S., & Jackendoff, R. (2005). The faculty of language: what's special about it? *Cognition, 95*, 201–236.

Platek, S. M., Critton, S. R., Myers, T. E., & Gallup, G. G., Jr. (2003). Contagious yawning: the role of self-awareness and mental state attribution. *Cognitive Brain Research, 17*, 223–227.

Polak, A., & Harris, P. L. (1999). Deception by young children following noncompliance. *Developmental Psychology, 35*, 561–568.

Posner, E. A. (2000). *Law and Social Norms.* Cambridge, MA: Harvard University Press.

Posner, R. A. (1992). *Sex and Reason*. Cambridge, MA: Harvard University Press.

Posner, R. A. (1998). The problematics of moral and legal theory. *Harvard Law Review, 111*, 1637–1717.

Povinelli, D. (2000). *Folk Physics for Apes*. New York: Oxford University Press.

Povinelli, D. J., & Eddy, T. J. (1996). What young chimpanzees know about seeing. *Monographs of the Society for Research in Child Development, 1–247*.

Povinelli, D. J., Rulf, A. B., Landau, K. R., & Bierschwale, D. T. (1993). Self-recognition in chimpanzees (*Pan troglodytes*): Distribution, ontogeny, and patterns of emergence. *Journal of Comparative Psychology, 107*, 347–372.

Premack, D. (1976). *Intelligence in Ape and Man*. Hillsdale, NJ: Lawrence Erlbaum Associates.

Premack, D. (1990). The infant's theory of self-propelled objects. *Cognition, 36*, 1–16.

Premack, D., & Premack, A. (1983). *The Mind of an Ape*. New York: Norton.

Premack, D., & Premack, A. (2002). *Original Intelligence*. New York: McGraw Hill.

Premack, D., & Premack, A. J. (1995). Origins of human social competence. In M. Gazzaniga (Ed.), *The Cognitive Neurosciences* (pp. 205–218). Cambridge, MA: MIT Press.

Premack, D., & Premack, A. J. (1997). Infants attribute value+/- to the goal-directed actions of self-propelled objects. *Journal of Cognitive Neuroscience, 9*, 848–856.

Preston, S., & de Waal, F. B. M. (2002). Empathy: It's ultimate and proximate bases. *Behavioral and Brain Sciences, 25*, 1–72.

Preuschoft, S. (1999). Are primates behaviorists? formal dominance, cognition and free-floating rationales. *Journal of Comparative Psychology, 113*, 91–95.

Prinz, J. J. (2004). *Gut Reactions*. New York: Oxford University Press.

Prinz, J. J. (in press). Against moral nativism. In P. Carruthers & S. Laurence & S. Stich (Eds.), *Innateness and the Structure of the Mind, Vol. II*. Oxford: Oxford University Press.

Pylyshyn, Z. W., & Storm, R. W. (1998). Tracking multiple independent targets: Evidence for a parallel tracking mechanism. *Spatial Vision, 3*, 179–197.

Rachels, J. (1975). Active and passive euthanasia. *New England Journal of Medicine, 292*, 78–80.

Rachels, J. (2000). Naturalism. In H. LaFollette (Ed.), *The Blackwell Guide to Ethical Theory* (pp. 74–91). Malden, NJ: Blackwell Publishing.

Rachels, J. (2003). *The Elements of Moral Philosophy*. Boston: McGraw Hill.

Rachlin, H. (2000). *The Science of Self-Control*. Cambridge, MA: Harvard University Press.

Raine, A., Ishikawa, S. S., Arce, E., Lencz, T., Knuth, K. H., Bihrle, S., LaCasse, L., & Colletti, P. (2004). Hippocampal structural asymmetry in unsuccessful psychopaths. *Biological Psychiatry, 55*, 185–191.

Rakison, D. H., & Poulin-Dubois, D. (2001). Developmental origins of the animate-inanimate distinction. *Psychological Bulletin, 127*, 209–228.

Ratnieks, F. L. W., & Visscher, P. K. (1989). Worker policing in the honeybee. *Nature, 342*, 796–797.

Ratnieks, F. L. W., & Wenseleers, T. (2005). Policing insect societies. *Science, 7*, 54–56.

Rawls, J. (1950). *A study in the grounds of ethical knowledge: Considered with reference to judgments on the moral worth of character*. Unpublished Ph.D., Princeton University, Princeton, NJ.

Rawls, J. (1971). *A Theory of Justice*. Cambridge, MA: Harvard University Press.

Rawls, J. (2001). *Justice as Fairness*. Cambridge, MA: Belknap, Harvard University Press.

Reynolds, B., & Schiffbauer, R. (2004). Measuring state changes in human delay discounting: an experimental discounting task. *Behavioural Processes, 67*, 343–356.

Richerson, P. J., & Boyd, R. (2005). *Not by Genes Alone: How culture transformed human evolution.* Chicago: University of Chicago Press.

Ridley, M. (1996). *The Origins of Virtue.* New York: Viking Press/Penguin Books.

Riling, J., Gutman, D., Zeh, T., Pagnoni, G., Berns, G., & Kilts, C. (2002). A neural basis for social cooperation. *Neuron, 35*, 395–405.

Ristau, C. (1991). Aspects of the cognitive ethology of an injury-feigning bird, the piping plover. In C. Ristau (Ed.), *Cognitive ethology: the minds of other animals* (pp. 91–126). Hillsdale, NJ: Erlbaum.

Rizzolatti, G., Fadiga, L., Fogassi, L., & Gallese, V. (1999). Resonance behaviors and mirror neurons. *Archives Italiennes de Biologie, 137*, 83–99.

Rizzolatti, G., Fadiga, L., Matelli, M., Bettinardi, V., Perani, D., & Fazio, F. (1996). Localization of grasp representation in humans by positron emission tomography: 1. Observation versus execution. *Experimental Brain Research, 111*, 246–252.

Rochat, P., & Striano, T. (1999). Emerging self-exploration by 2-month-old infants. *Developmental Science, 2*, 206–218.

Rogers, A. R. (1994). Evolution of time preference by natural selection. *American Economics Review, 84*, 460–481.

Rogers, A. R. (1997). The evolutionary theory of time preference. In G. Bock & G. Cardew (Eds.), *Characterizing Human Psychological Adapatations* (pp. 231–252). London: Wiley.

Rolls, E. T. (1999). *Brain and Emotion.* Oxford: Oxford University Press.

Rorty, R. (1998). *Truth and Progress: Philosophical Papers, Volume 3.* Cambridge: Cambridge University Press.

Rosati, A. (2005). *Discounting in tamarins and marmosets.* Unpublished Undergraduate Honors Thesis, Harvard University, Cambridge, MA.

Roskies, A. (2003). Are ethical judgments intrinsically motivational? Lessons from "acquired sociopathy." *Philosophical Psychology, 16*, 51–66.

Rozin, P. (1997). Moralization. In A. Brandt & P. Rozin (Eds.), *Morality and Health* (pp. 379–401). New York: Routledge.

Rozin, P., & Fallon, A. E. (1987). A perspective on disgust. *Psychological Review, 94*, 23–41.

Rozin, P., Haidt, J., & McCauley, C. R. (2000). Disgust. In M. Lewis & J. M. Haviland-Jones (Eds.), *Handbook of Emotions, 2nd Edition* (pp. 637–653). New York: Guilford Press.

Rozin, P., Millman, L., & Nemeroff, C. (1986). Operation of the laws of sympathetic magic in disgust and other domains. *Journal of Personality and Social Psychology, 50*, 703–712.

Rozin, P., Nemeroff, C., Wane, M., & Sherrod, A. (1989). Operation of the sympathetic magical law of contagion in interpersonal attitudes among Americans. *Bulletin of the Psychonomic Society, 27*, 367–370.

Rubin, D. (2002). *Darwinian Politics: The evolutionary origins of freedom.* Newark, NJ: Rutgers University Press.

Rubin, P. (2002). *Darwinian Politics.* New Brunswick, NJ: Rutgers University Press.

Ruffman, T., Garnham, W., Import, A., & Connolly, D. (2001). Does eye gaze indicate implicit knowledge of false belief? Charting transitions in knowledge. *Journal of Experimental Child Psychology, 80*, 201–224.

Ruffman, T., Olson, D. R., Ash, T., & Keenan, T. (1993). The ABCs of deception: do young children understand deception in the same way as adults? *Developmental Psychology, 29*, 74–87.

Russell, E. (1965). Scapegoating and social control. *Journal of Psychology, 61*, 203–209.

Russell, J. (1997). How executive disorders can bring about an inadequate 'theory of mind.' In J. Russell (Ed.), *Autism as an Executive Disorder* (pp. 256–304). Oxford: Oxford University Press.

Russell, J., Jarrold, C., & Potel, D. (1994). What makes strategic deception difficult for children—the deception or the strategy? *British Journal of Developmental Psychology, 12*, 301–314.

Sachs, J. L., Mueller, U. G., Wilcox, T. P., & Bull, J. J. (2004). The evolution of cooperation. *Quarterly Review of Biology, 79*, 135–160.

Saffran, J., Johnson, E., Aslin, R. N., & Newport, E. (1999). Statistical learning of tone sequences by human infants and adults. *Cognition, 70*, 27–52.

Saffran, J. R., Aslin, R. N., & Newport, E. L. (1996). Statistical learning by 8-month-old infants. *Science, 274*, 1926–1928.

Sanfey, A. G., Rilling, J. K., Aronson, J. A., Nystrom, L. E., & Cohen, J. D. (2003). The neural basis of economic decision making in the ultimatum game. *Science, 300*, 1755–1758.

Santos, L., Flombaum, J., & Hauser, M. D. (in prep). *Rhesus monkey expectations about object motion.* Unpublished manuscript, Cambridge, MA.

Santos, L. R., & Hauser, M. D. (1999). How monkeys see the eyes: cotton-top tamarins' reaction to changes in visual attention and action. *Animal Cognition, 2*, 131–139.

Sapolsky, R. (1994). Individual differences and the stress response. *Seminars in Neurosciences, 6*, 261–269.

Sapolsky, R. (1998). *The Trouble with Testosterone.* New York: Scribner.

Sapolsky, R. (2001). *A Primate's Memoir.* New York: Scribner.

Sapolsky, R., & Share, L. (2004). *A pacific culture among wild baboons: Its emergence and transmission.* Public Library of Science Biology, 2: e106. doi: 10.1371/ journal.pbio. 0020106 [2.

Saporta, S. (1978). An interview with Noam Chomsky. *Linguistic Analysis, 4[4]*, 301–319.

Scheel, D., & Packer, C. (1991). Group hunting behaviour of lions: a search for cooperation. *Animal Behaviour, 41*, 697–709.

Schelling, T. (1960). *The Strategy of Conflict.* Cambridge, MA: Harvard University Press.

Schlottmann, A., Allen, D., Linderoth, C., & Hesketh, S. (2002). Perceptual causality in children. *Child Development, 73*, 1656–1677.

Schlottmann, A., & Surian, L. (1999). Do 9-month-olds perceive causation-at-a-distance? *Perception, 28*, 1105–1113.

Scholl, B. J., & Nakayama, K. (2002). Causal capture: contextual effects on the perception of collision events. *Psychological Science, 13*, 493–498.

Scholl, B. J., & Tremoulet, P. (2000). Perceptual causality and animacy. *Trends in Cognitive Science, 4*, 299–309.

Schultz, W., Dayan, P., & Montague, P. R. (1997). A neural substrate of prediction and reward. *Science, 275*, 1593–1599.

Schultz, W., & Dickinson, A. (2000). Neuronal coding of prediction errors. *Annual Review of Neuroscience, 23*, 473–500.

Sethi, A., Mischel, W., Aber, J. L., Shoda, Y., & Rodriguez, M. L. (2000). The role of strategic attention deployment in development of self-regulation: predicting preschoolers' delay of gratification from mother-toddler interactions. *Developmental Psychology, 36*, 767–777.

Seyfarth, R. M., & Cheney, D. (2003). Hierarchical social knowledge of monkeys. In F. B. M. d. Waal & P. L. Tyack (Eds.), *Animal Social Complexity: Intelligence, Culture, and Individualized Societies* (pp. 207–229). Cambridge, MA: Harvard University Press.

Seyfarth, R. M., & Cheney, D. L. (1984). Grooming alliances and reciprocal altruism in vervet monkeys. *Nature, 308*, 541–543.

Shakespeare, W. (1914). *Macbeth*. London: Oxford University Press.

Shallice, T. (1988). *From Neuropsychology to Mental Structure*. Cambridge: Cambridge University Press.

Shaw, G. B. (1916). *Pygmalion*. Bartleby. Available: http://www.bartleby.com/138/.

Shepher, J. (1971). Mate selection among second generation kibbutz adolescents and adults: incest avoidance and negative imprinting. *Archives of Sexual Behaviour, 1*, 293–307.

Sherman, N. (1989). *The Fabric of Character: Aristotle's Theory of Virtue*. New York: Oxford University Press.

Shimizu, Y. A., & Johnson, S. C. (2004). Infants' attribution of a goal to a morphologically unfamiliar agent. *Developmental Science, 7*, 425–430.

Siegal, M., & Peterson, C. C. (1998). Preschoolers' understanding of lies and innocent and negligent mistakes. *Developmental Psychology, 34*, 332–341.

Sih, A., & Mateo, J. (2001). Punishment and persistence pay: a new model of territory establishment and space use. *Trends in Ecology and Evolution, 16*, 477–479.

Silk, J. (2002a). The form and function of reconciliation in primates. *Annual Review of Anthropology, 31*, 21–44.

Silk, J. (2002b). Practice random acts of aggression and senseless acts of intimidation: the logic of status contests in social groups. *Evolutionary Anthropology, 11*, 221–225.

Silk, J. B. (2003). Cooperation without counting: the puzzle of friendship. In P. Hammerstein (Ed.), *Genetic and Cultural Evolution of Cooperation* (pp. 37–54). Cambridge, MA: MIT Press.

Simon, H. A. (1990). Invariants of human behavior. *Annual Review of Psychology, 41*, 1–19.

Singer, P. (1972). Famine, affluence, and morality. *Philosophy and Public Affairs, 1/3*, 229–243.

Singer, P. (1993). *Practical Ethics*. Cambridge: Cambridge University Press.

Singer, P. (2000). *A Darwinian Left*. New Haven, CT: Yale University Press.

Smetana, J. (1995). Morality in context: abstractions, ambiguities and applications. *Annals of Child Development, 10*, 83–130.

Smetana, J. (2005). Social-cognitive domain theory: consistencies and variations in children's moral and social judgments. In M. Killen & J. G. Smetana (Eds.), *Handbook of Moral Development*. Mahwah, NJ: Lawrence Erlbaum Publishers.

Smuts, B. B. (1987). Gender, aggession, and influence. In B. B. Smuts & T. S. Struhsaker & R. Seyfarth & D. Cheney & R. W. Wrangham (Eds.), *Primate Societies* (pp. 400–412). Chicago: University of Chicago Press.

Sober, E. (1994). *From a Biological Point of View*. Cambridge: Cambridge University Press.

Sober, E., & Wilson, D. S. (1998). *Unto Others*. Cambridge, MA: Harvard University Press.

Sodian, B., Taylor, C., Harris, P. L., & Perner, J. (1991). Early deception and the child's theory of mind: false trains and genuine markers. *Child Development, 62*, 468–483.

Solomon, R. C. (2002). Back to basics: on the very idea of "basic emotions." *Journal for the Theory of Social Behavior, 32*, 115–144.

Sorensen, R. (1991). Thought experiments. *American Scientist, May–June*, 250–263.

Spelke, E. (2000). Core knowledge. *American Psychologist, 55*, 1233–1243.

Spelke, E. S. (1994). Initial knowledge: six suggestions. *Cognition, 50*, 431–445.

Sperber, D., & Girotto, V. (in press). Does the selection task detect cheater-detection? In J. Fitness & K. Sterelny (Eds.), *New Directions in Evolutionary Psychology*. Abingdon, UK: Taylor & Francis.

Spinka, M., Newberry, R. C., & Bekoff, M. (2001). Mammalian pay: training for the unexpected. *Quarterly Review of Biology, 76*, 141–168.

Sprengelmeyer, R., Young, A. W., Calder, A. J., Karnat, A., Lange, H., Homberg, V., Perrett, D. I., & Rowland, D. (1996). Loss of disgust: perception of faces and emotions in Huntington's disease. *Brain, 119*, 1647–1665.

Sprengelmeyer, R., Young, A. W., Pundt, I., Sprengelmeyer, A., Calder, A. J., Berrios, G., Wingkel, R., Vollmoeller, W., Kuhn, W., Sartory, G., & Przuntek, H. (1997). Disgust implicated in obsessive-compulsive disorder. *Proceedings of the Royal Society of London, B, B264*, 1767–1773.

Stammbach, E. (1988). Group responses to specially skilled individuals in a *Macaca fascicularis. Behaviour, 107*, 241–266.

Stamps, J. A., & Krishnan, V. V. (1999). A learning-based model of territory establishment. *Quarterly Review of Biology, 74*, 291–318.

Stephens, D. W., McLinn, C. M., & Stevens, J. R. (2002). Discounting and reciprocity in an iterated prisoner's dilemma. *Science, 298*, 2216–2218.

Stevens, D., Charman, T., & Blair, R. J. R. (2001). Recognition of emotion in facial expressions and vocal tones in children with psychopathic tendencies. *Journal of Genetic Psychology, 162*, 201–211.

Stevens, J. R. (2004). The selfish nature of generosity: harassment and food sharing in primates. *Proceedings of the Royal Society, London, B, 271*, 451–456.

Stevens, J. R., Cushman, F. A., & Hauser, M. D. (2005). Evolving the psychological mechanisms for cooperation. *Annual Review of Ecology and Systematics, 36*, 499–518.

Stevens, J. R., Hallinan, E. V., & Hauser, M. D. (2005). The ecology and evolution of patience in two New World primates. *Biology Letters, 1*, 223–226.

Stevenson-Hinde, J., Stillwel-Barnes, R., & Zunz, M. (1980). Subjective assessment of rhesus monkeys over four sucessive years. *Primates, 21*, 223–233.

Stingl, M., & Colier, J. (2005). Reasonable partiality from a biological point of view. *Ethical Theory and Moral Practice, 8*, 11–24.

Stone, V. E., Cosmides, L., Tooby, J., Kroll, N., & Knight, R. T. (2002). Selective impairment of reasoning about social exchange in a patient with bilateral limbic system damage. *Proceedings of the National Academy of Sciences, USA, 99*, 11531–11536.

Sunstein, C. R., Hastie, R., Payne, J. W., Schkade, D. A., & Viscusi, W. K. (2002). *Punitive Damages. How Juries Decide*. Chicago: University of Chicago Press.

Sweetser, E. (1987). The definition of "lie": an examination of the folk models underlying a semantic prototype. In D. Hollard & N. Quinn (Eds.), *Cultural Models in Language and Thought*. New York: Cambridge University Press.

Talwar, V., & Lee, K. (2002). Development of lying to conceal a transgression: children's control of expressive behaviour during verbal deception. *International Journal of Behavioral Development, 26*, 436–444.

Talwar, V., Lee, K., Bala, N., & Lindsay, R. C. L. (2002). Children's conceptual knowledge of lying and its relation to their actual behaviors: implications for court competence examinations. *Law and Human Behavior, 26*, 395–415.

Tarr, M. J., & Gauthier, I. (2000). FFA: A flexible fusiform area for subordinate-level visual processing automatized by expertise. *Nature Neuroscience, 3*, 764–769.

Tetlock, P. E. (2003). Thinking the unthinkable: sacred values and taboo cognitions. *Trends in Cognitive Science, 7*, 320–324.

Tetlock, P. E., Kristel, O. V., Elson, S. B., & Lerner, J. S. (2000). The psychology of the unthinkable: taboo trade-offs, forbidden base rates, and heretical counterfactuals. *Journal of Personality and Social Psychology, 78*, 853–870.

Thompson, C., Barresi, J., & Moore, C. (1997). The development of future-oriented prudence and altruism in preschoolers. *Cognitive Development, 12*, 199–212.

Thomson, J. J. (1970). Individuating actions. *Journal of Philosophy, 68*, 774–781.

Thomson, J. J. (1971). A defense of abortion. *Philosophy and Public Affairs, 1*, 47–66.

Tinkelpaugh, O. L. (1928). An experimental study of representative factors in monkeys. *Journal of Comparative Psychology, 8*, 197–236.

Tinkelpaugh, O. L. (1932). Multiple delayed reaction with chimpanzees and monkeys. *Journal of Comparative Physiological Psychology, 13*, 207–224.

Tobin, H., Longue, A. W., Chelonis, J. J., & Ackerman, K. T. (1996). Self-control in the monkey *Macaca fascicularis. Animal Learning and Behavior, 24*, 168–174.

Todd, P. M., & Gigerenzer, G. (2003). Bounding rationality to the world. *Journal of Economic Psychology, 24*, 143–165.

Tomasello, M., & Call, J. (1997). *Primate Cognition.* Oxford: Oxford University Press.

Tomasello, M., Call, J., & Hare, B. (2003). Chimpanzees understand psychological states—the question is which ones and to what extent. *Trends in Cognitive Science, 7*, 153–156.

Tomasello, M., Carpenter, M., Call, J., Behne, T., & Moll, H. (2005). Understanding and sharing intentions: the origins of cultural cognition. *Behavioral and Brain Research, 28*, 795–855.

Tooby, J., & Cosimdes, L. (1998). Friendship and the Banker's Paradox: Other pathways to the evolution of adapations for altruism. In W. G. Runciman & J. M. Smith & R. I. M. Dunbar (Eds.), *Evolution of Social Behaviour Patterns in Primates and Man* (pp. 299–323). Oxford: Oxford University Press.

Tranel, D., Bechara, A., & Damasio, A. (2000). Decision making and the somatic marker hypothesis. In M. Gazzaniga (Ed.), *The New Cognitive Neurosciences* (pp. 1047–1061). Cambridge, MA: MIT Press.

Tranel, D., & Damasio, A. R. (1985). Knowledge without awareness: An autonomic index of recognition of prosapagnosics. *Science, 228*, 1453–1454.

Tranel, D., & Damasio, A. R. (1993). The covert learning of affective valence does not require structures in hippocampal system of amygdala. *Journal of Cognitive Neuroscience, 5*, 79–88.

Tranel, D., Damasio, A. R., & Damasio, H. (1988). Intact recognition of facial expression, gender, and age in patients with impaired recognition of face identity. *Neurology, 38*, 690–696.

Tranel, D., Damasio, H., & Damasio, A. (1995). Double dissociation between overt and covert recognition. *Journal of Cognitive Neuroscience, 7*, 425–532.

Tremblay, L., & Schultz, W. (1999). Relative reward preference in primate orbito-frontal cortex. *Nature, 398*, 704–708.

Trivers, R. L. (1972). Parental investment and sexual selection. In B. Campbell (Ed.), *Sexual Selection and the Descent of Man* (pp. 136–179). Chicago: Aldine Press.

Trivers, R. L. (1974). Parent-offspring conflict. *American Zoologist, 14*, 249–264.

Trut, L. N. (1999). Early canid domestication: the farm-fox experiment. *American Scientist, 87*, 160–169.

Turiel, E. (1983). *The Development of Social Knowledge: Morality and Convention*. Cambridge: Cambridge University Press.

Turiel, E. (1998). The development of morality. In W. Damon (Ed.), *Handbook of Child Psychology* (pp. 863–932). New York: Wiley Press.

Turiel, E. (2005). Thought, emotions, and social interactional processes in moral development. In M. Killen & J. G. Smetana (Eds.), *Handbook of Moral Development*. Mahwah, NJ: Lawrence Erlbaum Publishers.

Uller, C. (2004). Disposition to recognize goals in infant chimpanzees. *Animal Cognition, 7*, 154–161.

Unger, P. K. (1996). *Living High and Letting Die*. New York: Oxford University Press.

van den Berghe, P. L. (1983). Human inbreeding avoidance: culture in nature. *Behavioral and Brain Sciences, 6*, 91–123.

Vandello, J. A., & Cohen, D. (2004). When believing is seeing: sustaining norms of violence in cultures of honor. In M. Schaller & C. Crandall (Eds.), *Psychological Foundations of Culture*. Mawhah, NJ: Lawrence Erlbaum.

Wainryb, C. (2004). "Is" and "Ought": moral judgments about the world as understood. *New Directions for Child and Adolescent Development, 103*, 3–18.

Walzer, M. (2000). *Just and Unjust Wars: a moral argument with historical illustrations*. New York: Basic Books.

Watanabe, M. (1996). Reward expectancy in primate prefrontal neurons. *Nature, 382*, 629–632.

Wattles, J. (1996). *The Golden Rule*. Oxford: Oxford University Press.

Wedekind, C., & Milinski, M. (2000). Cooperation through image scoring in humans. *Science, 288*, 850–852.

Wegner, D. (2002). *The Illusion of Conscious Will*. Cambridge, MA: MIT Press.

Weinberg, S. (1976). The forces of nature. *Bulletin of the American Society of Arts and Sciences, 29*, 28–29.

Wells, J. C. K. (2003). Parent-offspring conflict theory, signaling of need, and weight gain in early life. *Quarterly Review of Biology, 78*, 169–202.

West, M. D. (2003). *Losers: recovering lost property in Japan and the United States*. Ann Arbor, MI: University of Michigan School of Law.

Westermark, E. (1891). *The History of Human Marriage*. London: Macmillan.

White, P. A. (1995). *The Understanding of Causation and the Production of Action*. Hillsdale, NJ: Lawrence Erlbaum Associates.

Whitehead, H., & Connor, R. C. (2005). Alliances. I. How large should alliances be? *Animal Behaviour, 69*, 117–126.

Whiten, A., & Byrne, R. W. (1988). Tactical deception in primates. *Behavioral and Brain Sciences, 11*, 233–273.

Whitman, J. Q. (1998). What is wrong with inflicting shame sanctions? *Yale Law Journal, 107*, 1055–1092.

Whitman, J. Q. (2003). *Harsh Justice.* New York: Oxford University Press.

Wilson, D. S. (2002a). *Darwin's Cathedral.* Chicago: University of Chicago Press.

Wilson, D. S. (2002b). Group selection. In M. Pagel (Ed.), *Oxford Encyclopedia of Evolution* (pp. 450–454). Oxford: Oxford University Press.

Wilson, D. S., Diertrich, E., & Clark, A. B. (2003). On the inappropriate use of the naturalistic fallacy in evolutionary psychology. *Biology and Philosophy, 18*, 669–682.

Wilson, E. O. (1998). The biological basis of morality. *The Atlantic Monthly, April*, 53–70.

Wilson, J. Q. (1997). *The Moral Sense.* New York: Free Press.

Wilson, J. Q., & Herrnstein, R. J. (1985). *Crime and Human Nature.* New York: Free Press.

Wilson, M., Daly, M., & Pound, N. (2002). An evolutionary psychological perspective on the modulation of competitive confrontation and risk-taking. *Hormones and Behavior, 5*, 381–408.

Wimmer, H., & Perner, J. (1983). Beliefs about beliefs: Representation and constraining function of wrong beliefs in young children's understanding of deception. *Cognition, 13*, 103–128.

Wolf, A. P. (1995). *Sexual Attraction and Childhood Associations.* Stanford, CA: Stanford University Press.

Woodward, A. L., Sommerville, J. A., & Guajardo, J. J. (2001). How infants make sense of intentional action. In B. F. Malle & L. J. Moses & D. A. Baldwin (Eds.), *Intentions and Intentionality: Foundations of Social Cognition* (pp. 149–171). Cambridge, MA: MIT Press.

Wrangham, R. W., & Peterson, D. (1996). *Demonic Males: Apes and the Origins of Human Violence.* Boston: Houghton Mifflin.

Wrangham, R. W., Pilbeam, D., & Hare, B. (in press). Convergent cranial paedomorphosis in bonobos, domesticated animals and humans? The role of selection for reduced aggression. *Evolutionary Anthropology*, still in press.

Wright, R. (1995). *The Moral Animal.* New York: Vintage.

Yang, C. D. (2004). Universal Grammar, statistics or both? *Trends in Cognitive Sciences, 8(10)*, 451–456.

Young, A. W., & DeHaan, E. H. F. (1988). Boundaries of covert recognition in prosopagnosia. *Cognitive Neuropsychology, 5*, 317–336.

Zacks, J. M., Braver, T. S., Sheridan, M. A., Donaldson, D. I., Snyder, A. Z., Ollinger, J. M., Buckner, R., & Raichle, M. E. (2001). Human brain activity time-locked to perceptual event boundaries. *Nature Neuroscience, 4*, 651–655.

Zacks, J. M., & Tversky, B. (2001). Event structure in perception and conception. *Psychological Bulletin, 127*, 3–21.

Zahavi, A. (1975). Mate selection: a selection for a handicap. *Journal of Theoretical Biology, 53*, 205–214.

Zahavi, A. (1987). The theory of signal selection and some of its implications. In V. P. Delfino (Ed.), *International Symposium of Biological Evolution* (pp. 305–327). Bari, Italy: Adriatica Editrice.

Zak, P. J., Kurzban, R., & Matzner, W. T. (2003). The neurobiology of trust. *Working Paper, Center for Neuroeconomic Studies.*

Zelazo, P. D., & Frye, D. (1997). Cognitive complexity and control: a theory of the development of deliberate reasoning and intentional action. In M. Stamenov (Ed.), *Language Structure, Discourse, and the Access to Consciousness* (pp. 113–153). Amsterdam: John Benjamins.

INDEX